# Surface Mount Technology

## Principles and Practice

# Surface Mount Technology

## Principles and Practice

Ray P. Prasad

*SMT Program Manager*
*Systems Group*
*Intel Corporation*
*Hillsboro, Oregon*

**VNR** VAN NOSTRAND REINHOLD
New York

Copyright © 1989 by Van Nostrand Reinhold
Library of Congress Catalog Card Number: 88-25885
ISBN 0-442-20527-9

Printed in the United States of America

Van Nostrand Reinhold
115 Fifth Avenue
New York, New York 10003

Van Nostrand Reinhold (International) Limited
11 New Fetter Lane
London EC4P 4EE, England

Van Nostrand Reinhold
480 La Trobe Street
Melbourne, Victoria 3000, Australia

Macmillan of Canada
Division of Canada Publishing Corporation
164 Commander Boulevard
Agincourt, Ontario M1S 3C7, Canada

16  15  14  13  12  11  10  9  8  7  6  5  4  3  2  1

**Library of Congress Cataloging in Publication Data**

Prasad, Ray,
   Surface mount technology: principles and practice/Ray Prasad.
      p.   cm.

Bibliography: p. 632
Includes index.
ISBN 0-442-20527-9
1. Printed circuits—Design and construction.   2. Electronic
packaging.   I. Title.
TK7868.P7P7 1989
621.381′74—dc19                                          88-25885
                                                              CIP

*To my wife Peggy
and my children Geeta, Joey, and Kevin
the ones who really made it possible.*

# Brief Contents

# Contents

**Chapter 6   Surface Mount Land Pattern Design      209**

**Chapter 7   Design for Manufacturability, Testing, and Repair      235**

# Preface

Surface Mount Technology is not a technology of tommorrow but a technology of today. It provides a quantum jump in the packaging technology to produce state-of-the-art miniaturized electronic products. However, in order to take advantage of this technology, a complete infrastructure must be put in place. This requires considerable investment in human and capital resources. Intel corporation has made these investments to keep its customers for components and systems on the leading edge of technology. Based on the experience of putting this infrastructure in place for system products, this book is written for managers who need to manage the risk during its implementation, and the practicing engineers who need to improve the design and manufacturing processes for improved yield and cost reduction. To accomplish this task, I have not only culled the information from published materials, but have also depended on input from both my colleagues in Intel and such outside organizations as the Institute of interconnecting and Packaging electronic Circuits (IPC), the Electronics Industries Association (EIA), and the Surface Mount Council. But the underlying basis for this book has been my first-hand experience in implementing this technology for Intel Systems Group and my experience at Boeing, my previous employer.

In a fast-changing technology like SMT, it is very easy to have obsolete information even before the book is published. For this reason, I have concentrated on the basic principles and practice of the technology. Formulas and theories are discussed when necessary to give a better understanding of the subject. Very little emphasis is given on the details of manufacturing equipment which are constantly changing. Instead, emphasis is placed on important equipment features and selection criteria.

This book is divided into three parts. The first part gives the overview of the technology and discusses the inherent technical and risk management issues that must be addressed for an effective implementation of the technology. The next two parts of the book provide the details of design and manufacturing solutions.

The second part of the book focuses on the component, substrates, and design considerations, such as land pattern design and design for manufacturability, that must be considered for designing a cost-effective surface mount assembly. The reader will find much new and practical information on design considerations and design for manufacturability. The basic concepts discussed will be valid even if a great change in the offerings of components occurs in the future.

In the third part of the book, the details of materials, processes (including workmanship standards), repair, and process control are discussed. No matter how good a design, the process must be adequately characterized for improving the manufacturing yield. The appendixes provide additional information on surface mount standards, a questionnaire for selection of pick and place equipment, and a glossary of terms.

# Acknowledgments

To the extent that I may have been successful in my stated objective, it could not have been possible without the support of many of my friends. However, I am solely responsible for any shortcomings. There are many colleagues from my SMT team at Intel who have made significant contribution to this book. The list is too long to mention them all. So I thank them collectively for their contribution. However, there are some individuals whose contributions have been significant. For example, Dr. Raiyomand Aspandiar, contributed extensively, especially in the areas of adhesive, soldering, and cleaning. George Arrigotti, also a key member of the SMT team, read the entire text and made good suggestions for improvement. Mark Christensen read Chapters 4 and 11, and provided good input. Joe Floren, Al Liske, and Sarah McDonald made good suggestions in Chapter 1. And Alan Donaldson, who conducted many of the experiments and produced many charts and photographs for use in this book. Dave Love and Oscar Vaz of Intel component group also provided useful suggestions.

Many of my colleagues in the IPC and Surface Mount Council also made valuable contributions. For example, Joseph Spalik of IBM read the entire text. Foster Gray of Texas Instruments, John DeVore of General Electric, and Werner Engelmaier and Bernard Wargotz of AT&T read different chapters in their areas of expertise. They provided valuable suggestions from their perspectives for improvement.

I want to thank my editor, Marjan Bace at VNR for his continued support and seeking outside reviewers and editors who played an important role improving the quality of this book.

I am thankful to Melinda Crawford for producing the majority of the illustrations in this book. I also thank Diane McKinnon, Allyson McKenna, Pam Noga, Carol Warren, and Ann Schoffstoll for producing many of the illustrations. Ann also spent considerable amount of her time in making corrections to the text. I also want to thank Gary Orehovec and various equipment manufacturers for supplying many of the photographs.

I would also like to thank the Intel management in general and my boss Fred Burris in particular for their support. It was Fred who encouraged me to write this book. Without his encouragement and support, I could not have completed this book.

And most important of all I am very appreciative of my wife and three children who provided moral support in completing this book. I especially appreciate my two-year-old son Joey, who was always eager to help me type the manuscript on my personal computer. He wanted to join in the fun and forced me to take breaks.

Writing this book has been more of a challenge than I originally imagined, but it has been a very useful educational experience as well. It gave me an opportunity to look at SMT from a broader perspective than only the job at hand required. It is my hope that this book makes a useful contribution for a better and balanced understanding of this emerging technology and in its proliferation.

November, 1988                          *Ray P. Prasad*
                                        Systems Group, Intel Corporation
                                        Hillsboro, Oregon

# Surface Mount Technology

## Principles and Practice

# Part One
# Introduction to Surface Mounting

# Chapter 1

# Introduction to Surface Mount Technology

## 1.0 INTRODUCTION

The present methods of manufacturing conventional electronic assemblies have essentially reached their limits as far as cost, weight, volume, and reliability are concerned. Surface mount technology (SMT) makes it possible to produce more reliable assemblies at reduced weight, volume, and cost. SMT is used to mount electronic components on the surface of printed circuit boards or substrates. Conventional technology, by contrast, inserts components through holes in the board. This deceptively simple difference changes virtually every aspect of electronics: design, materials, processes, and assembly of component packages and substrates.

The surface mount concept isn't new. Surface mounting has its roots in relatively old technologies such as flat packs and hybrids. But the design and manufacturing technologies used previously generally are not applicable to the surface mounting done today. The current vision of SMT requires complete rethinking of design and manufacturing, along with a new SMT infrastructure to develop and sustain it.

Electronics assembly originally used point-to-point wiring and no substrate at all. The first semiconductor packages used radial leads, which were inserted in through holes of the single-sides circuit boards already used for resistors and capacitors. Then Fairchild invented dual in-line packages (DIPs) with two rows of leads. This allowed wave soldering of DIPs along with other components such as resistors, capacitors, vacuum tubes, or vacuum tube socket leads.

In the 1950s, surface mount devices called flat packs were used for high reliability military applications. They can be considered the first surface mount packages to be mounted on printed circuit boards. However, the flat-pack devices had to be mounted too close to the board surface, required gold-plated leads, were very costly, and required discrete soldering (hot bar). They did not become popular, but they are still used in mostly military and some commercial applications. In the 1960s, flat packs were replaced with DIPs, which were easier to insert and could be wave soldered en masse.

Another significant contribution to today's SMT lies in hybrid technology. Hybrid components, which became very popular in the 1960s, are still widely used. They feature surface mount devices soldered inside a through-hole or surface mount ceramic package body. Hybrid substrates are generally ceramic, but plastic hybrids have been used to reduce cost in some commercial applications. The ceramic and plastic hybrids shown in Figure 1.1 have surface mount devices mounted on the outside. Hybrids may have devices mounted inside the package as well. Also, most of the passive components (resistors and capacitors) used today were originally made for the hybrid industry. Placement and soldering techniques developed for the hybrid industry have also become a part of today's SMT. While the hybrid industry has made a signficant contribution to SMT, hybrid substrates are so small compared with the substrates used for surface

**Figure 1.1 Some examples of ceramic and plastic hybrids. (Courtesy of the Systems Group, Intel Corporation.)**

mounting that the design, manufacturing, and test methodologies are drastically different. But there is no doubt that SMT has profited immensely from the developmental work in hybrids.

Today's SMT also has its roots in extensive development work done in the United States for military applications. For example, to shrink package size for larger pin counts, the military needed hermetically sealed devices with leads on all four sides. This led to the development of leadless ceramic chip carriers (LCCCs) in the 1970s.

LCCCs proved to have their own problems, however: They require expensive substrates with matching coefficients of thermal expansion (CTE) to prevent solder joint cracking due to CTE mismatch between the ceramic components and the glass epoxy substrates. The military has spent considerable human and financial resources in developing acceptable substrates for the LCCCs, but the results have been less than satisfactory.

While the electronics industry in the United States was preoccupied with developing substrates for LCCCs for a very limited military market, the Japanese and Europeans were responding to a much wider consumer market. In the 1960s, N.V. Philips, a European company, invented surface-mountable Swiss outline packages for the watch industry. These packages are known today as small outline (SO) or small outline integrated circuit (SOIC) packages. In the early 1970s the Japanese started building calculators using plastic quad packs, which are similar to flat packs but have leads on all four sides. The term "quad" is used for packages that have surface-mountable leads on all four sides, bent down and out like gull wings. This lead configuration is susceptible to bending and damage during handling and shipping. Even though the quad packs provided space savings because of their smaller size compared with their through-hole counterparts, the fragile gull wing leads required special handling to prevent lead bending. Package style has a big impact on manufacturing technologies such as handling, soldering, cleaning, and inspection.

The Japanese also perfected the current version of surface mount technology on high volume consumer products, such as radios, television receivers, and video cassette recorders. For the most part, these products require passive surface mount devices for their analog circuits and are wave soldered. Passive components have terminations instead of leads, and they are considerably smaller. Figure 1.2 compares commonly used surface mount passive components with their conventional leaded equivalents.

Whereas the 1970s saw widespread usage of passive devices in consumer products in Japan, the present decade is characterized by the use of active devices worldwide in mass applications such as telecommunications and computers. In the United States, the electronics industry is driven by digital circuits for telecommunications, computer, and military markets. The electronics assemblies for these markets primarily use LCCCs for the

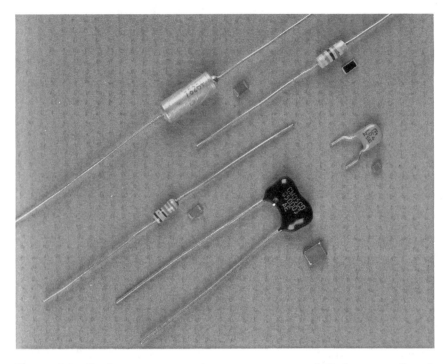

**Figure 1.2   Surface mount passive components and their conventional equivalents.**

**Figure 1.3   Surface mount active component [a plastic leaded chip carrier-(PLCC)] and its conventional equivalent.**

military applications, and SOICs and plastic leaded chip carriers (PLCCs) for the industrial markets.

PLCCs have leads on all four sides, like quad packs, but each lead is bent down and under like a "J"; thus PLCCs are also referred to as the J-lead packages. The J leads are less prone to damage during handling and provide better real estate efficiency on a substrate. A PLCC package and its conventional leaded equivalent are compared in Figure 1.3. PLCCs were developed to solve the problem of CTE mismatch between the package and the substrates by providing J leads to take up any strain during the component's use. They are the most commonly used packages in the United States. The Japanese have also adopted the PLCC packages, but generally for meeting the needs of U.S. markets.

Having briefly reviewed the history of SMT, let us discuss the advantages and disadvantages, the technical concerns, and the future of this emerging technology.

## 1.1 TYPES OF SURFACE MOUNTING

Since many components aren't yet available for surface mounting, SMT must accommodate some through-hole insertion. For this reason, "a surface mount assembly" is an incomplete description. The surface mount components, actives and passives, when attached to the substrate form three major types of SMT assembly—commonly referred to as Type I, Type II, and Type III, as shown in Figure 1.4. The process sequences are different for each type, and all the types require different equipment.

The Type III SMT assembly contains only discrete surface mount components (resistors, capacitors, and transistors) glued to the bottom side. The Type I assembly contains only surface mount components. The assembly can be either single-sided or double-sided. Type II assembly is a combination of Type III and Type I. It generally does not contain any active surface mount devices on the bottom side but may contain discrete surface mount components glued to the bottom side. This description of SMT types I, II, and III is by no means universal, but it is the most commonly used in the industry. For simplicity and consistency, these definitions for types of SMT have been used throughout this book.

The process sequence for Type III SMT is shown in Figure 1.5. First the through-hole components are autoinserted and clinched using existing through-hole insertion equipment. Next the assembly is turned over and adhesive is applied. Since the leads of the autoinserted components are clinched, they do not fall off the board. Then the surface mount components are placed by "pick-and-place" machine(s), the adhesive is cured in a convection or infrared oven, the assembly is turned over, and both leaded

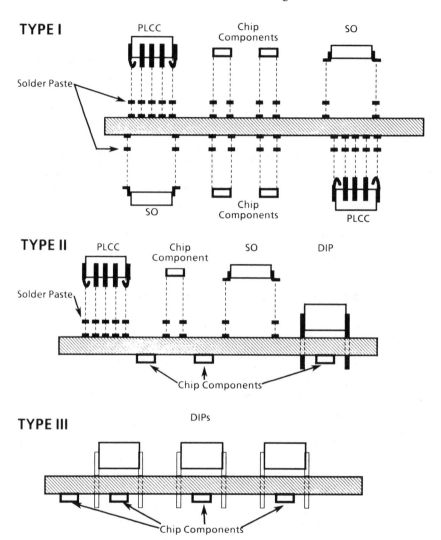

**Figure 1.4   The three principal surface mount assemblies in surface mount technology.**

and surface mount components are wave soldered in a single operation. The discrete components on the bottom side of the board are held in place by the adhesive during wave soldering. If autoinsertion equipment is not used, the leads are not clinched and the process sequence must be reversed: the adhesive is dispensed first; the discrete surface mount components are set in place; the adhesive is cured; the assembly is turned over and all the through-hole components manually inserted; and the assembly is wave soldered, cleaned, and tested.

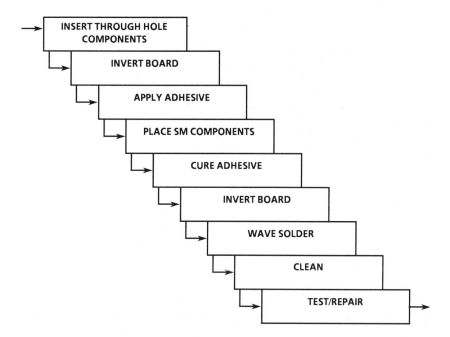

**Figure 1.5 Typical process flow for underside attachment (Type III SMT).**

I have assumed in this book that SOIC is not used in Type III assembly. Some companies do wave solder SOICs, but this is not a recommended process because flux may seep through the lead frame material during wave soldering, such seepage, in turn, will cause die corrosion, thus compromising the reliability of the device.

The process sequence for Type I SMT is shown in Figure 1.6. Type I assembly uses no through-hole components. First solder paste is screened, components are placed, and the assembly is baked in a convection or infrared oven to drive off volatiles from the solder paste. (Pastes are also available that do not require baking.) Finally the assembly is reflow soldered (vapor phase or infrared) and solvent cleaned. For double-sided Type I SMT assemblies, the board is turned over and the process sequence just described is repeated. The solder joints on the top side of the board are reflowed again. During the second reflow, the components are held in place by the surface tension of previously reflowed solder. Note that the assembly must be first reflowed on the top side of the board for the surface tension of solder to prevent components from falling off when the assembly is turned over during the second reflow.

The process sequence for Type II SMT is shown in Figure 1.7. Since a Type II assembly is a combination of Type I and Type III, it uses all the

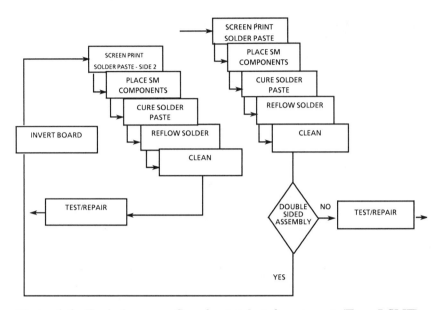

**Figure 1.6 Typical process flow for total surface mount (Type I SMT).**

processes needed for both. It is the most difficult assembly to manufacture because it has the most process steps. Type II assembly goes through the process sequence of Type I followed by the process sequence for Type III. Note that a Type II assembly may not have discrete surface mount components on the bottom side. In that case, the Type III process sequence of dispensing and curing adhesive and placing surface mount components on the bottom side will be omitted, but assembly will still use two soldering operations: reflow soldering for the active surface mount components on the top and wave soldering for the through-hole components also on the top side. The process details for all three types of SMT assembly—adhesive, solder paste, placement, soldering, and cleaning—are covered in Chapters 8, 9, 11, 12, and 13 respectively. And finally, every type of assembly is inspected, repaired if necessary, and tested. The details of inspection, test, and repair are discussed in Chapter 14.

## 1.2   BENEFITS OF SURFACE MOUNTING

Because the surface mount components are small and can be mounted on either side of the boards, they have achieved widespread usage. The benefits of SMT are available in both design and manufacturing.

Among the most important design-related benefits are significant savings in weight and real estate, and electrical noise reduction. As shown in Figure 1.8, surface mount components can weigh as little as one-tenth of

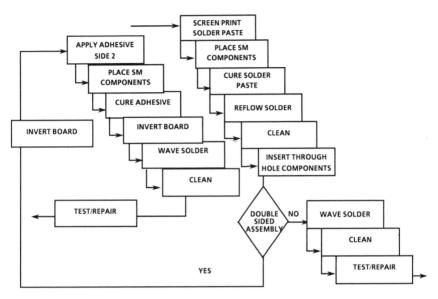

**Figure 1.7 Typical process flow for mixed technology (Type II SMT).**

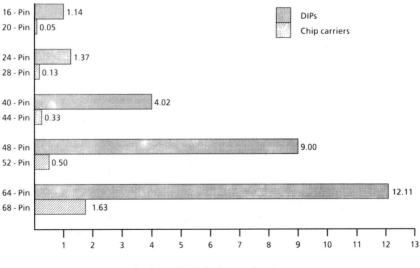

Package Weight (grams)

**Figure 1.8 The weight of various lead count chip carriers for SMT compared to the weight of an equivalent DIP [1].**

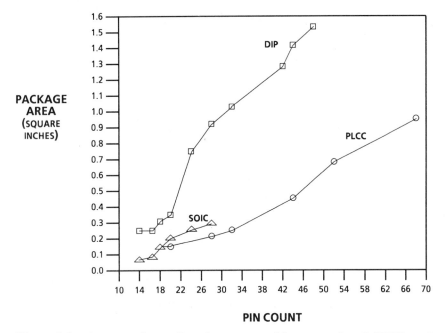

**Figure 1.9** **A comparison of package area with conventional (DIP) and surface mount (PLCCs and SOIC) packages.**

their conventional counterparts [1]. This causes a significant reduction in the weight of surface mount assemblies. Weight savings is especially important in aircraft and aerospace applications. Because of their smaller size, surface mount components occupy only about one-half to one-third of the space on the circuit board. Figure 1.9 compares the areas occupied by DIP, PLCC, and SOIC packages of different pin counts.

The size reduction in active components is a direct function of lead pitch, a term that describes the distance between the centers of adjacent leads. The active surface mount components have leads spaced at 0.050 inch (50 mil) centers or less. In other words, their lead pitch is at 50 mil centers or less. The packages that have lead pitches less than 50 mils are generally referred to as fine pitch packages. The lead pitches of commonly used fine-pitch packages are 33 and 25 mils, although 20 mil lead pitches are also used. For comparison, the lead pitch for DIPs is 100 mils. As the lead pitch gets smaller, the reduction in real estate requirement becomes more significant.

Size reduction is also a function of surface mount and through-hole component mix on the board. Since all components are not available in surface mount, the actual area saving on a board will depend on the percentage of through-hole mount components replaced by surface mount components. Depending on the component mix, the three types of surface mount assembly discussed earlier provide different levels of benefits. For example, the real estate savings are much smaller in a Type III SMT board, since only discrete components are replaced with surface mount counterparts. The real estate savings in a Type III board typically varies from 5 to 10%.

At the other extreme, the greatest real estate saving is accomplished in a Type I SMT board, which contains all components in surface mount. Since surface mounting also allows double-sided mounting, the real estate savings will be about twice as much in a double-sided Type I assembly. For example, half a megabyte (0.5 MB) of 256K dynamic random access memory (DRAM) with conventional DIP components will require about 4 × 4 inches of printed circuit board area. The same board with 100% surface mount components can accommodate 1 MB in a single-sided board or 2 MB in a double-sided memory board.

Figure 1.10 shows a 4 MB memory board (256K DRAM) in through hole and in surface mount. The surface mount boards, each with 2 MB of

**Figure 1.10   A comparison of board size for a 4 MB memory in through-hole and in surface mount using 256K DRAMs. (Courtesy of the Systems Group, Intel Corporation.)**

**Figure 1.11   An example of area savings is an experimental Type II SMT board in 1984. (Courtesy of the Systems Group, Intel Corporation.)**

memory, have components on both sides. If mounted on top of each using surface mount connector, as is the case in many Intel systems, they occupy only one-quarter the space. The surface mount connectors are gold plated and require selective gold plating on both the base and memory boards. The gold-plated pads are visible on both surface mount boards shown in Figure 1.10.

Type II SMT is a mixutre of through-hole and surface mount active (integrated circuits) components and passive components (resistors and capacitors). Since the availability of surface mount components has been constantly improving, so has the real estate savings for Type II assemblies. For example, Figure 1.11 (Intel SBC 286/10 board) shows that in a single-sided, mix-and-match board with approximately 50% surface mount components, the real estate savings possible in 1984 was about 20%. When the component availability in surface mount improved to 80%, the real estate savings jumped to 35% in 1986, as shown in Figure 1.12.

The two single-sided computer boards shown in Figure 1.11 and 1.12 were used for feasibility studies by the Systems Group of Intel Corporation.

Real estate savings for these two experimental Type II SMT boards vary from 20 to 35%, as summarized in Figure 1.13.

SMT also provides improved shock and vibration resistance as a result of the lower mass of components. Also, because of their shorter lead lengths, the surface mount packages have lower parasitics (undesirable inductive and capacitive elements) than through-hole counterparts, as shown in Table 1.1. This results in reduced propagation delays as shown in Figure 1.14 [1]. A reduction in parasitics also reduces package noise. For example, in a memory board with DIP memory devices on the top side and surface mount capacitors on the bottom side, about 25% noise reduction can be expected when compared with its conventional version with all through-hole mount components. As a result, only about half as many decoupling capacitors may be necessary to provide effective decoupling. Also refer to chapter 5, section 5.9.

In addition to the foregoing examples of design benefits of a surface mount assembly, SMT provides many manufacturing benefits. These include reduced board cost, reduced material handling cost, and a controlled manufacturing process. Routing of traces is also improved because there

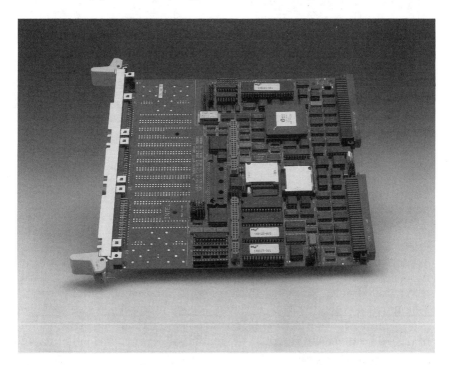

**Figure 1.12   An example of area savings in an experimental Type II SMT board in 1986. (Courtesy of the Systems Group, Intel Corporaiton.)**

THMT (THRU HOLE MOUNT TECHNOLOGY)

TYPE II SMT

1984 = *50*% SMT COMPONENTS IN
SURFACE MOUNT FOR iSBC 286/10

1986 = *81*% COMPONENTS IN SURFACE
MOUNT FOR iSBC 286/100

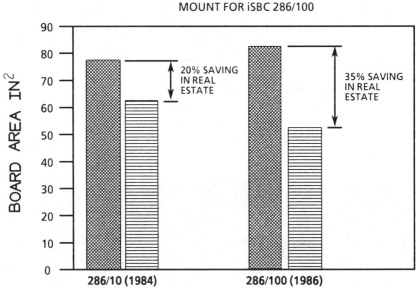

**BOARD TYPE**

**Figure 1.13   Area savings by different percentages of surface mount components in a mixed assembly.**

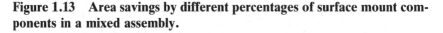

are fewer drilled holes, and these holes are smaller. The size of a board is reduced and so is the number of drilled holes. A smaller board with fewer drilled holes will naturally cost less. If the functions on the surface mount board are not increased, the increased interpackage spacings made possible by smaller surface mount components and reduction in the number of drilled holes may also reduce the number of layer counts in the printed circuit board, thus further lowering board cost.

Surface mounting is more amenable to automation than is through-hole mounting. It even reduces some of the process steps . For example,

**Table 1.1  Reduced inductance in surface mount devices**

| TYPE OF PACKAGE | INDUCTANCE (nH) |
| --- | --- |
| DIP, 14 pin | 3.2 − 10.2 |
| SOIC, 14 pin | 2.6 − 3.8 |
| PLCC, 20 pin | 4.2 − 5.0 |

different autoinsertion equipment (such as DIP inserter, radial inserter, axial inserter, and sequencer) are used for different through-hole mount components. This requires setup time for each piece of equipment. When SMT is used, the same pick-and-place equipment can place ALL types of surface mount components, thus reducing setup time. In addition, stock can be kept on the placement machine, further decreasing setup time.

What do all these manufacturing benefits mean in terms of cost? Depending on the application, the cost savings is 30% or better in placement time alone. Additional savings can be realized through the reductions in

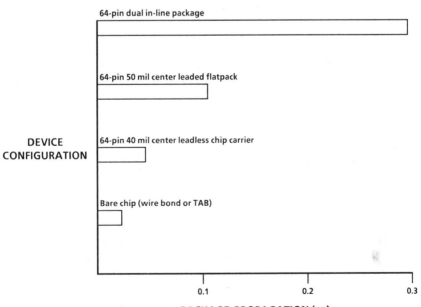

**Figure 1.14  Comparison of the propagation delays of surface mount and DIP packages [1].**

material and labor costs that are associated with automated assembly. Moreover, because repairs are fewer, less scrap is generated.

In through-hole assemblies, damage during repair is a major concern if the process variables are not properly controlled [2]. As discussed in Chapter 14, the repair of surface mount assemblies, contrary to popular belief, is easier and less damaging to the assembly than the repair of through-hole assemblies.

In SMT there are some less tangible benefits as well. For example, the pick-and-place machines tend to be extremely quiet compared with autoinsertion equipment, thus providing a less noisy work environment. Also, because surface mount components are smaller, the need for warehouse space can be significantly reduced. Smaller size and weight of components and finished assemblies also mean lower shipping and handling costs.

Does this mean that SMT assemblies always cost less? No. Actually, quite the reverse may sometimes be true—for a very good reason, as explained in Section 1.4 (When to Use SMT).

## 1.3   SMT EQUIPMENT REQUIRING MAJOR CAPITAL INVESTMENT

Through-hole and surface mount assemblies (Types I, II, and III) have three major process steps in common: component mounting, soldering, and cleaning. To understand the major differences in the requirements for equipment for through-hole and surface mount assemblies, let us refer to Figure 1.15, which lists the major pieces of equipment required for conventional through-hole mount (THM) as well as all types of surface mount assemblies. It is not the intent of Figure 1.15 to show the specific process sequence in different types of SMT, but only to highlight the major equipment requirement for different assemblies. We supplement the legend to Figure 1.15 by noting that for through-hole assemblies, insertion, wave soldering, and aqueous cleaning equipment are shown. The cleaning equipment may be solvent type as well. For Type III, only two additional types of surface mount equipment are required: the component placement machine and an adhesive curing oven. For the assembly using Type I SMT, none of the through-hole equipment shown in the double box is required.

A completely through-hole process and a surface mount process have nothing in common. For a Type I assembly, all the equipment shown in the single boxes of Figure 1.15 is required. If, however, the placement equipment for Type III SMT already exists, the additional equipment needed for Type I will be a solder paste screen printer, a reflow soldering

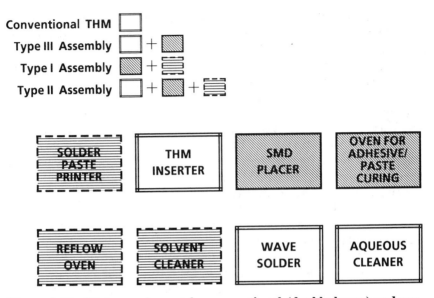

**Figure 1.15** **Major equipment for conventional (double boxes) and surface mount (single boxes) assemblies (Type II SMT).**

(vapor phase or infrared), and a solvent cleaner. If an infrared oven is used for the curing of adhesive for Type III SMT, it can also be used for the reflow soldering.

For Type II SMT, it is a different matter, however. All the equipment (single and double box) shown in Figure 1.15 is required. This means that if one has the capability for a Type II SMT, the capabilities for through-hole, Type III, and Type I automatically exist.

Let us briefly discuss the surface mount equipment shown in Figure 1.15. The details on this equipment are provided in later chapters. The through-hole equipment is also required for Type II and III assemblies, but their discussion is beyond the scope of this book.

## 1.3.1　Pick-and-Place Equipment

The pick-and-place machine is one of the most critical and expensive pieces of equipment in an SMT manufacturing line. The cost of this equipment can vary from $20,000 to millions of dollars. The most important, at least in the early phases of implementation, are the cheaper models for the lab or prototype line. They can cost from $20,000 to $100,000. They have limited capacity and are not very accurate (placement tolerances vary from 0.004 to 0.012 inch). The midrange pick-and-place equipment costs from \$150,000 to \$500,000. This is production equipment for good placement accuracy. It can be either flexible or dedicated.

Flexibility, however, comes at a cost; flexible equipment is generally slower in placement rate (1000–3000 components per hour). The placement rate specified by manufacturers may be higher if they do not take into account downtime and setup time. Since dedicated placement equipment can place only a limited variety of components, it is faster. The lack of versatility is compensated for by higher placement rates (5,000–10,000 placements/hour). Finally, there is the expensive equipment for very high speed placement. It can cost \$1 million to \$3 million and is intended for very high volume manufacturing lines. In each of these categories numerous pieces of equipment are available, and new models are continually being introduced. In selecting a particular type of equipment, current and future needs for volume and flexibility should be the deciding factors.

Some companies also use dedicated robots for placement. The equipment is generally very inexpensive; but the software and hardware development costs can be considerable. Robotic placement should be considered for components for which pick-and-place equipment is not widely available. Also, robotic placement implies that a company wants the ultimate in flexibility and has sufficient engineering resources to develop the necessary hardware and software. Examples of the placement equipment discussed in this section are shown in Chapter 11.

## 1.3.2   Solder Paste Screen Printer

A screen printer is needed to screen solder paste onto the printed circuit board (PCB) before placement of surface mount components. Screen printers have been widely used by the PCB industry for screening solder mask. This equipment has also been extensively used by the hybrid industry for screening solder paste. Different equipment, however, is used for the screening of solder mask and solder paste. The cost of screen printers can vary widely, depending on degree of automation and the size of board they can handle. Process and equipment for screen printing are discussed in Chapter 9.

## 1.3.3   Curing/Baking Oven

For Type III surface mount, components glued to the bottom of the board require curing before wave soldering. In addition, components placed on solder paste in Type I and Type II assemblies achieve reduced solder defects (such as solder balls) by being baked before reflow soldering. An oven is required for both adhesive curing and solder paste baking operations. The same or different ovens may be used for adhesive curing and baking of solder paste. If an infrared (IR) oven is used for reflow soldering, it is not necessary to have separate adhesive and solder paste baking ovens. The same IR oven can be used for adhesive curing, solder paste baking, and reflow soldering. The decision to combine solder paste baking, solder reflow, and adhesive curing in the same oven such as IR depends on volume requirements. It also means that a decision has been made to use IR instead of other reflow soldering processes such as vapor phase, laser soldering, or convection belt soldering.

## 1.3.4    Reflow Soldering Equipment

The selection of reflow equipment is the second major decision, after pick-and-place equipment has been chosen. As discussed in Chapter 12, there are many reflow soldering methods. Each has advantages and disadvantages, and none of them is perfect for every application. Equipment cost, maintenance cost, and yield are some of the leading deciding factors. The most widely used reflow processes are vapor phase and infrared. The vapor phase process is very versatile in that it can be applied to assemblies of any shape, and various models are available in both batch and in-line types. The capital costs of the vapor phase and infrared processes are comparable, but the operation cost may be higher for vapor phase. In addition to cost, many other technical issues that have an impact on frequency of solder defects should be considered in selecting a particular reflow soldering process.

Infrared equipment has been widely used in the hybrid industry, but at much higher temperature, for firing thick film deposits on ceramic substrates. The same equipment was initially tried for reflow soldering on glass epoxy substrates—with terrible results: burning, charring, and warpage of boards were fairly common. Now the industry has made tremendous progress in redesigning heating elements to eliminate earlier problems. Effective IR equipment generally applies nonfocused heat (more convection energy than radiation energy). Along with elimination of burning and warpage, this mode of heat application reduces the sensitivity of heat energy to color of components and boards. The near-IR process, which relies more on infrared radiation than on convection energy, also has been used very successfully for reflow soldering. The pros and cons of convection IR versus near-IR reflow soldering are discussed in detail in Chapter 12.

## 1.3.5    Solvent Cleaning

After soldering, the electronic assemblies must be cleaned to remove the flux. The subjects of flux and cleaning are closely intertwined. In reflow soldering, solder paste consisting of rosin flux requires solvent cleaners for effective cleaning. Companies that use rosin fluxes generally have either a batch or an in-line solvent cleaner. In such cases, purchase of solvent cleaner for surface mounting may not be required. However, if an aqueous cleaner is used for through-hole mount assembly, solvent cleaner must be purchased. The selection of solvent and cleaning equipment depends on the flux used and the cleanliness requirement. In addition, an international protocol signed recently by various nations including the United States establishes restrictions on the use of chlorinated fluorocarbons (CFCs) to prevent depletion of the ozone layer. For a detailed discussion on flux and cleaning, refer to Chapter 13.

## 1.3.6    Wave Soldering Equipment

As mentioned earlier, the discussion of equipment used for conventional through-hole technology is beyond the scope of this book. Components glued to the bottom of a mixed-technology board do require wave soldering, however, although the required wave geometry is different from that for through-hole components. In the most common wave geometry, the dual-wave type, one wave is turbulent and the other wave is smooth, as in conventional technology soldering.

The need for wave soldering equipment can also be met by simply retrofitting the conventional solder pot with the dual pot—at a small fraction of the cost of a new machine. A completely new machine is not necessary.

Another wave geometry that is becoming popular for surface mounting is a vibrating single wave known as the omega wave. Like the dual-wave geometry, it helps in reducing solder defects, as discussed in Chapter 12.

## 1.3.7    Repair and Inspection Equipment

Equipment unique to surface mounting includes inspection and repair/ rework equipment. Inspection equipment for surface mounting has not matured yet. There are two main types of automated inspection equipment on the market: optical, x-ray and laser. The effectiveness of these devices is questionable, however, since they are mostly still under development.

For this reason, most companies depend on visual inspection at 2 to 10X, using either a microscope or magnification lamp.

Repair/rework equipment uses conductive tips or hot air for removal and replacement of components. Conductive tooling, such as soldering iron tip attachments, is very inexpensive. The tips come in different configurations for different components but have limited applications. At a slightly higher cost, there are other conductive tips with built-in heating elements. At a much higher cost, hot air repair/rework tools are available that use nozzles of different sizes for components of different types and sizes. The hot air repair tooling is most commonly used for surface mounting because it is ideal for damage-free repair.

Even in-line (and expensive) repair systems are available on the market. However, when selecting repair/rework equipment, it must be kept in mind that repair/rework tooling does not need to be an automated, expensive in-line system designed for high volume repair. If the number of boards to be repaired is high, automated repair equipment is not the answer. Instead, the process should be controlled to keep the defect rate low. See Chapter 14 for details on repair, inspection, and process control.

## 1.4  WHEN TO USE SURFACE MOUNTING

With all the benefits SMT has to offer, it might seem foolish not to use it in every product. For many technical reasons, however, it is not appropriate for every application; even when it is, SMT requires a complete new infrastructure to implement it. The issue of infrastructure is discussed in Chapter 2. The major technical issues that must be overcome are discussed briefly in Section 1.5 of this chapter. The rest of the book is devoted to resolving those technical issues. Since the surface mount assembly processes cannot be performed manually, and since the processes are relatively high speed, thousands of assemblies can be built before the problem is detected. It has the potential to produce more scrap than the conventional through hole technology for the following reasons:

- The processes are incredibly high speed.
- They must be performed by machines.
- The equipment must be thoroughly characterized. Characterization can be defined as understanding all parameters that affect the equipment's performance.
- Vendors may say it is easy—it is not.
- Most large companies have assigned engineers to optimize; small companies "learn as they go."

- Learn as they go is not a real option, since revenue or product schedule or both may be adversely impacted.
- Expect to invest considerable engineering resources in process characterization, training, and documentation.

It is stated at the outset that SMT should not be considerd for an application if the performance, weight, functionality, and real estate requirements can be met by the conventional technology. One must not use SMT just for the sake of using a new technology. The SMT infrastructure requires considerable investment in human resources and capital equipment. One might ask, however, whether companies that have already invested in an SMT infrastructure, and happen to have conventional design and manufacture capability, should switch to SMT for all new designs. The answer is no; conventional boards of lower functionality will be cheaper than SMT boards of higher functionality. Thus if the needs of the marketplace can be met without jeopardizing the future, one should stick to conventional technology for such products.

Since the customer is buying the functionality, not the technology, one must carefully analyze the marketplace before deciding to use SMT. Some of the important questions that should be asked are: How much real estate is needed, and which type of SMT best meets those needs? Should Type I, II, or III SMT be used? Are all the components available?

Many components are not available in surface mount format. For example, many very large scale integrated (VLSI) and electromechanical devices, such as sockets and connectors, delay lines, crystals, and oscillators, are not widely available. Some connectors and sockets are available in surface mount but tend to take up more real estate than conventional devices, thus defeating one of the purposes of surface mounting. However, surface mount connectors do provide better routing efficiency, since they help reduce the number of via holes. In addition, some components have a cost premium, although this situation is improving rapidly. Because of cost premiums and the newness of the technology, SMT boards tend to be more expensive, especially in the beginning stages, when yields generally are much lower.

A surface mount board costs more than a through-hole board in most applications because through-hole boards are rarely redesigned just to achieve cost reduction. In most new designs, instead of reducing board size, designers put added functions on the same board. This increases cost per board, since the same board will contain more components, which constitute a significant portion of its cost. The board cost will also be higher, since a denser board with more functions than a through-hole board will require finer lines and spaces while maintaining or even increasing the layer count. If the technology requires the use of blind and buried via holes, the cost may be further increased.

However, the cost per function will be significantly less, since instead of designing and building two assemblies in through hole, only one surface mount board is needed. See Chapter 5 (Section 5.5: Cost Considerations) for more details on cost.

## 1.5 TECHNICAL ISSUES IN SURFACE MOUNTING

Since SMT is a new technology, the complete internal and external infrastructure needed to develop and sustain it does not yet exist. Part of the external infrastructure is the supply of engineers; but in addition the technology requires trained technical personnel, a mature vendor base for materials and components and equipment, and, above all, industry standards for these items.

Development of external infrastructure takes time, since the resources of a single company, no matter how large, are not sufficient. For example, few process, manufacturing, and design engineers are experienced in SMT. Learning about it has been, for the most part, a self-teaching process in larger companies with in-house R&D capability. Also, surface mount standards for components are lacking, although various organizations, such as the Electronics Industries Association (EIA), the Institute of Interconnecting and Packaging Electronics Circuits (IPC), and the Surface Mount Council are addressing the standards issues. It should be noted that the Surface Mount Council does not replace either the IPC or the EIA. It does not even write any specifications as the IPC or the EIA do. It is simply a management body initially sponsored by these two organizations and now supported by many others. Its main function is to coordinate the activities of industry and governmental organizations to promote SMT. In addition, every company needs to develop its own internal infrastructure such as documentation, training, and a manufacturing line. See Chapter 2 for details on developing internal and external infrastructure for an effective implementation of SMT.

Many technical issues in SMT also must be resolved before this technology can be implemented in a product. There are only two very simple differences between surface mount technology and through-hole technology. The surface mount components are smaller and they mount on the surface of the board. These simple differences, however, create technical concerns that affect every aspect of the design and process of building an electronic assembly. These technical issues in SMT are summarized in

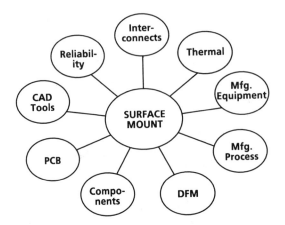

**Figure 1.16    Key technical issues in Surface
Mount Technology.**

Figure 1.16. Let us briefly discuss them, not to provide an answer but
simply to gauge the extent of the problem.

Let us start with a discussion of components. As far as the components
are concerned, the main issues are price, availability, and standardization.
These matters are interrelated and are addressed in Chapter 3. Another
major issue associated with components is package reliability.

Like conventional through-hole dual in-line packages (DIPs), surface
mount plastic components absorb moisture during storage. However, when
surface mount packages are reflow soldered at 200 to 240°C, the moisture
expands and may crack the package. The same plastic molding compound
is used for DIPs, but DIP packages do not crack during soldering simply
because during wave soldering, the plastic bodies of DIP packages do not
experience the high temperatures that surface mount packages undergo
during reflow soldering (vapor phase or infrared). In wave soldering only
DIP leads are heated to high temperatures.

Not all surface mount plastic packages are susceptible to package
cracking. Susceptibility here depends on package design, die size, strength
and thickness of the molding compound, and moisture content of the package. Baking of components before reflow solves this problem. See Chapter
5 for details. The ceramic packages used in military applications are not
plagued by such cracking because ceramic does not absorb moisture. However, the ceramic packages have their own problems.

Military applications require hermetically sealed ceramic devices
(leadless ceramic chip carriers or LCCCs), but these cannot be used on
glass epoxy substrates. To prevent cracking of solder joints, the coefficients
of thermal expansion of surface mount packages and the substrates should

match. Even when the X and Y CTEs of the substrates match, the CTE in the Z direction may be different. This may cause cracking in the via or plated through hole in the substrates. The danger is eliminated by selecting an appropriate substrate material. However, cracking of a plated through hole in the substrate is a matter for concern. Commercial applications eliminate solder joint cracking problems by using surface mount packages with compliant leads. See Chapters 4 and 5.

Let us refer again to the technical issues presented in Figure 1.16. Solder joint reliability, computer-assisted design (CAD) tools, thermal management, and interconnect design rules are also affected when SMT is used. For example, in conventional assemblies, the plated through hole provides a strong mechanical connection with plenty of safety margin. In surface mounting, the safety margin for the mechanical strength of a solder joint is not as great because there is no plated through hole.

CAD tooling for SMT is also one of the critical issues in SMT. The CAD system should have the capability for autorouting and on-line design rule checking. Most CAD systems are adequate for the Type III layout and for single- or double-sided memory boards, but not for the design of double-sided logic boards with or without blind/buried vias. By laying out pilot boards, one can determine whether the existing CAD system will be adequate or whether it must be replaced.

Since SMT makes it possible to increase the function per square unit of area, the power density per square unit also increases. If a thermal problem already exists on a board, SMT will compound it. For the increased functional density, SMT will also require finer line width and spaces for routing of boards. This increases the potential of cross coupling of signals. These issues are addressed in Chapter 5.

The land pattern design, which is unique to SMT is covered in Chapter 6. Here we note simply that in addition to proper selection of components and substrates, proper land pattern design is essential if reliable solder joints are to be provided.

SMT also requires good design for manufacturability (DFM) in Figure 1.16). This technology is not really suited for manual placement and soldering. As explained in detail in Chapter 7, designing for manufacturability is very critical for good yield.

SMT also has an impact on the manufacturing processes and equipment. In our earlier discussion of major equipment, it was noted that the technology is still evolving. Thus users are hesitant to invest in capital equipment for fear of obsolescence. For example, as late as 1986, vapor phase was the predominant method of reflow soldering. By 1987, infrared was widely used and is today considered superior to vapor phase for many applications. To further complicate the situation, there is a compatibility problem among different pieces of equipment in mechanical and electrical

interface. The industry group known as Surface Mount Equipment Manufacturers Association is developing standards to resolve this problem,
however, and the situation is improving.

Another major issue in manufacturing is testing. The straightforward
solution to testing problems—namely, to provide access to all critical test
points—will offset some of the increased density benefits of SMT.

In SMT manufacturing, issues of materials (adhesive, solder paste)
and processes (solderability, placement, soldering, cleaning, and inspec-
tion/repair) are new and unique to the technology. Among manufacturing
issues, adhesive, solderability, and cleaning are more critical than technical
problems in placement, solder paste, or soldering. For example, if adhesive
is cured too rapidly, voids generated during cure may absorb flux when
the assembly is wave soldered. This can cause serious cleaning problems,
especially if aggressive fluxes are used. The solutions to materials and
process problems are discussed in Chapters 8 to 14.

Reading this section, one may get the feeling that we are almost in a
development phase in SMT. This is not true. None of the technical issues
discussed above poses a serious impediment to using SMT as a production-
worthy manufacturing process. It is probably more accurate to say that
SMT is in a transition mode. Considerable progress in each of the issues
discussed above has been made. We are essentially talking about improvement and fine-tuning of the technology rather than any fundamental concern. To keep things in proper perspective, it should be noted that even
in conventional through-hole technology, not all the technical problems
have been totally resolved. For example, the industry is still struggling with
solderability, repair/rework, inspection, and solder defect problems even
after four decades of widespread use of through-hole technology.

## 1.6   TREND IN SURFACE MOUNTING

There are many benefits and concerns in surface mount technology,
but the benefits far exceed the concerns. Even though many of the issues
have not been totally resolved, the electronics industry, having recognized
the importance of SMT, is moving toward solutions. For example, various
industry organizations are addressing the surface mount standards issue.
Nor should the role played by through-hole technology in the future of
surface mount be neglected. For example, the through-hole technology
has matured to the point at which no further significant improvements in
cost effectiveness are possible. Moreover, its component tooling costs have
been amortized by the component suppliers, and new investments generally
are made in the SMT areas.

The suppliers of surface mount components have started to recover their investment in tooling for SMT, although this process has taken longer than forecasted. In 1986 Intel shipped 9% of its volume in PLCC. By 1990, the PLCCs' share of Intel's volume is projected to reach at least 50%. Even the conservative forecasters are predicting at least 10% annual growth in SMT.

The cost structure of SMT has nearly caught up with conventional technology. For example, most passive and memory devices are available at no extra cost. This is not surprising, because these devices are typically used in large numbers on electronics assemblies of all types. But logic devices also will catch up with conventional products very soon. Indeed, many vendors do not charge a premium for surface mount components over through-hole components.

Component availability in surface mount is improving. For example, as illustrated earlier in Figure 1.13, the space savings was only 20% in 1984. But by 1986 component availability had reached such a level that the space savings became 35% for a similar board. If a study were undertaken today, the real estate savings would be considerably more.

Other developments in surface mount technology are under way. For example, certain high pin devices are now available only in surface mount. Examples would be Intel 386SX microprocessor and many application specific integrated circuits (ASIC) devices. When the count exceeds 84 pins, they must be in fine pitch (lead pitch at 33 mils and under) as well as in surface mount. This trend is going to continue, resulting in finer and finer lead pitches, until eventually the industry will get rid of the package entirely and mount the die directly on the board, as discussed in Section 1.7 (The Future).

This is not to suggest that through-hole technology will disappear all together. It won't, just as hand soldering did not disappear, even after wave soldering became commonplace. Surface mount technology will complement, not replace, the conventional technology. Still, the decision to convert to SMT is a gigantic step. Many companies wonder when and how they should convert. This problem resembles the question faced by the industry when wave soldering began to replace hand soldering. Because of technical problems of icicles and bridging in wave soldering in the 1950s, the industry was really debating whether to replace hand soldering with wave soldering. Today it would be unthinkable to go back to hand soldering for volume production. Similarly, in the mid-1990s, the use of SMT will be widespread, and the process will be universally accepted for almost all applications.

The industry is in a transition phase. By 1990s the SMT infrastructure will be firmly in place and will be used by more than half of electronics production firms. More importantly, companies using SMT will improve their productivity and responsiveness to changes in the business cycle and will be involved in interconnection of high pin count devices that are more difficult to design, assemble, and test. State-of-the-art, cost-effective miniature products are easier to market. It is safe to say that SMT is not a fad; it is here to stay. For many companies, such as Intel SMT is not a technology of tomorrow; it is the technology of today. It keeps our customers on the leading edge of their business.

## 1.7    THE FUTURE

Surface mount technology is the current frontier of packaging high pin count components. Even though it provides substantial design and manufacturing benefits, its potential is far from being fully realized. Surface mount packages, as small as they may be compared with their conventional through-hole counterparts, can be shrunk in size considerably by decreasing their lead pitches (lead center to lead center spacing). The lead pitch of current surface mount packages is at 50 mil (0.05 inch) centers. Fine pitch packages (33, 25, and 20 mil pitches) are also available for better real estate efficiency. This trend could lead us to 10 mil center packages. Real estate requirements on printed circuit boards can be further reduced only by mounting the die (singular) or dice (plural) directly on the substrate: the so-called chip on board (COB) technology.

Since the die is considerably smaller than the packages of any lead pitch, the real estate savings provided by COB can be substantial. Table 1.2 summarizes the real estate occupied by packages of four types.

The COB technology has been widely used in consumer products, especially in Japan [3]. The Japanese have essentially dominated this type of consumer market, but COB is also being used in this country (for example, in calculators made by Hewlett Packard). The driving force for COB is the need for miniaturization (size and weight) in products such as smart cards, wristwatches, cameras, calculators, and thermometers, just to name a few. One of the requirements for this technology is high volume, because very few devices are used per board. Since specialized tooling and

**Table 1.2    Comparison of areas occupied by various packages [3]**

| PACKAGE TYPE | DIP 24 | SOP 24 | CC 24 | TAB 24 | BARE CHIP |
|---|---|---|---|---|---|
| Area (mm²) | 472.44 | 157.7 | 124.99 | 50.12 | 37.12 |
| Relative area | 12.73 | 4.25 | 3.37 | 1.35 | 1 |

equipment are required, volume must be high to justify the capital expenditures.

There are two main concerns about COB technology: the availability and reliability of dice. Unless a company is vertically integrated, it must rely on the very limited supply of bare chips that is available from outside semiconductor suppliers. Semiconductor suppliers are reluctant to ship bare dice. One reason is that they do not like to disclose the yield data on their wafers. Also, the loss of profit margins in selling packaged dice is a deterrent. Thus close cooperation between the supplier and the user is necessary for the success of COB technology.

Since a bare chip cannot be tested before placement and attachment to the substrate, enhancing the yield of an assembled board can pose a serious problem. The yield of the assembly decreases dramatically as the number of chips per board increases. This can be explained by using a simple probability theory. When $n$ chips are mounted on a board when the yield of assembling an individual chip is $Y_p$, the total yield ($Y_t$) of the assembly can be expressed as follows:

$$Y_t = (Y_p)^n$$

Figure 1.17 plots total yield ($Y_t$) against total number of chips ($n$) on the substrate. As is obvious from Figure 1.17, for a small decrease in the assembly yield of an individual chip (from 95% to 90%), the final yield of the assembly drops rapidly (from 60% to 35%) for 10 chips per board. If the number of chips per board is higher, the decrease in total yield is drastic. Improved passivation technology has increased chip reliability, but they can be damaged easily during shipping and handling, which can affect the final yeild. The above formula can also be used to predict the yield during placement or soldering where $n$ is the total number components to be placed or soldered.

There is also some concern about the long-term reliability of the assemblies because they no longer enjoy the protection provided by the packages. Silicon or epoxy resin encapsulation is applied and then cured over the chips after assembly. But these blob-type encapsulants are not adequate for high performance and high wattage devices (>1 W per chip). For this reason, their application should be considered only where field reliability requirements are not stringent. Repair of bare chips is also a matter of concern.

The COB technology can be divided into three main categories: chip and wire, tape automated bonding (TAB), and flip chip. These are discussed next.

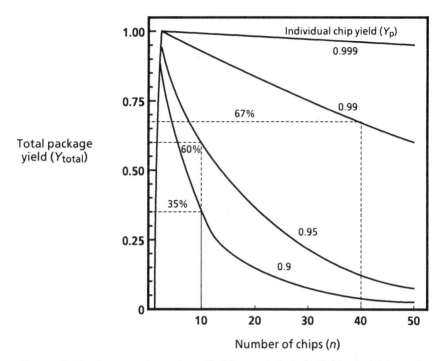

**Figure 1.17 Impact of number of chips and their individual yield on the total yield of the assembly in chip-on-board (COB) technology.**

## 1.7.1 Chip-and-Wire (Chip-on-Board) Technology

In chip-and-wire technology, also referred to as chip on board (COB), a bare integrated circuit chip is attached to any substrate, including glass epoxy (FR 4) substrate. The stress generated by the coefficient of thermal expansion mismatch between the bare silicon chip (CTE of nearly 2.6 ppm/°C) and the FR 4 (CTE of about 14 ppm/°C) is not of concern. The ductile wire leads are compliant and can take up the stress without breaking.

In the schematic view of COB and tape automated bonding (TAB) presented in Figure 1.18, note the difference in height requirement for COB and TAB. TAB is discussed in the next section. The assembly process for COB is as follows.

Before IC placement, a conductive adhesive (generally silver epoxy paste) is screened on the substrate; it is cured after IC placement. After adhesive cure at about 150°C, die bond pads are wire bonded to the pads on the substrate.

The die is placed using a pick-and-place machine, as in hybrids. It is presented to the machine either in waffle packs (plastic trays segregated into squares or rectangular cavities) or in wafer form. In waffle packs, each cavity holds one die. Only good dice are packed in the waffle packs. Sometimes the entire silicon wafer is supplied, and the user cuts it into individual dice and presents them to the pick-and-place machine. The advantage of this method is minimization of scratches and other forms of damage to the silicon chips during shipping. When using wafers, the user or the supplier must mark the bad dice. This means that the supplier must agree to divulge his wafer yield and the user must have dicing or sawing capability.

Chip-and-wire technology affords reductions in cost and in real estate.

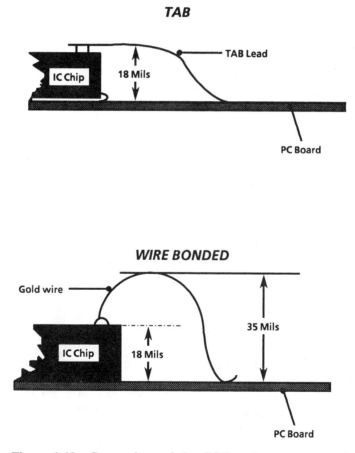

**Figure 1.18   Comparison of the COB and tape automated bonding (TAB) technologies.**

The basic cost reduction is realized because bare IC chips on substrates are used. There is no cost for packaging of the die. Since the silver paste used in chip and wire is very expensive, solder paste is substituted for silver paste to reduce cost in some applications. For using solder paste, however, a film of gold is required on the backside of the die, which tends to offset any cost saving. This also puts an extra burden on board manufacturer, who is not accustomed to plating dense, wire-bondable gold on a printed circuit board.

### 1.7.2   Tape-Automated Bonding (TAB)

The main advantage of TAB over COB technology is that TAB provides lower profile on the PC board as shown in Figure 1.18. Moreover, since wire lengths are longer in COB, the lead inductance tends to be about 20% higher [4]. This makes TAB better than COB for good electrical properties, especially at high frequencies. TAB is also a faster process, since inner (or outer) leads can be gang bonded.

Gang bonding is typical of the processes used to bond inner leads. However, outer leads are generally bonded either individually, one side at a time, or all four sides simultaneously. Gang bonding decreases bonding time and cost. TAB also makes possible denser interconnection, since it allows bonding to 2 mil pads on 4 mil centers as compared to 4 mil pads on 8 mil centers in COB [4]. Because of this feature, if TAB instead of a standard wire bonding process is used, even the die size can be shrunk for many devices. Most important of all, TAB when packaged in individual carriers, allows pretesting and burn-in of devices before assembly. This increases product reliability and reduces cost, since there is less rework.

TAB can be packaged in either individual carriers like camera slides (to allow pretesting) or on a reel (no pretesting possible). A tape is used in either case which acts as the transport medium for the die. The typical TAB tape in 35 mm format shown in Figure 1.19 resembles a movie tape. Tapes come in single, double, and triple layers. The conductors are etched on the tape to match the inner and outer lead pads. The tape is unsupported in its center to accommodate the die. For the inner lead bonding (attaching the chip to the tape), two methods are used: bumped tape automated bonding (BTAB) or bumped die. The bumps elevate the leads from the die and prevent shorting to the edge of the die.

Copper, solder, and tin are used for bumping. In bumping a die, die pads are plated with gold. Bumps on tape are formed by etching a thicker layer of copper so that a bump is left on the lead tip. Then the tip is coated with other metal layers. If a thermocompression bonding process is scheduled, gold is used on the tape and on the die pads.

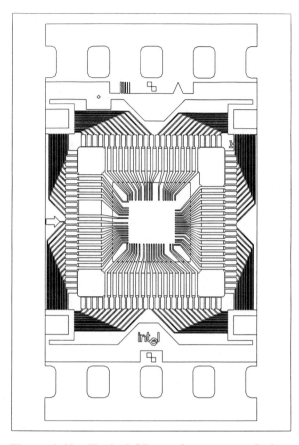

**Figure 1.19    Typical 35 mm format tape design for TAB. (Courtesy of Intel Corporation.)**

If bumped dice are used, the chip supplier electroplates gold pads on the die before sawing the wafer. The gold bumps on IC chips compensate for some nonplanarity during inner lead bonding. Bumped chips require extra steps during chip fabrication, provide additional hermeticity and reliability, and hence command higher prices. If the bumped tape (BTAB) process is used, bumping of tape is done by the tape supplier; the die resembles the die used in chip and wire.

The inner lead bonding is either high speed solder or a thermocompression bonding process. In the solder process, an intermetallic tin-gold bond is achieved between the gold bumps on the chip and tin-plated tape leads at relatively low temperature (284°C) and pressure (40–75 g per lead) [5]. A typical thermocompression cycle includes preheating of the chip at 200°C before inner lead bonding at about 500°C for 1 to 3 seconds at about 100 g per lead force, for a lead 3 mil wide and a 4 by 4 mil bump. The pressure is very critical at the interface area.

The outer lead bonding process in TAB is similar to the reflow soldering operation in SMT. For outer lead bonding, the thermocompression bonding or conductive adhesive bonding process can also be used. In thermocompression bonding, the functions of the placement and soldering machines are combined in one piece of equipment. Such equipment excises the outer leads from the tape, aligns the tape visually on the PC board, and then solders it to the board by thermocompression bonding.

## 1.7.3    Flip Chip or Controlled Collapse Bonding

The flip chip process provides the ultimate densification possible, since it eliminates both the bond wires and the packages. The chip is turned or flipped over (hence the name) and mounted directly on the substrate. The interconnection medium is solder pads on the die and the substrate. Figure 1.20 is a schematic representation of a flip chip on a substrate. The most widely known application of the flip chip process is the IBM's thermal conduction module containing up to 133 chips. The flip chip process requires very tight control, but it can be significantly faster than any other COB technology. Of course, the placement speed depends on the type of placement equipment, as is the case in SMT.

There are other side benefits to the flip chip process. Since the solder pads can be provided at any location on the die, a die design that is very efficient with respect to die size can be used. The flip chip process is a better reworkable process than either TAB or COB, but flip chip is not a truly chip-on-board technology, since the "board" must have a coefficient of thermal expansion value compatible with that of the silicon chip to prevent solder joint cracking. This problem is similar to those faced by the military when using leadless ceramic chip carriers on incompatible substrates. The CTE mismatch problem in flip chip is less critical in comparison, however, since the silicon die is much smaller than an LCCC. The smaller the die size, the less severe is the stress on the solder joint due to CTE mismatch between the die and the substrate. Solder joint stress due to CTE mismatch can be futher reduced by using a silicon substrate.

The flip chip process has some similarities to the SMT processes and also some differences. The placement, soldering, and cleaning processes are identical; thus one is concerned about process issues similar to those found in SMT—but at a miniature level. The differences lie in the fact that the manufacturing environment resembles a semiconductor operation rather than a surface mount line. So one is likely to see this process only in a semiconductor company.

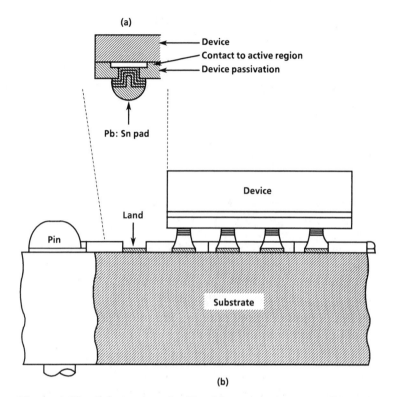

**Figure 1.20    Schematics of a flip chip on a substrate. (Courtesy of Intel Corporation.)**

There are six major process steps in a flip chip process. First is preparation of solder bumps on dice and substrate. This is a critical operation, and the solder bump size and diameter must be tightly controlled. It is because of this preparation step that the process is also referred to as the controlled collapse process. That is, the name reflects the need to maintain an appropriate gap between the chip and the substrate. This is done by depositing in vacuum a controlled amount and geometry of solder on dice before the wafer is sawed. The composition of the solder is also tightly controlled.

The next step is to do a wafer sort (reject bad dice), saw the dice, and package them in a transport medium (generally waffle pack) that can be presented to the pick-and-place machine. The placement machine is very similar to the ones used in SMT, but it has a vision system and tighter placement accuracy. Before the placement of chips, a rosin (Type R with no halides) flux is applied on the substrate to provide tackiness and to prevent chip movement. The assembly is reflow soldered. It is cleaned, inspected (usually by process control and destructive pull testing on a sampling basis since solder joints cannot be seen), and encapsulated. Thus the manufacturing issues in flip chip (tackiness of flux, accuracy of placement, part movement and reflow soldering profiles, and cleanliness) are very familiar to people working in the SMT field.

## 1.8   SUMMARY

Surface mount is a packaging technology that mounts components on the surface of a printed wiring board instead of inserting them through holes in the board. It provides state-of-the-art, miniature electronics products at reduced weight, volume, and cost.

SMT is rooted in the technology of flat packs and hybrids of the 1950s and 1960s. But for all practical purposes, today's SMT can be considered to be a revolutionary technology, and an entirely new internal and external infrastructure is necessary to develop and sustain it.

It is prudent for almost every electronics company to start developing an SMT infrastructure. Companies that do not do so will be the laggards of the 1990s. SMT has provided a quantum jump in the manufacturing technology, much as occurred in the early 1950s, when the industry was debating the switch from hand soldering to wave soldering. Another change of this magnitude in the manufacturing technology from SMT to the next leading technology—such as chip on board (COB), tape automated bonding (TAB), or flip chip technologies for mass application—is unlikely to occur very soon.

Certainly many technical issues must be taken into account and thoroughly understood before converting to SMT. The latter chapters will discuss the details of design, materials, processes, and equipment issues for both military and commercial applications to take advantage of this emerging technology.

## REFERENCES

1.   Technical Staff, ICE Corporation, Surface Mount Packaging Report, *Semiconductor International,* June 1986, pp. 72–77.

2.   Prasad, P. "Contributing factors for thermal damage in PWB assemblies during hand soldering." *IPC Technical Review,* January 1982, pp. 9–17.

3. Nakahara, H. "Japan's swing to chip-on-board." *Electronic Packaging and Production,* December 1986, pp. 38–41.

4. Schwartz, W.H. "Chip-on-board: Shrinking surface mount." *Assembly Engineering,* March 1987, pp. 32–34.

5. IPC-SM-784. Guidelines for Chip-on-Board Technology Implementation. March 1987, IPC, Lincolnwood, IL 60646.

# Chapter 2

# Implementing Surface Mount Technology

## 2.0 INTRODUCTION

Insofar as the internal and external infrastructures needed to sustain surface mount technology are not yet fully developed, SMT is in a transition phase. Clearly a good internal infrastructure (process and design rule development, validation, documentation, and training and development of manufacturing capability) is necessary to develop and sustain the technology. The internal infrastructure is especially important, and it is management's task to understand the technical issues and put the resources in place to develop it.

In this chapter we briefly discuss the major process and design issues in SMT and explore ways to resolve them. We highlight the technical SMT issues from the managerial standpoint necessary to put together a plan for successful development and implementation of SMT with a minimum amount of risk. The focus of this chapter is on the management apsects of SMT. The details of the technical issues are addressed in various chapters in the rest of the book.

With the technology in a transition phase, is use of an outside assembly house a better option? When the technology matures, the "right" equipment can be purchased, thus avoiding obsolescence. What are the pros and cons of this approach for the specific needs of a particular market? What is a feasible time frame for full implementation of SMT based on business condition, equipment lead time, and the level of maturity of internal and external SMT infrastructure?

Finally, management must decide whether to implement this technology all in a single phase or in various phases depending on the type of SMT (I, II, or III), the stability of the technology, and the available capital. In this chapter we take these issues into account in discussing the effective implementation of SMT.

## 2.1   SETTING THE IMPLEMENTATION STRATEGY

SMT is relatively new. Indeed, it is commonly referred to as a leading edge technology, and almost every aspect of design and process in building an electronic module with SMT is pushed to its very limits. SMT requires new design rules, materials, processes, and equipment. So it is highly critical that the surface mount board be packaged in such a manner that it is not only electrically, thermally, mechanically, and environmentally reliable, but producible with zero defects. This is definitely a difficult goal, considering the reality that many companies are still struggling with excessive amounts of defects in conventional through-hole mount assemblies that have had the benefits of design and process refinement for decades. Therefore, achievement of denser, more reliable SMT assembly with zero defects is a difficult goal, but certainly worth aiming for.

Many engineering issues (see Figure 1.16 in Chapter 1) in both design and manufacturing SMT assemblies need to be resolved. The technical issues fall into two major categories: design and manufacturing, as briefly discussed in Chapter 1 (Section 1.5). Here we briefly review some of the design and manufacturing problems that the management must understand and overcome by putting the appropriate resources in place.

Because of differences in types of packages, processes, and solder joints for SMT, the reliability window through which conventional boards pass is inadequate for SMT. Stress in solder joints is introduced by mechanical handling, vibration, soldering processes, and temperature fluctuations in service. Thermal and mechanical cycling tests for solder joint reliability should be defined. This will help determine optimum pad sizes of active and passive components that allow a maximum number of traces between pads without compromising solder joint reliability.

Increased interconnect density in surface mount boards, resulting from using finer lines and spaces, may potentially increase signal cross-talk levels. Plans should be made to evaluate the placement of critical traces in the ground and power planes. In addition to providing the design rules for reducing cross-talk, one should also qualify vendors with capability for producing fine line boards. Design rules should be established and implemented in work station tools for circuit simulation capability. Refer to Chapters 3 to 7 for details on the resolution of design issues.

Manufacturing and quality assurance organizations should be concerned with the issues of design and process compatibility, and control of machine and nonmachine variables, to achieve zero defects in SMT assemblies. There are standards for some processes and none for others. Even where standards exist, they may not be adequate, as in the areas of solderability, cleanliness test methods and requirements, and shipping container specifications. Also, new SMT equipment and processes are continually being introduced in the marketplace. Other conditions to be considered include the following: (a) the design of manufacturing equipment is constantly changing, (b) the "best" soldering methods are still being debated, (c) the equipment vendor base is slowly maturing, and (d) it is hard to predict which vendor will still be around few years from now to service the equipment. It is not easy to weave through these issues and prevent equipment obsolescence.

Most of the materials and processes for SMT differ also from those of conventional technology, even as they play a critical role in SMT. The first goal should be to establish the generic materials and process requirements that should not depend on any specific equipment purchased and installed by manufacturing. The requirements should be based on determining the critical variables in adhesives, fluxes, pastes, solderability, soldering, cleaning, and repair/rework, and their impact on reliability and manufacturability. The technical details of materials, process, and equipment are the focus of Chapters 8 to 14.

Any materials and process requirements should be equally valid for in-house manufacturing and an outside assembly house. The issues related to in-house manufacturing versus outside assembly house are discussed in detail in this chapter.

Given the technical issues involved in the design and manufacturing of SMT assemblies, where should one begin? The most complex type of board has the biggest payoff in terms of real estate and cost per function savings, but it is not the place to start. For example, the easiest types of surface mount board should be introduced first, before moving on to more difficult types. It becomes very hard to come up with answers if one tries to resolve all the issues for the most difficult type of SMT, namely, double-sided logic board. For example, component availability, CAD layout, thermal management, and test issues make it very difficult to design and build a double-sided logic board as the first product. It will become even more difficult if fine pitch packages (25 or 20 mil center lead pitch) are added on the board. The strategy for implementation of SMT should satisfy four major requirements:

1.   The product requires the inherent advantages of SMT.
2.   The packages are available.
3.   The assembly and testing capabilities are ready.
4.   SMT is introduced only to the degree necessary.

That is, SMT should be implemented in a logical manner in various stages. It should be an evolutionary and not a revolutionary process. A gradual introduction of technology allows development of coordinated approach for design guidelines and CAD tools, packages (availability, style, reliability), assembly (process, equipment), and testing.

It is fortunate that SMT is modular, hence can be implemented in various phases. The suggested phases are Type III SMT, Type I memory, Type II logic, single-sided Type I logic, Type II/Type I with fine pitch packages, and finally double-sided Type I logic. Once a decision has been made to install SMT capability, Type III SMT is the logical first step for a company before making a full commitment to SMT. Type III entails the lowest possible risks in both technology and capital investment. Capital investment is minimum because the pick-and-place machine and the adhesive dispensing and curing apparatus are the major pieces of new equipment required for a Type III SMT line. It impacts only components and their placement in a significant way.

If the need for high density memory exists, it is also relatively easier to implement this technology, especially if the capability for Type III exists. It is a different matter for Type II SMT, however. The design and assembly of a Type II logic board is more complex than that of either Type III or Type I memory. The Type II logic boards also tend to have many programmable logic arrays. They require manual handling for programming, which tends to cause lead bending (coplanarity). They also require two soldering processes (reflow and wave soldering) and two cleanings.

Using Type I double-sided logic calls for the most complex technology. Even though the manufacturing technologies for both memory and logic products are the same in Type I, the design and test technologies are very different. Also, due to increased interconnect routing requirements, Type I double-sided logic boards generally require blind and buried via technologies for the printed circuit board. The vendor base for such a technology is not mature at this time. In addition, since not all components are available in SMT, a full surface mount logic board is generally not feasible. For this reason, most new designs tend to be Type II SMT for most companies that have access to full SMT capability.

One certainly has the option of introducing all three types of SMT products simultaneously, provided the resources exist to resolve all the SMT issues and build a stable infrastructure. This not only requires substantial investment in capital and personnel but poses a higher risk.

Chip-on-board, tape automated bonding, and flip chip technologies should be attempted last. These approaches have many similarities to surface mount but require entirely different design and manufacturing strategies, as indicated in Chapter 1.

A gradual progression into different levels of surface mount technology allows the advantage of closely managing its introduction, while spreading the costs of such a large program over several years. This approach also helps build a storng infrastructure to support the technology.

## 2.2 BUILDING THE SMT INFRASTRUCTURE

Both external and internal infrastructure are necessary for successful implementation of SMT. There is not too much an organization can do about the external infrastructure, such as components, design, and manufacturing equipment availability. These factors are mostly driven by the market conditions and will take time to change. Other external infrastructure issues, such as component, process, and design rule standardization, are set by industry organizations. Active participation in standards organizations by both user and supplier member companies helps develop meaningful standards that benefit all concerned.

## 2.2.1 Developing Internal SMT Infrastructure

The key to a successful SMT implementation, above all, is formation of an SMT team with the full support of top management. The team member should be drawn from functional organizations such as packaging, R&D, CAD, manufacturing, testing, quality, components, reliability, and design engineering. The team should be headed by a program manager with both technical and program management experience.

The first task of the program manager should be selection of team members with specific technical skills to address the issues summarized previously (Figure 1.16). The skill set needed may not be specifically in SMT. The first task of the team should be to write an SMT program plan that addresses various technical issues and establishes major milestones. The plan should be approved and supported by management. The program manager needs the full support of the managers of technical personnel that are part of the team.

Before writing the plan, the team members should visit other companies and attend conferences to familiarize themselves with SMT issues. There are some professional conferences and even societies that are targeted at SMT. Private (and expensive) SMT courses are available in the industry. Such courses tend to be general and may not address the specific needs of a company, but they may be a good place to start to learn about the technology.

Reading published material (including this book) and attending outside training classes and conferences are no substitute for hands-on experience. As Dr. Charles Hutchins formerly at Texas Instruments has pointed out, learning about SMT is a lot like learning to ride a bike. You can read about it and talk about it, but unless you ride on it and fall a few times you are never going to learn to do it. The same is true for SMT, or any other new technology for that matter. An SMT plan is not realistic if it does not take into acocunt a few mistakes. The goal should be to prevent fatal mistakes, not all mistakes.

One of the key elements of a successful SMT implementation strategy should include, if possible, having an in-house SMT laboratory. This can be accomplished at a cost of less than $100,000. Whenever possible, the laboratory equipment and process should be compatible with the manufacturing equipment and processes to be selected for production. In the beginning stages, the lab serves many functions: it helps in minimizing the risk in SMT implementation, it provides prototyping capability for engineering and test boards, and it plays a very important role in providing hands-on training to in-house personnel.

SMT lab helps prevent expensive mistakes. For example, it is very useful in developing and characterizing the process that can help in the final selection of appropriate manufacturing equipment. The lab can also serve as a backup capability especially for prototyping actual products in the future. Even when a full-fledged manufacturing line has been set up, the lab can meet short-term prototyping needs when, for example, production equipment is down.

The next step in the implementation plan should be to develop and characterize some SMT processes. Test boards with varying design rules (different size and patterns, varying interpackage spacing) should be designed and built. They can be used for developing appropriate design rules and characterizing processes such as screen printing, placement, soldering, cleaning, and repair. The test boards can also be used to characterize materials, such as solder paste, adhesive, and solder mask.

There are two main guiding factors in selecting the final process and design variables: solder joint reliability and solder defects. The materials, process, and design variables that give the best results in terms of yield and reliabiiity for the desired application should form the basis for process and design specification. In-house materials, process, and design specification are some of the key elements of the internal SMT infrastructure. The workmanship standards should define the requirements, and the design and process specifications should establish rules to ensure that those requirements are met. Four different categories of specifications are required to control the quality and reliability of the final products.

1.    Material specifications for such surface mount materials as adhesive, solder paste, components, and PCBs are generally intended to control the quality of incoming materials. See Chapters 3 (components), 4 (substrates), 5 (baking of surface mount plastic packages before reflow to prevent package cracking), 8 (adhesive), 9 (solder paste), and 14 (work-

manship standards) for guidelines in developing applicable specifications. These specifications are intended for vendor use.

2. Nonequipment-dependent process specifications. Non-equipment-dependent process specifications are for internal use. They establish procedures to meet the requirements for producing assemblies with low defect rates and high quality. Examples include guidelines and rules for the application of adhesive and solder paste, and process parameters for placement, soldering, cleaning, and repair.

The process specifications establish the process "windows." As long as the process is maintained within the specified "window," the solder joints will be metallurgically sound, providing reliable solder joints, no matter what equipment is used.

3. Equipment- and process-dependent specifications. The specifications in this group, though also for internal use, are specific to the equipment used in manufacturing. They depend very heavily on equipment and will play the key role in achieving zero defects in the final product. Each piece of equipment should be extensively characterized to develop this specification. Some examples are: pressure nozzle size, and temperature to control adhesive dot size in a pick-and-place machine and individual panel settings, and conveyor speed to achieve a desired adhesive cure or solder reflow profile in an IR oven. The equipment-dependent specifications must meet the windows defined under the materials and process specifications.

4. Design rules and guidelines. The design guidelines and rules are for internal use also, and for vendors. No matter how good the materials and the process parameters are, if the board is not designed for manufacturability, zero defects can never be achieved. Design rules and guidelines are discussed in connection with system design considerations (Chapter 5), land pattern design (Chapter 6), and design for manufacturability, testing, and repair (Chapter 7).

The four groups of specifications listed above form the key to internal infrastructure. The next step is to ensure that they are followed both in letter and spirit; otherwise all the time devoted to writing those specifications will have been simply wasted. Such compliance can be achieved only with the support of top management.

Development of specifications and top management support are not enough, however. Effective SMT implementation calls for the development of in-house training courses in SMT manufacturing and design. There is another reason for having in-house training classes. Generally the specifications do not include the details of why they were established. In the training class, the backup material and rationalization for the rules and guidelines can be explained. If people understand the reasons behind the rules, enforcement should not be a problem. In addition, in-house classes are a quick and cost-effective way to train newly hired personnel and provide updates on changes in the technology. In-house training classes are directly applicable to the specific needs of the organization. In most cases they cannot be substituted for general outside training classes even at higher expense.

For successful implementation of SMT, the existence of a core group of in-house "experts" in SMT is also necessary but it is only a beginning. Unless process and design rules exist and personnel across all functional organizations are properly trained, there will not be a strong infrastructure to support and sustain SMT in production volumes.

### 2.2.2  Influencing External SMT Infrastructure

The external SMT infrastructure refers to component availability and standardization; maturing of design, process, and equipment technology; availability of experienced technical personnel; existence of a mature vendor base for fabrication and assembly of boards; and existence of industry standards on boards, materials, and process and equipment specifications. Standardization is almost mandatory for surface mount technology because of the need for automation. Indeed, adaptation of SMT has been slower than expected partly because of the lack of standards.

Standards make the market grow at a faster rate than it would without them. The proliferation of various packages offers an example. Users generally think that only the component suppliers are responsible for the proliferation of standards, but the users have contributed to the proliferation of standards by accepting packages any way they can get them, with or without standards.

A good standard benefits both users and suppliers. For example, if the package size tolerances are tightly controlled (within the requirements of the standard), the user can properly design the land pattern and program his/her pick-and-place machine. The user can use the same design for all suppliers of that package. The supplier also benefits because, as long as packages meet the standard, he or she can meet the needs of all customers. A good standard is a win-win situation for everyone.

There is much to be done in the international arena of standards, but much progress has been made in the United States. As far as surface mounting is concerned, two main organizations set standards in this country: the Electronics Industries Association (EIA) and the Institute of Interconnecting and Packaging Electronic Circuits (IPC), a technical society.

The IPC draws its membership mostly from the users, and it sets standards for printed circuit boards, assemblies, and interconnect design and manufacturing processes. Various IPC committees undertake many activities on surface mount land patterns, design guidelines, component mounting, soldering and cleaning methods, and requirements to promote surface mounting.

EIA is a trade organization of electronics component manufacturers. Various parts committees of EIA establish outlines for passive and electromechanical components and the JC-11 committee of the Joint Electron Device Engineering Council (JEDEC), which is part of EIA, is chartered to establish a mechanical outline for packages for active devices. Similarly, various "P" or parts committees of EIA are responsible for setting standards for the mechanical outlines of passive devices.

How are the standards set by IPC and EIA? Refer to IPC and EIA standardization flow charts in Appendix A. Let us further discuss the standardization process of EIA to clarify their process better. EIA maintains a very definite distinction between standards and registration. Both EIA and JEDEC committees essentially approve standards in the same manner. We will use the JEDEC procedure to illustrate the standardization process.

For JEDEC there are two levels of standards: JEDEC registered and true standard. Any manufacturer can submit an outline for registration. If two-thirds of the committee members vote in favor, the package is registered but is not considered a standard. A committee member need not be in attendance to vote, but all "no" votes must be accompanied by an explanation so that the committee may make an informed decision. The committees welcome constructive input, and an attempt is made to resolve all "no" votes.

To become a true standard, a package must be registered and all conflicts resolved. After the package has been registered, the proposal is submitted to a second group called the JEDEC council. If two-thirds of the council members approve, the package becomes a standard. Very few packages are true standards: LCCC, PLCC, and SOIC are some examples.

One note of caution about JEDEC/EIA standards: these documents should not be treated as procurement specifications. The standards tend to have loose tolerances and may not meet the requirements of manufacturing or process specifications. Nevertheless, in practice, many people use JEDEC/EIA standards as procurement specifications. For this reason, the responsible committees are attempting to tighten tolerances.

Some very large companies, by their sheer size, tend to dictate standards. But the task is too big even for them, and it is very important for manufacturers at all levels to get involved in the committees of IPC and EIA to set the direction for the electronics industry. By actively participating on the EIA and IPC committees and subcommittees, member companies can influence standards. The committees generally welcome input from the various sectors of the industry to develop useful and balanced standards. Various surface mount standards applicable to SMT are shown in Appendix A.

It has been a difficult process to coordinate the standards set by EIA and IPC. The activity of one group directly affects the other as far as surface mounting is concerned. For example, the setting of land patterns for surface mount (IPC-SM-782) by IPC is like laying out railroad tracks without consulting the manufacturer of railway cars. In 1986 IPC and EIA joined forces and formed the Surface Mount Council, to coordinate various surface mount standards. The Surface Mount Council does not replace either EIA or IPC and does not set any standards. It brings the users and developers of standards together for better coordination and points out new SMT issues that need industry focus.

The membership of the council consists of chairmen of EIA and IPC committees, and representatives from users, suppliers, equipment manufacturers, the Department of Defense, and the army, navy, and air force. Any company can become an associate member of the Surface Mount Council, which has as its objective the promotion of surface mount technology by promoting development of coordinated standards.

Other organizations, such as the International Society for Hybrid and Microelectronics (ISHM) and the International Electronics Packaging Society (IEPS) are working to resolve technical issues. In addition, organizations solely dedicated to SMT have been formed. The Surface Mount Technology Association (SMTA) and the Surface Mount Equipment Manufacturers Association (SMEMA) are working with the Surface Mount Council to promote surface mount technology.

## 2.3  SETTING IN-HOUSE MANUFACTURING STRATEGY

Setting manufacturing strategy is one of the key items to be decided on in the early stages of conversion to SMT. There are various options depending on budget and manpower resource constraints. One may decide to have a full surface mount line in-house or depend on an outside assembly house at least during the initial stages of SMT implementation. Both approaches have their strengths and weaknesses. Going to an outside assembly house will delay the capital investment but will make close coordination and feedback between design and manufacturing difficult. Close coordination, though important in complex logic board designs, may be less significant in a memory product.

In-house SMT manufacturing capability is good for the long term but it requires considerable investment in capital equipment and personnel.

The cost can vary from a few hundred thousand dollars to millions. The cost of the pick-and-place machine constitutes about half of the total. Refer to chapter 1 (Section 1.3) for a brief description of various SMT equipment.

Once the decision has been made to go ahead with the installation of an in-house manufacturing line, there are many variables to consider. The most important involve the types of product to be built, the maximum board size, fixed or variable form factors, product mix and volumes, and types of surface mount assembly. These variables will determine the types of equipment to be purchased and their total cost, as well as the lead time and the manpower resources necessary to select, purchase, install, and debug the line.

Other important item to consider is whether the line will be brought up for all types of surface mount at one time or in stages. The lead time on some of the equipment can be from 4 to 9 months after placement of purchase order, excluding the time for capital budget authorization and capital equipment survey and selection. Also, time must be allowed for installation and debugging of the line and training of personnel. Thus at least a year should be allowed to set up and debug a manufacturing line and about 6 months to set up a surface mount lab.

The alternative to installing an in-house manufacturing line may be to go to an established subcontract assembly house to gain quicker access to a manufacturing line. However, this route is not as quick as is often believed. Proper selection of a subcontract assembly house takes about 3 months for a conditional qualification and additional few months, depending on volume, for dock to stock that requires no incoming inspection or approved qualification. The details of selection and qualification issues of a subcontract assembly house are discussed next.

## 2.4   SELECTING AN OUTSIDE SMT ASSEMBLY HOUSE

With so many contract assembly houses looking for business, it may seem very tempting to offload manufacturing to them. But this means losing control to an outside assembly house for a technology that will be very critical in the future. Thus there are pros and cons to offloading manufacturing to an outside assembly house.

As indicated previously, going to an outside assembly house has the advantage of delaying the purchase of capital equipment. In many cases, the equipment selected for the in-house manufacturing may not have a track record of performance because only a very few have been installed in the field. This may be a significant consideration, especially when the technology is in a transition phase, hence the chances of purchasing the "wrong" equipment may be high. In addition, for many companies, the best way to get into surface mounting may be to start with a subcontract assembly house. Developing an in-house SMT infrastructure (design, manufacturing, training, and documentation) can cost millions of dollars. An SMT subcontractor can either supplement or substitute in-house manufacturing capability.

Many subcontract houses have sprung up across the country. Sales and marketing personnel usually give the impression that their SMT assembly capability is excellent. However, due to relative infancy and lack of experienced personnel, some of these subcontractors may not have the maturity necessary to build all types of SMT product *in large volumes and with consistent quality over time*. And they may not have skilled and experienced technical personnel.

To build surface mount board products requires not only the right type of equipment but also a solid infrastructure of skilled personnel, as well as extensive documentation and process control, as we discussed earlier. For instance, possession of sophisticated repair/rework equipment is not a ticket to trouble-free surface mount component and board repair. Successful repair and rework depend critically on trained operators. Moreover, proper characterization of the equipment for time and temperature variables for different components is important to prevent board damage. Characterization of process and equipment, which is necessary for higher yield, requires an investment in engineering resources that has not been made by all subcontractors.

## 2.4.1 Evaluation and Qualification of Subcontractors

Contrary to one common conception, a proper evaluation of a subcontractor during the qualification phase does indeed require considerable resources of the user. Sometimes even more human resources may be necessary than would be the case if the product were built in-house. Based on a personal experience it can be stated that some suppliers can be qualified very rapidly, but for others the process can take up to 2 years.

A thorough evaluation of the subcontractor will prevent product crashes at the outside facility when one can least afford them during the production phase. At the same time, an extensive evaluation puts the subcontractor on notice that he can enjoy the continued business only if he adheres to the user's quality requirements.

Selection and qualification of subcontracted assembly is a major task, and at least 6 months should be allowed for selection and qualification. First a strategy to assess and qualify the SMT assembly subcontractor must be put in place. The evaluation strategy, stressing the importance of the "partnership" between the subcontractor and the user, should also hinge on a methodology that rates the ability and maturity of the subcontractor in critical areas after specific periods of time and quantities of SMT boards manufactured. Based on the results of this rating, a qualification status can be granted to the successful candidate. Such a strategy will weed out immature subcontractors at the prototype stage rather than at the production stage, when the manufacturing capability is critical.

## 2.4.2 The Various Stages of Subcontractor Qualification

It is advisable for the user to form a team before embarking on subcontractor qualification. The procedure for subcontractor qualification should be divided into various stages, such as Preliminary Survey: Evaluation status, Conditional status, and Approved status.

The subcontractor qualification team should be staffed by personnel who have expertise in critical areas, such as technology, manufacturing, procurement, and quality assurance. All subcontractors under consideration should be surveyed over the phone to get a good basic idea about each one's capability and maturity. If a subcontractor is located in close proximity, the preliminary survey may involve a visit to the premises by one or two members of the team. Also, a visit to overseas subcontractors may be necessary.

On the basis of the data obtained over the phone or the initial visit, the qualification team should narrow the list of candidates. Any subcontractor not exhibiting sufficient capability and maturity to warrant an Evaluation status should be dropped from further consideration.

During the Evaluation stage, it is absolutely essential that the qualification team visit potential assembly contractors with prepared questionnaires. (Section 2.4.3 suggests questions.) The type of equipment the subcontractor has plays an important role in selection: if the equipment matches the current or planned in-house equipment, it may be easier to transfer tools and fixtures later on, if necessary.

The next step should be to order the building of some prototype boards to assess the quality of the board and process yield. This is very important because the design and process may not be compatible. The results will indicate where corrective action is necessary. A contractor who meets the yield and quality requirements should be put on the Conditional qualified status.

Now the list of potential subcontractors can be narrowed to one or two final candidates, depending on the subcontractor's capacity and the customer's volume requirements. Purchase orders for production lot quantities can be placed, and the defect rate and first-pass acceptance should be closely monitored.

A subcontractor who demonstrates consistently good results for various production lots should be put on Approved status for continued business. A subcontractor's stability is also essential for approved status. It is absolutely necessary to work very closely with the vendor to bring the firms to an Approved status. As mentioned earlier, an "Approved Status" implies no or minimum incoming inspection of the assembly.

## 2.4.3 Questionnaires for Rating of Subcontractors

During the various qualification stages, the subcontractor should be rated in four critical categories: technology, manufacturing, quality assurance, and business issues. The qualification team prepares the questionnaires and a checklist for each of these categories, to form the basis of the ratings given during each phase of qualification. In each area the types of questions should be such that the answers will permit a clear understanding of the subcontractor's capability, maturity, and viability. (Examples of questions for each area are discussed in the next four sections.) The minimum points required and the maximum points possible for each category during the qualification stages are established by the team.

Points can be assigned for various questions in each category. For example, during the Evaluation stage, technology should be of paramount importance. Even the best of manufacturing equipment or a strong financial position will not compensate for a subcontractor's failure to demonstrate sound technological know-how. Hence, the initial Evaluation stage should stress technological maturity to the fullest.

The qualification requirements get stiffer as the vendor moves from Evaluation to Approved status. For example, to get an Approved status, the subcontractor should not only demonstrate continuous improvement during the first few lots (minimum of three) but also should maintain the higher rating for several lots thereafter. A sudden decrease in any of the given areas should flag an impending problem and may require careful reevaluation by the user.

### 2.4.3.1 *Technology Questions*

To gauge the engineering experience of a potential subcontractor's personnel, the following questions may be asked:

- How many engineers are working on SMT process development?
- How many engineers are working on SMT production?
- What is the experience of SMT personnel?
- What is the average monthly volume of product being built for each SMT type?

The following questions could be used to gauge the subcontractor's understanding in materials and process development:

- What are the properties (compositions, metal content, particle size range, etc.) of the solder paste used?
- What is the solder application method (stencil or screen)? Why was this method chosen?
- What is the paste deposit thickness?
- What is the reflow soldering method used (vapor phase or infrared), and what are typical time-temperature profiles on the board surface and at a solder joint?
- Have the reflow processes been compared for manufacturing yields?
- What cleaning method and solvent are used?
- How is the cleanliness of product boards monitored?
- What sort of repair/rework equipment is used, and what are the time-temperature profiles for repairing each surface mount device type?

The idea here is to establish whether the subcontractor has done extensive process evaluation and whether he understands the importance of critical materials and process variables on product quality and reliability.

### 2.4.3.2    Manufacturing Questions

The questions that are to determine the manufacturing capability of a subcontractor should relate to the capacity and utilization of the manufacturing equipment, material control and handling methods, equipment operating procedures, equipment maintenance and calibration schedules, and operator training. Questions on these and similar matters should cover each type of equipment in use at the subcontractor's facility.

The type of equipment used by a subcontractor can give a good idea about the firm's capability. The manner in which the equipment is organized is a good indicator of the contractor's awareness of SMT. Good quality, in-line equipment is prefarable to batch equipment for automation.

Ask what other equipment was considered and why the firm bought a particular brand and model. Observe the performance of the equipment in operation. How often is it down for repair? How accurate is the pick-and-place equipment? How closely can the solder profile be controlled? Does the screen printer provide good and consistent print? How effective is the soldering and cleaning equipment in providing good and clean boards? Batch versus in-line cleaning for surface mounting should be closely investigated. With tighter spacings among components and lower standoff heights above board, in-line cleaning systems with sufficient nozzle pressure are very effective in providing boards free of solder balls and flux residues.

And finally, one must note that having the best equipment is no guarantee for excellent yields. Does the subcontractor know what the yields are? Is process control used to monitor and control defects? Have all the equipment and processes been characterized? Does documentation exist for operation and maintenance of all equipment?

### 2.4.3.3 Business Questions

Questions about business are aimed at determining the subcontractor's financial stability, long-term viability as a business, pricing policy, and so on. One typical point of debate between the subcontractor and the user centers around the question of which workmanship standards should be followed to ascertain product quality and acceptability. It should be decided at the Evaluation stage if the workmanship standard used will be that of the subcontractor, the user, or an industry standard, such as the IPC-815 workmanship standards. Also see Appendix A for various standards applicable to SMT. Other pertinent questions include the following:

- Does the subcontractor guarantee quality?
- Is the subcontractor going to be responsible for damaged boards?
- What is the subcontractor's policy for compensating the user for damaged boards and their potential impact on the user's product schedule, and what is the procedure for implementing this policy?
- Will the subcontractor guarantee a certain percentage of capacity in the future, or is the user likely to be dropped; should more lucrative business opportunities materialize?

After the subcontractor has reached the Approved status, an agreement should be drawn to "freeze" the recipe used by the subcontractor to manufacture the user's product, so that consistent quality is maintained over time. Some subcontractors will object to providing and freezing of the "recipe," which they claim as proprietary "know-how." This is really not a valid objection, however, because a finely tuned process such as solder profile may be product specific, hence not proprietary at all. The idea here is not to induce the subcontractor to give away proprietary information but to discuss the technical issues in the beginning and have a clear understanding even before Conditional qualification status is granted.

### 2.4.3.4    Quality Assurance Questions

Generally most of the problems from the subcontractor are encountered in the quality assurance area. Some sample questions for assessing whether the subcontractor can deliver *consistent quality over time* and whether the firm has good process control as well as documentation in place to meet commitments are given below.

- Are areas sensitive to electrostatic discharge (ESD) clearly identified?
- Do all personnel and visitors wear approved ESD protective clothing and wrist straps while in the ESD-sensitive area?
- Are all work surfaces, dissipative or conductive, grounded through a resistor per specification?
- Are conductive or antistatic containers issued to transport ESD-sensitive materials from one location to another?
- Are ESD-sensitive areas and containers periodically audited for conformity to documented ESD guidelines?
- Are documented workmanship standards maintained? (This is not an issue if the user's workmanship standards are to be used.)
- Is a controlled document control procedure in place?
- Is a minimum distribution list for new specifications maintained?
- Is there a controlled method to assure removal of previous revisions of documents?
- How is defect logging/archiving and failure analysis accomplished?
- Are work instructions, production equipment, and appropriate work environments in place to accomplish the major SMT processes?
- Are the workmanship standards available to assemblers and inspectors?
- Is a corrective action system maintained, and does it provide preventive measures also?
- Is statistical quality control (SQC) used in the production area? This may be the most important questions of all. Refer to Chapter 14 for a discussion on SQC.

## 2.5 MANAGING THE RISK: PILOT TO PRODUCTION

Taking risk is an integral part of developing competitive products using new technology. Even when an old technology is used, not all new products make it to the marketplace. The chances of survival with new technology are even lower because of added risk. The point is not to avoid a new technology like surface mount but to carefully assess the inherent risk before committing a product to it. One should not use a new technology for the sake of technology; rather, the goal should be to provide the functionality and quality to the customer at a low cost within the available market window. So the driving force in any new product should not be a new technology, but functionality or cost or both. In other words, the new technology must fulfill the customer's need.

Once it has been decided that surface mount is the most desirable way to meet the customer's needs and gain a competitive advantage, steps should be taken to carefully assess and manage the risks. Yet before the use of the new technology can be considered, the infrastructure for the new technology must be in place. What does this mean?

Certain basic questions must be asked. For example, are the components available and qualified? Has the design tool been validated? Are the manufacturing locations ready? (They could either be in-house or at a subcontract assembly house.) Has the documentation for process and design been released? Are trained personnel available? And finally, has any product been designed and assembled using this technology?

The risk will be the greatest if the product is committed before the technology has been validated on a nonproduct board, because the technology, manufacturing, and reliability problems will be unknown. For example, the following circumstances must be anticipated: (a) the first product may have to go through multiple redesigns, (b) the materials and process problems in manufacturing may cause very poor yield, and (c) the product may not be testable. Under any of these conditions, there may be an adverse impact on schedule and revenue.

To minimize the risk of introduicng a new technology, to manufacture a new product, development and validation of the technology should be conducted in phases. First, the material, process, and design issues should be resolved on nonproduct boards or technology pilot boards. The use of the nonproduct board serves to flush out technical problems in design, manufacturing, and testing and to assess the risk associated with the technology itself. The process and design variables that are found to affect the quality, reliability, and manufacturability of the board should be documented and incorporated in the process and design specifications.

Next the technology should be validated in a manufacturing environment on an actual product. Only a minimum number of products, preferably one or two, should be committed at this time because not all problems in a manufacturing environment can be foreseen. One limits the number of products at this time simply to validate the process and design rules in a produciton environment without incurring significant risks. Thus the company's revenue will not suffer significantly if serious problems are discovered along the way.

Although some technical problems are bound to emerge no matter how carefully the technology is implemented, the key concern should be to avoid major problems. For example, even when the technology is sound, the vendor base may be unstable, and there may be problems in the areas of receiving components and boards on time. This is very common. Thus time in the project schedule should be allowed for startup problems in manufacturing and vendor deliveries. After successful completion of the first round of real products, the technology can be released for additional new products.

A successful release of technology is defined as an actual product being built in a manufacturing environment at a very low rate of defects per unit and a high first-pass acceptance. There are no absolute numbers on defect rate and first-pass acceptance. A decision here depends on the product complexity and the sophistication level of the company. The defects per unit or the first-pass acceptance numbers should be comparable to those for products using conventional technologies on which the company has considerable experience.

If no problems are encountered, additional products should be introduced. The risk on the third and fourth products will be almost negligible. After a certain volume of products has been built at a predefined quality level consistently, widespread usage of the new technology can begin.

The approach just discussed should be repeated for each category of new technology boards. For surface mount, the technology should be divided into four categories for the purpose of minimizing risk. In ascending order of risk incurred, these are Type III logic, Type I memory, Type II logic (or Type I single-sided logic), and Type I double-sided logic boards. For manufacturing, if the capability exists for Type II logic board assembly, there may be no additional risk for any other category. For design, testing, thermal management, and component availability, the distinction of SMT in various categories is very important.

After some initial experience with successful releases, the level of difficulty in implementing the additional phases should be lower even for the more complex technologies. Two to three years should be allowed to develop and validate the full surface mount process and design capability.

An even longer period may be needed for Type I double-sided surface mount logic boards because of issues such as testability, CAD capability, thermal management, and above all availablility of all components (including crystals, oscillators, delay lines, switches, sockets, connectors, and peripheral devices) in surface mount.

Double-sided Type I logic products, depending on product complexity and density, may also require blind and buried vias or even resistive layers to satisfy increased interconnect routing requirements. The vendor base for such state-of-the-art printed circuit board technology is not mature today. These products are examples of external infrastructure-related issues, their development will take time. As the usage of the technology grows, these issues will disappear. Active participation in the standards-setting bodies discussed earlier will expedite the availability of these low usage devices in surface mount and will make Type I single- and double-sided surface mount assemblies possible.

Most important of all, good working relations with vendors will play a critical role in the management of risk and the successful implementation of new technologies.

## 2.6  SUMMARY

For the successful implementation of surface mount technology, the first order of business for management consideration is to obtain a clear outline of the outstanding technical issues. Then an SMT team consisting of a selected group of individuals from all functional organizations is assembled to address those issues. The team is headed by a program manager who not only is qualified technically but also has the full support of top management.

Because of such constraints imposed by the external infrastructure as component availability and the evolutionary nature of the technology, SMT should be developed and introduced in stages. This allows for gradual building of the internal infrastructure and the spreading out of capital investment over a longer period of time.

In some cases even subcontract assembly houses may meet the interim need and help postpone capital investment. But extreme care should be taken in the evaluation and seleciton of a subcontractor. The successful candidate must meet stringent technology, manufacturing, business, and quality requirements to prevent "vendor crashes" at a critical moment. Appropriate selection of a subcontractor requires considerable investment in human resources.

Even when the technology has been evaluated and debugged, it should be implemented on a few selected products first. Only when the technology has been validated on some "shared" technology and product pilots should it be used across product lines. Such caution is important in reducing risk and preventing multiple failures across product lines, which could adversely affect the company's revenue.

# Chapter 3
# Surface Mount Components

## 3.0 INTRODUCTION

Surface mount devices, active or passive, are functionally no different from their conventional through-hole counterparts. Thus the design and electrical function of an internal device is not unique to surface mounting, hence beyond the scope of this book. What is different in surface mounting is the packaging of those devices. Surface-mounted devices (SMDs) or components (SMCs) provide greater packaging density because of their smaller size.

Among many other benefits of surface mounting, the real estate saving is of paramount importance. The surface mount packages affect not only the real estate on the board but also the electrical performance of the device and the assembly. Moreover, due to component packaging differences, the parasitic losses such as capacitance and inductance in surface-mounted devices are considerably less than those for the through-hole technology.

The component packages, in addition to saving real estate and providing better electrical performance, serve many other functions. They protect the devices within them from the environment, provide communication links, remove heat, and offer a means for handling and testing.

Full benefits of surface mounting are not available today because of lack component availability and standardization. This lack of availability also tends to increase the cost of some surface mount components. SMT is in a transition phase, and new packages are continually evolving. Increased user demand will eventually improve component availability, cost, and standardization. Standardization of surface mount components has been achieved for components of some types but is lacking in others. In addition, standards for different package configurations for other devices such as crystals, oscillators, connectors, and sockets are still evolving.

It is safe to say that when it comes to component packaging, the world of surface mounting is much more complex than that of conventional through-hole mount technology. In this chapter we discuss commonly used passive and active components. We discuss packaging for shipping and handling and procurement specifications for surface mount packages.

## 3.1 SURFACE MOUNT COMPONENT CHARACTERISTICS

Certain characteristics are common to all the many surface mount package types. For example, they all mount on the surface of the substrate instead of protruding through the plated through hole, as is the case in through-hole mount technology (THMT) boards. This means that the solder joint, which provides both mechanical and electrical connections, is very important for reliability of the assembly.

The surface mount packages are designed to meet the requirements of two major types of application: commercial and military. The commercial applications with benign environments can use nonhermetic packages. The operating temperature requirements are generally from 0 to +70°C.

Military or high reliability applications designed for severe environments require the use of hermetic packages in the −55C to +125°C range. Of course there are other environments between these extremes for which the designer must choose appropriate packages to meet the reliability requirements.

The hermetic chip carrier is expensive and is intended for high reliability products. It requires substrates of matching coefficient of thermal expansion and is still prone to solder joint failure during thermal cycling.

The surface mount packages also see much higher temperatures during soldering. They must be designed with this requirement in mind. Because of their smaller size, it is very difficult to provide part markings on them, especially on passive components. If the devices do get mixed, they must be positively identified or thrown away. This almost mandates packaging of these devices to facilitate automated placement.

The packages also mount very close to the surface of the substrate, hence have relatively less clearance off the substrate. To achieve the required cleanliness, therefore, very good process control is necessary.

Since these devices do not have leads in the conventional sense, and since they experience lower temperature in soldering, the solderability test methods and requirements need to be different. (The surface mount packages undergo higher temperatures, as mentioned earlier, but the surface mount leads or terminations see lower soldering temperatures as compared with DIP leads.)

## 3.2 PASSIVE SURFACE MOUNT COMPONENTS

The world of passive surface mounting is somewhat simpler. Monolithic ceramic capacitors, tantalum capacitors, and thick film resistors form the core group of passive devices. The shapes are generally rectangular and cylindrical. The surface mount versions of these devices have gained wide acceptance because they occupy half the space when mounted on the top of the substrate. When mounted on the bottom of the substrate, as is the case in Types II and III SMT boards, they utilize the space otherwise not occupied at all. (See Chapter 1 for definitions of Type I, II, and III SMT assemblies.) The mass of these devices is about 10 times lower than that of leaded devices, and the existence of termination instead of leads provides design benefits of better shock and vibration resistance and reduced inductance and capacitance losses.

The use of surface mount resistors and capacitors has been very extensive in Japan ( consumer electronics) and in the automotive electronics industry in the United States. Because of this extensive usage, surface mount resistors and capacitors are less expensive than through-hole axial components.

The surface mount resistors and capacitors come in various case sizes to meet the needs of various applications. One should avoid both the smaller case sizes ($<$ 0.12 inch by 0.060 inch), to minimize pick-and-place difficulties, and the larger sizes ($>$ 0.25 inch by 0.12 inch), to avoid CTE mismatch problems on glass epoxy boards.

These devices come in both rectangular and tubular (MELF: metal electrode leadless face) shapes, with latter at lower cost but also some what lower reliability. Passive components generally do not have part markings and are very difficult to use for manual handling. Part markings are available but at a premium price. Thus automated placement using tape and reel packaging is highly recommended.

Wraparound terminations are preferred over flat terminations (no solderable terminations on the side) for good solderability, and plated terminations are preferred over solder-dipped terminations for accurate placement. Nickel barrier underplating, especially for wave soldering applications, should be required to prevent dissolution (leaching) of solderable coating in the solder bath.

Problems such as misalignment and tombstoning of parts during reflow soldering and leaching during wave soldering are of concern when passive surface mount devices are used. Solutions to these problems are addressed in Chapters 6 and 7 (design), 10 (metallurgy of soldering), and 12 (soldering). In the sections that follow, we discuss in further detail passive components of various types.

## 3.2.1 Surface Mount Discrete Resistors

There are two main types of surface mount resistors: thick film and thin film. Thick film surface mount resistors are constructed by screening resistive film (ruthenium dioxide based paste or similar material) on a flat, high purity alumina substrate surface, as opposed to depositing resistive film on a round core as in axial resistors. The resistance value is obtained by varying the composition of resistive paste before screening and laser trimming the film after screening.

In thin film resistors the resistive element is nichrome film that is sputtered on the substrate instead of being screened on. The construction details of a surface mount resistor is shown in Figure 3.1.

Figure 3.1 shows a resistive element on a ceramic substrate with protective coating (glass passivation) on the top and solderable terminations (tin-lead) on the sides. The terminations have an adhesion layer (silver deposited as thick film paste) on the ceramic substrate, and nickel barrier underplating followed by either dipped or plated solder coating. The nickel barrier is very important in preserving the solderability of terminations because it prevents leaching (dissolution) of the silver or gold electrode during soldering.

The resistive layer on the top surface dissipates the heat and should always face away from the substrate surface. The passivation layer is very brittle, however, and should not be probed with hard points such as test probe points during testing. Damaging the passivation layer could expose the resistive layer to the environment, thus degrading the resistor.

Thick film surface mount resistors are available in various tolerances (1, 5, 10, and 20%). The thin film resistors are made for very high precision circuits that require very close tolerances (<1%). Depending on the tolerance requirement, there can be substantial difference in cost. For ex-

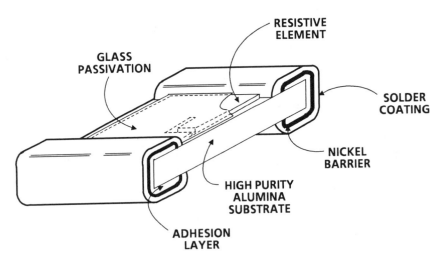

**Figure 3.1   The construction details of a surface mount resistor.**

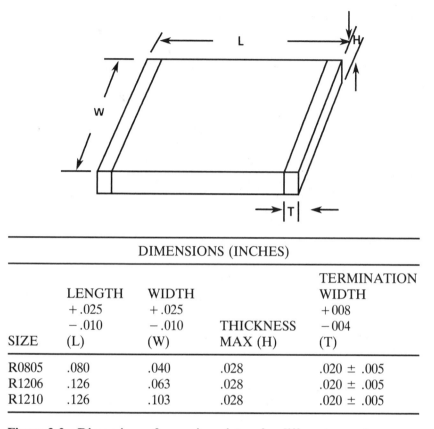

| DIMENSIONS (INCHES) | | | |
|---|---|---|---|
| LENGTH +.025 −.010 | WIDTH +.025 −.010 | THICKNESS | TERMINATION WIDTH +008 −004 |
| SIZE (L) | (W) | MAX (H) | (T) |
| R0805 .080 | .040 | .028 | .020 ± .005 |
| R1206 .126 | .063 | .028 | .020 ± .005 |
| R1210 .126 | .103 | .028 | .020 ± .005 |

**Figure 3.2   Dimensions of ceramic resistors for different case sizes.**

ample, resistors with 1% tolerance generally cost twice as much as 5% tolerance resistors. Tighter tolerance components, which are expensive and rarely necessary, should be avoided.

Resistors come in 1/16, 1/8, and 1/4 watt ratings in 1 ohm to 100 meg ohm resistance in various sizes. The commonly used designations for sizes (0805, 1206, 1210 etc.) are established by EIA specification IS-30. The EIA size refers to dimensions in hundredths of an inch. Dimensions vary from manufacturer to manufacturer, but generally speaking, 1/16, 1/8, and 1/4 watt resistors come in EIA sizes 0805 (0.08 inch by 0.05 inch), 1206 (0.12 inch by 0.06 inch), and 1210 (0.12 inch by 0.100 inch). The 1206 size with an 1/8 watt rating is the most commonly used size. Zero ohm resistors (jumpers) are also available in the 1206 size.

A surface mount resistor has some form of colored resistive layer with protective coating on one side and generally a white base material on the other side. Thus the outside appearance offers a simple way to distinguish between resistors and capacitors. Capacitors have the same color (generally brown) on both sides. It must be emphasized that there is no standard on color. However, there is a height difference between these components, ceramic resistors being about half as thick as ceramic capacitors. The Electronics Industries Association is considering standardization on the same mechanical outline for both resistors and capacitors. As discussed in Chapter 6, resistors and capacitors require different land pattern designs because of height differences even when they have the same widths and lengths. The other critical dimension for both resistors and capacitors is the width of terminations. The EIA dimensions for resistors are shown in Figure 3.2.

There generally are no part markings for resistance values on surface mount resistors. Some vendors will supply part markings at additional cost. There is generally not an absolute need for part markings, however, because these devices are most widely used on a tape and reel. Also, some pick-and-place equipment, such as Zevatech from Switzerland, can test capacitors and resistors before placement without compromising placement speed especially if only the first few parts on a reel are tested for correct values. This is very helpful when wrong tape and reel is used in the specified feeder slot.

## 3.2.2    Surface Mount Resistor Networks

The surface mount resistor networks or R-packs are going through the process of standardization. The currently available styles are based on the popular SOIC (small outline integrated circuits) packages, but the body dimensions vary. SOICs are discussed later (Section 3.4.2). But unlike SOICs, the standards for R-pack outlines have not been finalized yet.

The most commonly available body dimensions are as follows: 0.150 inch body, known as the SO package, with 8, 14, and 16 pins; a 0.220 inch body width known as SOMC, with 14 and 16 pins; and a 0.295 inch version, the SOL, with 16 and 20 pins. Some vendors are even supplying body widths of 0.410 inchs. The dimensions of SO and SOL are based on SOIC packages and have the same dimensions. The body dimensions based on SOIC package dimensions are likely to gain more user acceptance because of the popularity of SO packages and commonality in land pattern design and feeders for the placement machines.

The establishment of standards will help stimulate demand as well. Because R-packs are used relatively less often, they tend to be much more expensive than regular through-hole mount network packages. As is the case with other components, price parity is expected as SMT becomes widely used. In addition to price, the surface mount R-packs can have lead coplanarity problem (Section 3.7.1). Less than 0.04 inch coplanarity is now generally accepted by most users.

## 3.2.3 Ceramic Capacitors

In a high frequency circuit application, it is important to physically locate the capacitor as close to the high speed device as possible and to keep the lead length to a minimum, to minimize circuit inductance. The surface mount capacitor is ideal for such a purpose because it does not have any leads and can be placed underneath the package on the opposite side of the board. In very critical and high speed applications, capacitors can be placed directly underneath the surface mount packages on the same side. There are capacitors available only 18 mil thick that can fit underneath a leaded surface mount package. Since the solder joints cannot be inspected, the process yield must be 100% for such applications.

Surface mount capacitors are mostly used for decoupling applications rather than for frequency control, as is the case in the hybrid industry. This requires smaller sizes for hybrid applications and larger sizes for surface mounting. Smaller sizes are more expensive and difficult to place on the board. Luckily, they are not as much needed for surface mounting applications as they are for hybrid applications.

Single-layer disc capacitors form the basis for today's multilayer monolithic ceramic capacitors because of improved volumetric efficiency. In the multilayer ceramic capacitor, the electrodes are internal and are interleaved with ceramic dielectric. The alternate electrodes are exposed at the ends and connected to the end termination.

The construction of a multilayer ceramic capacitor is shown in Figure 3.3. The end terminations of the capacitors are similar in construction to those of resistors, with a silver adhesion layer and a nickel barrier to prevent leaching. This kind of ceramic capacitor construction results in a rugged block that can withstand a harsh environment and such treatments associated with the surface mount processes as immersion in the solder.

Monolithic surface mount capacitors are available in three different dielectric types per EIA RS-198, namely COG or NPO, X7R, and Z5U. They have different capacitance ranges as shown in Table 3.1. The COG or NPO dielectric capacitors are used when high stability is needed over a wide range of temperatures, voltages, and frequencies. The X7R and Z5U dielectric capacitors have poorer temperature and voltage character-

1. Termination
2. Dielectric
3. Electrode
4. Chip length
5. "A" electrode print
6. Electrode print
7. Cap (Topping layer)

8. End margin
9. Base layer
10. Shim (Active dielectric layer)
11. Side margin
12. Chip thickness
13. Chip width
14. Termination width

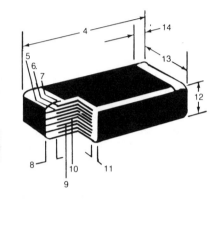

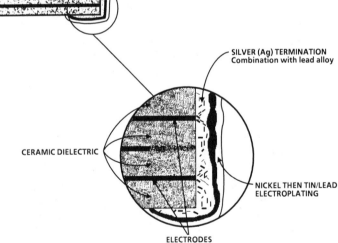

SILVER (Ag) TERMINATION
Combination with lead alloy

CERAMIC DIELECTRIC

NICKEL THEN TIN/LEAD
ELECTROPLATING

ELECTRODES

**Figure 3.3   Construction detail of ceramic capacitor with exposed alternate electrodes at the ends and connected to the end termination (top figures). The cross-sectional views (bottom figures) show the details of end termination materials.**

**Table 3.1    Capacitance range for different dielectric materials**

| SIZE | COG | X7R | Z5U |
|------|-----|-----|-----|
| C0805 | 10–560 pF | 120 pF–0.012 μF | — |
| C1206 | 608–1500 pF | 0.015–0.033 μF | 0.033–0.10 μF |
| C1812 | 1800–5600 pF | 0.039–0.12 μF | 0.12–0.47 μF |

istics, but since they are mostly used for bypass and decoupling applications, stability of capacitance is not very important.

The surface mount capacitor is highly reliable and has been used in high volumes in under-the-hood automotive applications. It also has a

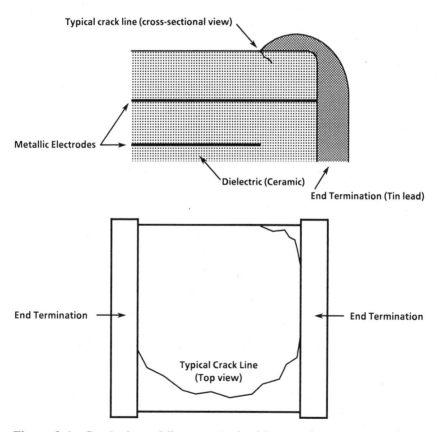

**Figure 3.4    Cracks in multilayer ceramic chip capacitors: cross-sectional view (top) and top view (bottom).**

proven history in military and aerospace applications. However, ceramic capacitors are prone to cracking (see Figure 3.4) during wave soldering. The cracks, which are generally difficult to see because they are very small, can grow in service and cause failures. There are various causes of cracking, including excessive or uneven solder fillet due either to poor land pattern design or to improper orientation of components. See Chapter 6 for a discussion of land pattern design and Chapter 7 for recommended component orientation.

The primary causes of cracking in ceramic capacitors are thermal shock during soldering and poor quality control by the vendor. As shown in Figure 3.3, a ceramic capacitor contains a metallic electrode, termination material, and ceramic dielectric. Each has a different coefficient of thermal expansion. For example, the CTEs of the internal electrode, the tin-lead termination and nickel over silver, and the ceramic, respectively, are 16, 18, and 9.5 to 11.5 ppm/°C [where ppm = microinch (mil) per inch]. During preheat and soldering, the electrodes and terminations heat up more rapidly than the ceramic body. If the expansion differential is too rapid due to a steep thermal profile, the stresses generated by CTE mismatch can crack the component.

Thermal shock can be minimized by gradually preheating the board before the assembly goes over the solder wave. Refer to Chapter 14 (Section 14.3.2: Component-Related Defects).

If the design is correct, and if component quality is properly controlled by the vendor and the user maintains the *top side* preheat and solder pot temperatures at 220 to 240°F and 475°F (for eutectic solder), respectively, cracking should not be a problem. Also note that the Z5U ceramic capacitors are more susceptible to cracking than the X7R. Since their costs are comparable, X7R capacitors should be used if possible.

The capacitance values of the capacitors are related to their case sizes. Case sizes are going through standardization, but the most commonly used case size is EIA 1206 (0.120 inch by 0.060 inch nominal). The selection of either lower or higher case size should be considered when the required capitance value is not available in the desired dielectric type. Table 3.1 can be used as a guideline for selecting a capacitance range in the dielectric material.

Figure 3.5 shows the case sizes for ceramic capacitors. The component termination width is the most critical dimension for soldering and land pattern design. The components from vendors that provide closer component tolerances and meet other procurement requirements (discussed in Section 3.8) will provide assemblies with higher yield.

The most widely used packaging for ceramic capacitor is 8 mm tape and reel. When establishing procurement specifications, the objective

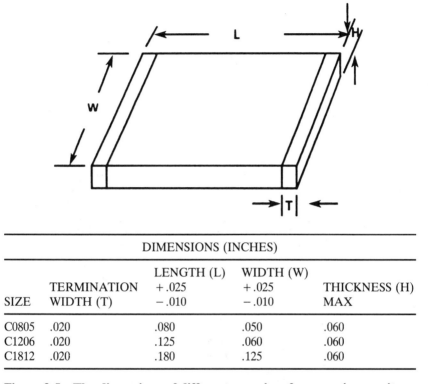

| DIMENSIONS (INCHES) | | | |
|---|---|---|---|
| SIZE | TERMINATION WIDTH (T) | LENGTH (L) + .025 − .010 | WIDTH (W) + .025 − .010 | THICKNESS (H) MAX |
| C0805 | .020 | .080 | .050 | .060 |
| C1206 | .020 | .125 | .060 | .060 |
| C1812 | .020 | .180 | .125 | .060 |

**Figure 3.5   The dimensions of different case sizes for ceramic capacitors.**

should be to minimize the case sizes and part numbers (capacitance values). This will provide increased purchasing leverage and will allow for reduced inventory. In addition, the number of feeders for the pick-and-place machine in manufacturing can be reduced.

## 3.2.4   Tantalum Capacitors

For capacitors, the dielectric can be either ceramic or tantalum. Surface mount tantalum capacitors offer very high volumetric efficiency or a high capacitance-voltage product per unit volume and high reliability. Standardization for the components is lacking. Indeed, even the designation for case sizes is not standardized. Both alphabetical and numerical designations for case sizes are used.

One of the most important considerations in selecting tantalum capacitors should be the construction of the end terminations, which come

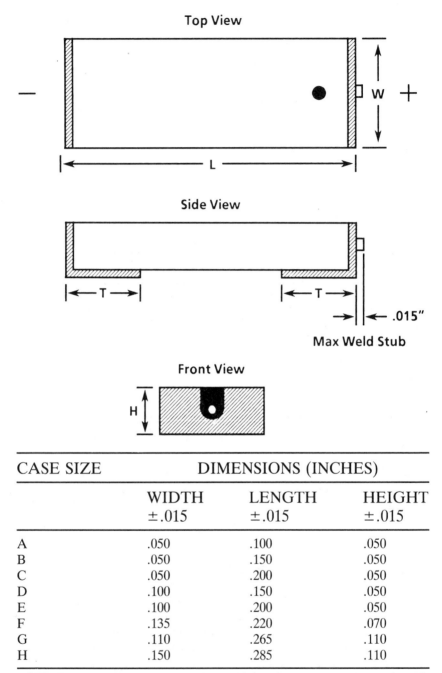

**Top View**

**Side View**

← .015″

**Max Weld Stub**

**Front View**

| CASE SIZE | DIMENSIONS (INCHES) | | |
|-----------|------------------|------------------|------------------|
| | WIDTH ±.015 | LENGTH ±.015 | HEIGHT ±.015 |
| A | .050 | .100 | .050 |
| B | .050 | .150 | .050 |
| C | .050 | .200 | .050 |
| D | .100 | .150 | .050 |
| E | .100 | .200 | .050 |
| F | .135 | .220 | .070 |
| G | .110 | .265 | .110 |
| H | .150 | .285 | .110 |

NOTE: The solderable and termination width for all components shown above is .030 inch except for components G and H for which end termination width is 0.050 inch.

**Figure 3.6** **The dimensions (inches) of different case sizes for tantalum capacitors with welded stub.**

in two major configurations-welded sub contacts (unmolded, Figure 3.6) and with wrap-under lead contacts (plastic molded, Figure 3.7). These capacitors are constructed using a sintered tantalum anode and a solid electrolyte. The welded stub causes problems during placement, and these capacitors come with gold terminations, which can embrittle solder joints. However, the stub is a good visible polarity indictor.

The wrap-under lead capacitors, commonly called plastic molded tantalum capacitors, have leads instead of terminations and a beveled top as a polarity indicator. There are no soldering or placement concerns when using the molded plastic tantalum capacitors. Their case sizes fall into two main categories, called standard and extended range. Each category has four different case sizes as shown in Figure 3.7 (bottom).

There is wide variation in dimensions from manufacturer to manufacturer because, as is the case with R-packs, the standards for tantalum capacitors are still evolving. The case sizes and designations shown in Figure 3.7 are taken from IPC-SM-782 [1]. The IPC-SM-782 committee adopted these dimensions from EIA RS-228 but modified them to conform to the land pattern design established in IPC-SM-782.

The capacitance values for tantalum capacitors vary from 0.1 to 100 $\mu F$ and from 4 to 50 V dc in different case sizes. Given the lack of a standard, mentioned earlier, one should refer to vendor literature for the capacitance range for a given voltage and case size. Tantalum capacitors should never be used above the rated voltage and should never be used in reversed polarity. If these precautions are ignored, the devices will blow up when powered. They can withstand immersion in the wave soldering process but are susceptible to thermal shock.

The tantalum capacitors have been found by some suppliers to be susceptible to moisture induced cracking during soldering (wave, vapor phase, or infrared). The susceptibility to cracking depends on the level of moisture absorbed, which in turn depends on the relative humidity, time and temperature of component storage. If we refer back to Figure 3.7 (top), moisture easily permeates through the encapsulant and the conductive coatings during storage. In addition, the core electrolyte (fine particles of tantalum surrounded by tantalum oxide and manganese oxide) is very porous and easily absorbs moisture. When the moisture absorption reaches a threshold limit, the moisture will convert to steam during soldering and cause the component to crack. The basic mechanism of moisture absorption and cracking is very similar to that of plastic package cracking discussed in detail in Chapter 5, section 5.7.1. The solution requires baking of components before soldering to drive of the moisture. For specific details on the susceptibility of the component to cracking and proposed solution, the users should contact the supplier. Also, like ceramic capacitors, they should be preheated gradually before wave or reflow soldering.

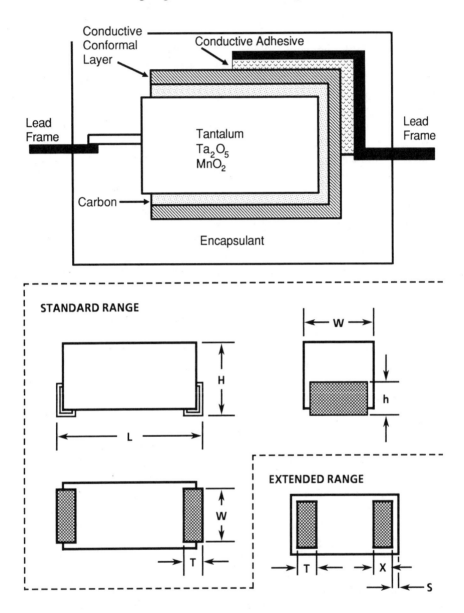

**Figure 3.7 The construction details of a molded tatalum capacitor (top) and the case sizes (inches) and designations for plastic molded tantalum capacitors with standard and extended capacitance range (bottom).**

## SIZE CODE/STANDARD CAPACITANCE RANGE

| | [A] 3216 | [B] 3528 | [C] 6032 | [D] 7243 |
|---|---|---|---|---|
| L | .118–.134 | .130–.146 | .224–.248 | .268–.299 |
| H | .055–.071 | .067–.083 | .087–.110 | .098–.122 |
| W | .05–.071 | .102–.118 | .114–.138 | .157–.181 |
| M Min | .028 | .028 | .040 | .040 |
| *Y | .043–.051 | .083–.09 | .083–.09 | .090–.098 |
| T* | .020–.043 | .020–.043 | .020–.043 | .020–.043 |

## SIZE CODE/EXTENDED CAPACITANCE RANGE

| DIM | 3518 | 3527 | 7227 | 7257 |
|---|---|---|---|---|
| L | .130–.146 | .130–.146 | .272–.295 | .272–.295 |
| H* | .067–.083 | .067–.083 | .098–.122 | .118–.146 |
| W | .063–.079 | .095–.118 | .095–.118 | .205–.244 |
| M Min | .028 | .028 | .040 | .047 |
| Y* | .063–.071 | .095–.102 | .095–.102 | .213–.228 |
| X* | .024–.040 | .024–.040 | .031–.047 | .031–.047 |
| T* | .028–.043 | .028–.043 | .051–.067 | .051–.067 |
| S* | .016–.024 | .016–.024 | .024–.032 | .024–.032 |

*Dimensions are tailored to facilitate standard land pattern design.

**Figure 3.7** (*Continued*)

Tantalum capacitors are available with or without marked capacitance values in bulk, in waffle packs, and on tape and reel. Tape and reel packaging is preferred for handling. Since tantalum capacitors tend to be expensive and are used in small numbers on a given board, tube feeders are commonly used to place them on the board.

## 3.2.5 Tubular Passive Components

The cylindrical devices known as metal electrode leadless faces (MELFs) are used for resistors, jumpers (zero ohm resistors), ceramic and tantalum capacitors, and diodes. They are cylindrical and have metal end caps for soldering. The construction detail of a MELF device is shown in Figure 3.8.

Since they are cylindrical, the resistors do not have to be placed with

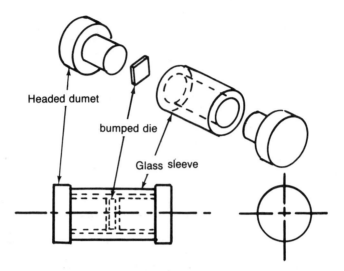

**Figure 3.8   Construction of a metal electrode leadless face (MELF) component [1].**

resistive elements away from the board surface as is the case with the rectangular resistors discussed earlier; in addition, MELFs are cheaper than the rectangular devices. Like the conventional axial devices, MELFs are color coded for values.

MELFs do have their problems—they may roll away during handling after placement in uncured adhesive or paste or even during reflow. To prevent movement during reflow, a notch is generally recommended in the land pattern (see Chapter 6). The devices are generally packaged on tape and reel. MELFs are identified as MLL41 and MLL34. Their dimensions are shown in Figure 3.9.

## 3.3   ACTIVE COMPONENTS: CERAMIC PACKAGES

Surface mounting offers considerably more types of active and passive packages than are available in through-hole mount technology. For example, in the DIP, there are only three major body widths: 300, 400, and 600 mils, in 100 mil center pitch. The size of the package and the lead configuration are the same for both ceramic and plastic bodies. The world of surface mounting is considerably more complex in comparison.

Figure 3.10 sets forth the hierarchy of chip carriers available for surface mounting. There are two main categories of chip carriers: ceramic and plastic. The plastic chip carriers are primarily used in commercial appli-

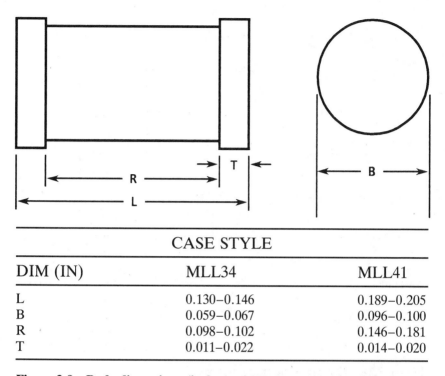

|  | CASE STYLE | |
|---|---|---|
| DIM (IN) | MLL34 | MLL41 |
| L | 0.130–0.146 | 0.189–0.205 |
| B | 0.059–0.067 | 0.096–0.100 |
| R | 0.098–0.102 | 0.146–0.181 |
| T | 0.011–0.022 | 0.014–0.020 |

**Figure 3.9   Body dimensions (inches) of MELF components [1].**

cations. Just as in dual in-line packages, the surface mount chip carriers are available in many constructions and materials. The ceramic packages provide hermeticity and are used primarily in military applications.

The most commonly used type of chip carrier in military applications is of leadless construction. Its main disadvantage, and it is a very significant one, is that any mismatch between the coefficients of thermal expansion of the package and the substrate will cause solder joint cracking because thered are no leads to take up the stresses. This requires use of expensive substrates with matching CTEs for leadless chip carriers. This issue is discussed in detail in Chapter 4.

Then the industry came up with plastic leaded chip carriers (PLCCs) with compliant J leads to solve the problem of solder joint cracking for nonhermetic commercial applications. The small outline integrated circuits (SOIC) packages also are used. They are more space efficient for pin counts of 20 and under. The Japanese, however, have not embraced the J-lead configuration for either lower or higher pin count packages.

Now the U.S. industry has turned around and is offering the high pin count, fine pitch packages in gull wing lead configurations, as the Japanese

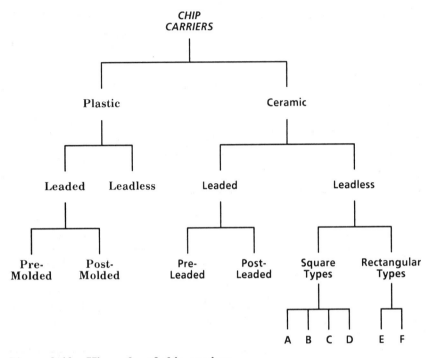

**Figure 3.10   Hierarchy of chip carriers.**

have done. However, as we will see later, the U.S. fine pitch packages have bumpers at the corners to protect the leads from damage. This design is superior to the Japanese fine pitch packages because, for example, there are fewer problems in manufacturing. See Section 3.4.5.

There are also discrete active devices known as small outline transistors (SOTs), having three gull wing leads. They are of two different configurations: SOT 23 and SOT 89. Other discrete diode devices generally have four leads, both gull wing and the J type. In the sections that follow we discuss various categories of active surface mount component packages in some detail.

### 3.3.1   Leadless Ceramic Chip Carriers

Ceramic chip carriers usually are constructed from a 90 to 96% alumina or beryllia base. Like DIPs, these carriers are constructed in single or multiple layers and can be used in the operating temperature range of −55 to 125°C. They are hermetically sealed and provide good environmental protection. The ceramic chip carriers come in leadless and leaded config-

**Table 3.2  The designation of ceramic packages and their pin counts** (*Source:* **JEDEC Standard, Publication 95). Also refer to Appendix A.**

| DESIGNATION | CHIP CARRIER | PIN COUNT |
|---|---|---|
| MS-002 | Leadless, Type A | 28, 44, 52, 68, 84, 100, 124, 156 |
| MS-003 | Leadless, Type B | 28, 44, 52, 68, 84, 100, 124, 156 |
| MS-004 | Leadless, Type C | 16, 20, 24, 28, 44, 52, 68, 84, 100, 124, 156 |
| MS-005 | Leadless, Type D | 28, 44, 52, 68, 84, 100, 124, 156 |
| MS-006 | Leaded, Type A | 24 only |
| MS-007 | Leaded, Type A | 28, 44, 52, 68, 84, 100, 124, 156 |
| None | Leadless, Type E | 28, 32 |
| None | Leadless, Type F | 18, 20 |
| MS-009 | Leaded, Type B | 16, 20, 24, 32, 40, 48, 64, 84, 96 |
| MS-014 | Leadless, 40 mil pitch | 16, 20, 24, 32, 40, 48, 64, 84, 96 |

uration. In this section we discuss leadless chip carriers. The leaded ceramic chip carriers are covered in the next section.

As the name indicates, leadless chip carriers have no leads. Instead they have gold-plated, groove-shaped terminations known as castellations. The castellations provide shorter signal paths, which allow higher operating frequencies because inductive and capacitive losses are lower. Reduced signal lead resistance and improved power dissipation also are provided by leadless chip carriers.

The list of hermetic leadless ceramic chip carrier types alone is very long. For example, in the leadless ceramic package, there are two pre-dominant pitches: 50 mil centers and 40 mil centers. This does not include the 20, 25, and 33 mil center packages for fine pitch. The variety does not end here. The 50 and 40 mil center packages break down further into Types A, B, C, D, E, and F. The ceramic packages of various types have unique designations (MS002, MS003, etc.) given by JEDEC. Table 3.2 shows the designation of packages and their pin counts for each package type.

The leadless ceramic chip carriers can be divided in different families depending on the pitch of the package. The most common is the 50 mil

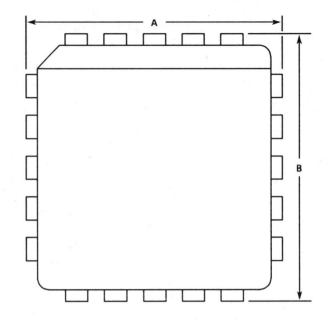

| PACKAGE DIMENSION (INCH) | | | | |
| --- | --- | --- | --- | --- |
| | LCCC PACKAGE | | PLCC PACKAGE | |
| PIN COUNTS LCCC/PLCC | WIDTH (A) MAX | LENGTH (B) MAX | WIDTH (A) MAX | LENGTH (B) MAX |
| Rectangular Package | | | | |
| 18-Short | .300 | .440 | .327 | .467 |
| 18-Long | | | .327 | .527 |
| 22 | | | .327 | .527 |
| 28 | .360 | .560 | .395 | .595 |
| 32 | .460 | .560 | .495 | .595 |
| Square Package: | | | | |
| 16 | .310 | .310 | | |
| 20 | .360 | .360 | .395 | .395 |
| 24 | .410 | .410 | | |
| 28 | .460 | .460 | .495 | .495 |
| 44 | .660 | .660 | .695 | .695 |
| 52 | .760 | .760 | .795 | .795 |
| 68 | .960 | .960 | .995 | .995 |
| 84 | 1.160 | 1.160 | 1.195 | 1.195 |
| 100 | 1.360 | 1.360 | 1.335 | 1.335 |
| 124 | 1.660 | 1.660 | 1.695 | 1.695 |
| 156 | 2.060 | 2.060 | | |

**Figure 3.11   Maximum package sizes for leadless ceramic and plastic leaded chip carriers: LCCCs and PLCCs. (Intel Corporation.)**

(1.27 mm) family. The castellations of these packages are at 50 mil centers. There are also 40, 25, and 20 mil families of leadless chip carriers. The 20 and 25 mil center families are discussed under Fine Pitch Packages in this chapter (Section 3.4.5). The 50 mil center family is most widely used, but we should keep in mind that the 40 mil family is also a part of the JEDEC standards. These packages are used by the military for high density applications. The 40 mil center family came into usage before the 50 mil center family.

The leadless ceramic chip carriers in the 50 mil center family come in square and rectangular configurations. Their dimensions are shown in Figure 3.11, which also shows the dimensions of plastic leaded chip carriers. It is worth noting that even though there are many types of plastic and ceramic chip carriers, they have common land patterns, as discussed in Chapter 6.

Why there are so many ceramic package types? Each type is designed for a different application. Some are designed for heat dissipation in air and others for heat dissipation through the substrate. Some can be directly mounted on the board and others are designed for socketing. The packages that can be mounted directly on the board can have their lid either up or down depending on the heat dissipation requirements (dissipation through the air or the substrate).

"Lid up" implies a cavity-up position and "lid down" implies a cavity-down orientation. These terms refer to the backside of the dies, which serves as the package's primary heat transfer or heat removal surface. When the lid is up, the backside of the die faces the substrate and the heat is dissipated through the substrate. This kind of package is not suitable for air-cooled systems or for the attachment of heat sinks. It does allow for a larger cavity size, and thus a larger die. Type B and C packages fall into this category. Figure 3.12 shows the lids of Type B and C packages on the opposite side of the seating or mounting plane.

The orientation in which the cavity or lid is down is suitable for air-cooled systems, since the backside of the die faces away from the substrate. Also, heat sinks can be attached to the backside of the die in cavity-down packages. However, the cavity size is reduced, hence the die size. Types A and D fall into this category (Figure 3.12 shows the lids of Type A and D packages on the same side as the seating plane). For socketing, Type A (lid-down position) and Type B (lid-up position) are used; for direct attachment, Type C (lid-up position) and Type D (lid-down position) are used. Thus these four packages meet the requirements of both socketing and direct attachment.

The leadless Type E and F packages are rectangular and are intended for memory devices. They are not high power devices, hence thermal properties are not important. They are generally used for direct attachment in the lid-up position, and the heat is dissipated through the substrate.

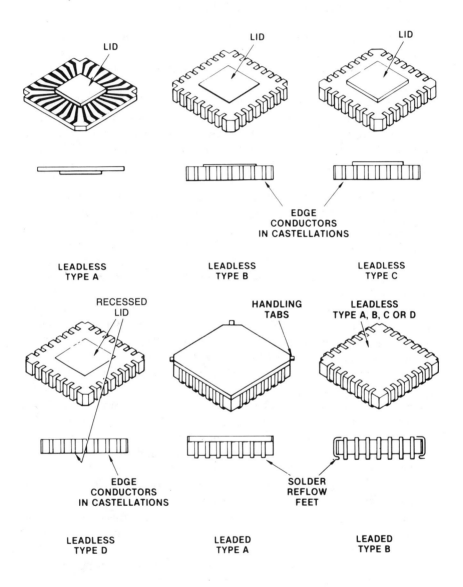

**Figure 3.12    Seating planes and lid orientations for leadless ceramic chip carriers [1].**

Refer again to Figure 3.10 for the hierarchy of chip carrier packages by type, A through F.

## 3.3.2   Ceramic Leaded Chip Carriers (Preleaded and Postleaded)

As discussed earlier, the leadless ceramic chip carriers may cause solder joint cracking if there is mismatch between the coefficients of thermal expansion of the package and the substrate. This problem may be prevented by attaching compliant leads to the leadless packages. (In Chapter 4, Section 4.3 discusses the impact of package type on substrate selection.)

The leaded ceramic carriers are available in both preleaded and postleaded formats. The preleaded ceramic chip carriers have copper alloy or Kovar leads that are attached by the manufacturer. The leads, which are either top brazed or attached to the castellations, can be formed to different configurations, such as J, L, S, or gull wing shape. The preleaded chip carriers are supplied with the leads straight and attached to a common strip. The user cuts the common strip and forms the lead in the desired configuration. This minimizes the bending of leads during shipping and handling.

In postleaded chip carriers, the user attaches the leads to the castellations of the leadless ceramic chip carriers. The leads are generally edge clips supplied on common carrier strips. The leads may be supplied unbent, or preformed. If unbent, the user can bend the leads in the desired configuration, as in the case of preleaded chip carriers. The edge clips are attached to the castellations and then reflowed to accomplish the bond.

When the leadless types discussed earlier (A, B, C, and D) are mounted with leads for direct soldering to the substrate, they are called leaded Type B. The leaded Type B packages cannot be socketed because of the nature of their lead configuration. They are intended for soldering directly to the board. The leaded Type A breaks down further into three different categories: leaded ceramic, premolded plastic, and postmolded plastic. All Type A leaded packages can either be socketed or soldered directly to the substrate.

When leaded ceramic packages are used, their dimensions are generally the same as in plastic leaded chip carriers. Example of this include JEDEC MS-044 leaded ceramic package with CERDIP/CERQUAD construction. It should be kept in mind that leaded ceramic chip carriers are generally not used. The most commonly used packages in the industry are leadless ceramic and the leaded plastic packages. The reason is very simple.

Very few users are set up for the tedious process of attaching leads, which is not suitable for volume production in any event.

Industry is getting ready to supply J or gull wing packages in the same way that plastic packages (PLCC and SOIC) are supplied. However, there is no clear direction in lead configurations for ceramic packages. This may change, however, because the main reason for using leaded ceramic packages is to avoid the expensive and exotic substrates that are necessary for the leadless ceramic chip carriers.

## 3.4 ACTIVE COMPONENTS: PLASTIC PACKAGES

As discussed earlier, ceramic packages are expensive and are used primarily for military applications. The plastic packages, on the other hand, are the most widely used packages for nonmilitary applications, where hermeticity is not required. Whereas the ceramic packages have solder joint cracking problems due to CTE mismatch between the package and the substrate, the plastic packages are by no means trouble free.

Some surface mount plastic packages may be moisture sensitive. For example, if they contain moisture above a threshold limit and are subjected to reflow soldering temperatures, the package may crack when the moisture expands. The susceptibility to cracking is dependent on the thickness of the plastic, the moisture content, and the die size. See Chapter 5 (Section 5.7.1) for a discussion of the mechanism of cracking and recommended solutions.

Other constraints for plastic packages include limited temperature range (0–70°C) and nominal environmental protection. The commercial sector of the electronics industry needs the plastic packages for cost effectiveness and reliable interconnection on glass epoxy substrates. The most common plastic packages used in commercial electronics are the discrete transistors known as small outline transistors (SOTs), small outline integrated circuits (SOICs) with gull wing leads, small outline devices with J leads (SOJ), plastic leaded chip carriers (PLCCs) with J leads, and fine pitch devices with gull wing leads known by various names such as mini-packs or plastic quad flat packs (PQFPs).

SOTs have unique lead spacing and construction as described later. The SOICs, SOJs, and PLCCs are 50 mil center pitch packages. In terms of real estate considerations, the most efficient packages are SOICs for pin counts under 20 and PLCCs for pin counts between 20 and 84. For pin counts above 84, the fine pitch packages with lead pitches of 20, 25, and 33 mils are most comon. For even higher pin count, tape automated bond-

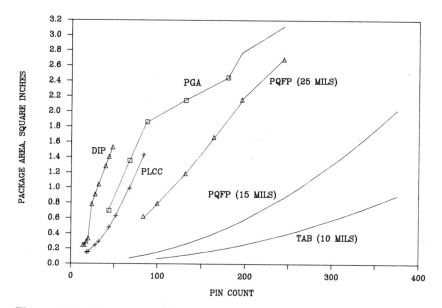

**Figure 3.13** **Real estate efficiencies of packages of various types for surface mounting. (Intel Corporation.)**

ing (TAB) devices are necessary. Figure 3.13 plots real estate efficiencies of packages of various types for surface mounting.

## 3.4.1 Small Outline Transistors

The small outline transistors are one of the forerunners of active devices in surface mounting. They are three- and four-lead devices. The three-lead SOTs are identified as SOT 23 (EIA TO-236) and SOT 89 (EIA TO-243). The four-lead device is known as SOT 143 (EIA to 253). These packages are generally used for diodes and transistors. The SOT 23 and SOT 89 packages have become almost universal for surface mounting small transistors.

SOT 23 is the most commonly used three-lead package, with internal construction as shown in Figure 3.14. This package can accommodate a maximum die size of 0.030 inch by 0.030 inch and can dissipate up to 200 mW in free air and up to 350 mW when attached to a ceramic substrate. It comes in low, medium, and high profiles as shown in Figure 3.15, to meet the differing needs of both hybrid and printed board applications.

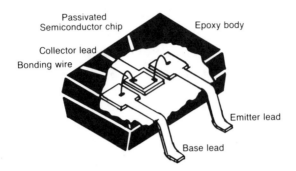

**Figure 3.14   Internal construction of SOT 23 [1].**

The higher profile package is more suitable for printed board application because it provides better cleaning. The dimensions of the package are given in Figure 3.16.

The SOT 89 is used for higher power devices. To be able to transfer heat to the substrate more efficiently, this package sits flush on the substrate surface. The outline dimensions of SOT 89 are shown in Figure 3.17. The SOT 89 package can accommodate a maximum die size of 0.060 inch by 0.060 inch. At 25°C it can dissipate 500 mW in free air and 1 W when mounted on an alumina substrate.

The SOT 143 packages come in the four-lead configuration (Figure 3.18) and are commonly used for RF transistor applications. They are like SOIC packages but with lower clearance off the board. The SOT 143

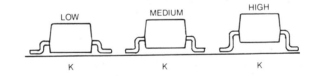

| PROFILE | MIN.-MAX. "K" DIMENSIONS (INCHES) |
|---|---|
| High Profile | .004–.010 |
| Medium Profile | .003–.005 |
| Low Profile | .0004–.004 |

**Figure 3.15   Clearance (in inches) from top of board to bottom of SOT 23 in low, medium, and high profiles [1].**

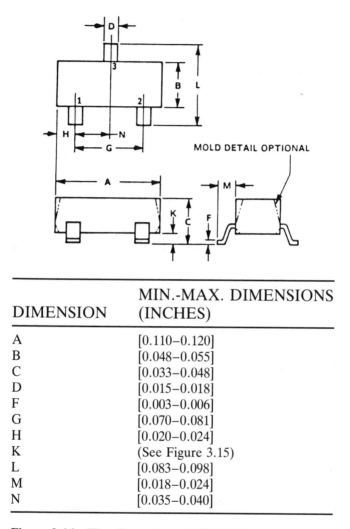

| DIMENSION | MIN.-MAX. DIMENSIONS (INCHES) |
|---|---|
| A | [0.110–0.120] |
| B | [0.048–0.055] |
| C | [0.033–0.048] |
| D | [0.015–0.018] |
| F | [0.003–0.006] |
| G | [0.070–0.081] |
| H | [0.020–0.024] |
| K | (See Figure 3.15) |
| L | [0.083–0.098] |
| M | [0.018–0.024] |
| N | [0.035–0.040] |

**Figure 3.16   The dimensions of SOT 23 [1].**

package can accommodate about the same die size (0.025 inch by 0.025 inch) as the SOT 23 package.

The SOT 23, SOT 89, and SOT 143 packages are most commonly supplied in tape and reel per EIA standard RS-481. The most popular are conductive tapes with embossed cavities for the packages. These packages are the only active devices that are soldered using both wave and reflow (vapor phase or infrared) methods. The other active devices, such as SOICs and PLCCs, are mostly soldered using only reflow soldering methods.

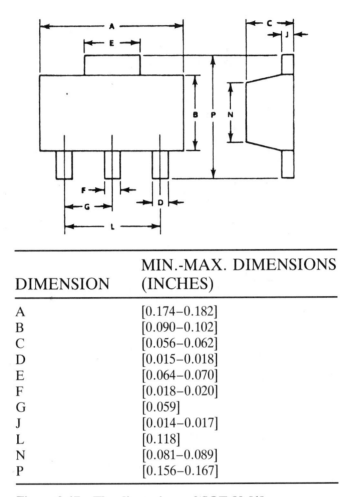

| DIMENSION | MIN.-MAX. DIMENSIONS (INCHES) |
|-----------|-------------------------------|
| A | [0.174–0.182] |
| B | [0.090–0.102] |
| C | [0.056–0.062] |
| D | [0.015–0.018] |
| E | [0.064–0.070] |
| F | [0.018–0.020] |
| G | [0.059] |
| J | [0.014–0.017] |
| L | [0.118] |
| N | [0.081–0.089] |
| P | [0.156–0.167] |

**Figure 3.17   The dimensions of SOT 89 [1].**

## 3.4.2   Small Outline Integrated Circuits

The small outline integrated circuit (SOIC, or SO for brevity) is basically a shrink DIP package with leads on 0.050 inch centers. It contains leads on two sides that are formed outward in what is generally called a gull wing lead. The Swiss watch industry is generally given the credit for originating this package style in the 1960s. However, for the modern electronics industry, N. V. Philips originated this package style in 1971.

In the United States, Signetics, a division of N. V. Philips, is considered to be the pioneer of this package style for digital devices. National Semiconductor committed to this package style after the automotive and te-

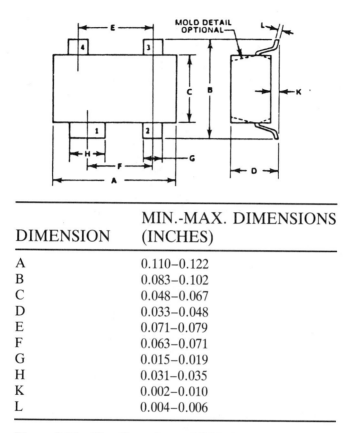

| DIMENSION | MIN.-MAX. DIMENSIONS (INCHES) |
|-----------|-------------------------------|
| A | 0.110–0.122 |
| B | 0.083–0.102 |
| C | 0.048–0.067 |
| D | 0.033–0.048 |
| E | 0.071–0.079 |
| F | 0.063–0.071 |
| G | 0.015–0.019 |
| H | 0.031–0.035 |
| K | 0.002–0.010 |
| L | 0.004–0.006 |

**Figure 3.18   The dimensions of SOT 143 [1].**

lecommunication industries provided support for it. The Japanese refer to this package as a mini-flat. The mini-flats have dimensions slightly different from the JEDEC dimensions.

Compared with J-lead packages, the SOs need to be handled more carefully to prevent lead damage. In the beginning, there was no lead coplanarity requirement for SOs. Now the JEDEC standard on SOICs requires 0.004 inch lead coplanarity, as is the case in PLCCs. The SOs prolvide better real estate savings than PLCCs for pin counts under 20, and the solder joints are easier to inspect. SOs are available in 8, 14, 16, 20, 24, and 28 lead counts.

The SOIC packages come in mainly two different body widths: 150 and 300 mils. The body width of packages having fewer than 16 leads is 150 mils; for more than 16 leads, 300 mil widths are used. The 16 lead package comes in both body widths. Figure 3.19 gives SOIC package dimensions. The 300 mil body width packages are referred to with suffix

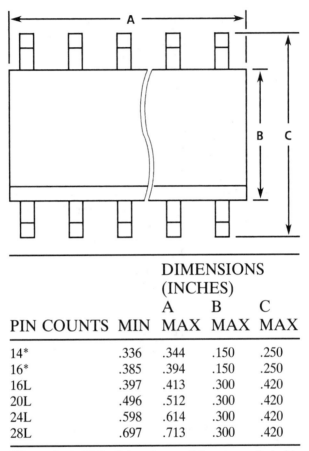

| | | DIMENSIONS (INCHES) | | |
| | | A | B | C |
| PIN COUNTS | MIN | MAX | MAX | MAX |
| --- | --- | --- | --- | --- |
| 14* | .336 | .344 | .150 | .250 |
| 16* | .385 | .394 | .150 | .250 |
| 16L | .397 | .413 | .300 | .420 |
| 20L | .496 | .512 | .300 | .420 |
| 24L | .598 | .614 | .300 | .420 |
| 28L | .697 | .713 | .300 | .420 |

*Narrow body (150 mils) packages. Others are called wide body (300 mils) packages.

**Figure 3.19   The dimension of SOIC packages. [1]**

"L", such as 16 SOL or 18 SOL. The narrow body packages are also referred to as JEDEC MS-012 AA-AC, and the wide body packages are referred to as JEDEC MS-013. There also SOICs with a body width of 330 mils. They are generally used for static RAMs, which are mostly Japanese packages.

At one time there was great concern about the reliability of SO packages. The concern had more to do with theoretical than with empirical data. Theoretically speaking, the amount of plastic around the die in an SOIC package is a fraction of that in a standard DIP. Because of this, the thermal problem was expected to be compounded in an SOIC package.

The actual reliability problem has not been encountered because the manufacturers have done such things as using copper lead frame material instead of Kovar or alloy 42 and improved plastic materials.

The wave soldering process is not very effective on PLCCs, but some companies do use wave soldering for SO packages. One should be careful in wave soldering any plastic package (SO, SOT, or PLCC) because package reliability may be compromised at the relatively higher temperatures in the wave. Also, the flux used in wave soldering may seep inside the package. This is possibly due to the different coefficients of thermal expansion between the plastic body and the lead frame. Any such seepage may have an adverse impact on the reliability of the device. If the SOIC happens to be moisture sensitive, the reliability problem may be even more serious. In Chapter 5, Section, 5.7.1 gives details on the moisture sensitivity of plastic packages and the mechanism of cracking during reflow soldering.

### 3.4.3    Plastic Leaded Chip Carriers

The plastic leaded chip carrier (PLCC) is a cheaper version of the ceramic chip carrier. The PLCCs are almost a mandatory replacement for plastic DIPs, which are not practical above 40 pins because of excessive real estate requirements. The PLCCs come in a lead pitch of 0.050 inch with J leads that are bent under the packages.

The packages with (higher pin count) and generally with the gull wing leads of the SOIC are defined as fine pitch quad packages and are discussed later on. The PLCC was introduced in 1976 as a premolded plastic package, but it did not become very popular. The postmolded PLCC, pioneered by Texas Instruments, was adopted by the major semiconductor companies such as Intel, Motorola, and National Semiconductor, and the major Japanese manufacturers for products to be sold in the United States.

This package was originally referred to as postmolded Type A leaded, but JEDEC abandoned this nomenclature in favor of PLCC. A JEDEC task force was organized in 1981 for the registration of PLCCs. The JEDEC outline known as MO-047 for square packages having 20, 28, 44, 52, 68, 84, 100, and 124 terminals was completed in 1984. These packages have an equal number of J leads on all four sides. See Figure 3.20 for dimensions.

A second registration, known as JEDEC outline MO-052, was completed in 1985 for rectangular packages with pin counts of 18, 22, 28, and 32. The 18 pin PLCCs have two sizes: a shorter one for 64K dynamic random access memory (DRAM) and the longer one for 256K DRAMs. The 28 and 32 pin packages are used for electrically programmable read only memories (EPROMs); see Figure 3.21 for dimensions.

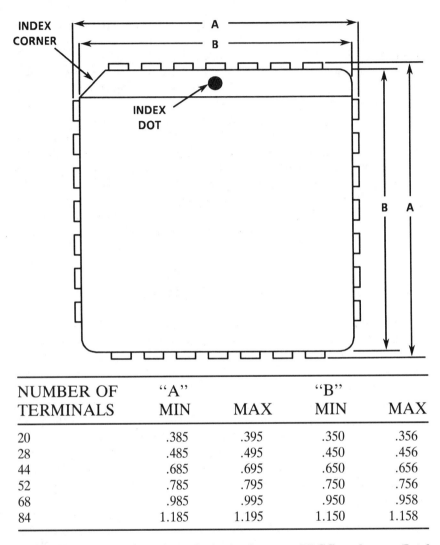

| NUMBER OF | "A" | | "B" | |
|-----------|-----|-----|-----|-----|
| TERMINALS | MIN | MAX | MIN | MAX |
| 20 | .385 | .395 | .350 | .356 |
| 28 | .485 | .495 | .450 | .456 |
| 44 | .685 | .695 | .650 | .656 |
| 52 | .785 | .795 | .750 | .756 |
| 68 | .985 | .995 | .950 | .958 |
| 84 | 1.185 | 1.195 | 1.150 | 1.158 |

**Figure 3.20  The dimensions (inches) of square PLCC packages. (Intel Corporation.)**

The pin numbering system for PLCC is different and should be noted to prevent problems of misorientation in manufacturing. The convention for the location of pin 1 in SOIC is the same as in DIP (starting at the top left-hand corner and progressing counterclockwise). In PLCC, however, pin 1 is the top center pin when there is an odd number of pins on each side and one to the left of center when there is an even number of pins

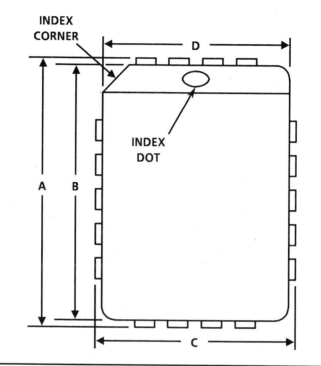

| NUMBER OF | "A" | | "B" | | "C" | | "D" | |
|---|---|---|---|---|---|---|---|---|
| TERMINALS | MIN | MAX | MIN | MAX | MIN | MAX | MIN | MAX |
| 18[1] | .317 | .327 | .457 | .467 | .282 | .293 | .422 | .428 |
| 18[2] | .317 | .327 | .517 | .527 | .282 | .293 | .487 | .493 |
| 28 | .385 | .395 | .585 | .595 | .347 | .353 | .547 | .553 |
| 32 | .485 | .495 | .585 | .595 | .447 | .453 | .547 | .553 |

1. Predominately used for 64K DRAM
2. Predominately used for 256K DRAM

**Figure 3.21   The dimensions (inches) of rectangular PLCC packages. (Intel Corporation.)**

on each side. In larger PLCC, pin 1 can also be at the top left-hand corner of the package; this is the corner with the beveled edge. The location of pin 1 is confirmed by a dot in the package, and the progression of the pin count continues counterclockwise.

The leads in PLCCs provide the compliance needed to take up the solder joint stresses and thus prevent solder joint cracking. Since the design of the J lead is intended to provide lead compliance, it is important to ensure that the ends and sides of the leads are not touching the plastic body. In PLCCs there is some concern that bent leads touching the plastic body will be constricted in their movement, and hence may be noncompliant. This may happen if the leads are bent during either shipping and handling or planarization.

Planarization, the process of forcing the finished package with formed leads through a die, is conducted by some suppliers to control lead coplanarity. Planarization is not a recommended method for controlling lead coplanarity, however. This subject is discussed later.

There is also some concern about the thermal resistance of PLCCs because the surface area for heat dissipation is smaller. This is a more valid concern in SOIC than in PLCC. For example, the junction-to-ambient thermal resistance ($\theta$ja) for a 20 pin DIP, a 20 pin SOIC, and a 20 pin PLCC are 68, 97, and 72°C/W, respectively.

This means that SOIC will run hotter than DIP, but DIP and PLCC have a minor difference in thermal resistance. It is worth noting that in SOIC and PLCC, the combination of the copper lead frame and the short distance form the die to the PC board interface readily conducts heat away from the source. In DIPs, the lead frame is made of alloy 42 or Kovar, neither of which is as thermally conductive as copper.

Why isn't copper used in DIPs to lower thermal resistance? The reason is simple. Although more conductive thermally, copper is not as stiff as Alloy 42 or Kovar. The DIP leads must be stiffer to withstand the insertion and withdrawal forces of a socket, which are not encountered by surface mountable packages. Also, in PLCC, the leads are bent and tucked under the plastic body, hence do not need the stiffness required by a straight DIP lead.

It should also be noted that some PLCCs with large die-to-package ratios may be susceptible to package cracking due to moisture absorption. Suppliers and users are still struggling to find an appropriate solution. This issue is discussed in some more detail in Chapter 5, (Section 5.7.1).

### 3.4.4    Small Outline J Packages

Small outline J (SOJ) surface mount packages are used for high density (1 and 4MB) DRAMs. The low density DRAMs (64 and 256K) come in PLCCs as discussed earlier and shown in Figure 3.21. The SOJ packages have J-bend leads like PLCCs, but they have pins on only two sides, as shown in Figure 3.22. This package is a hybrid of SOIC and PLCC and combines the handling benefits of PLCC packages with the routing space efficiency of SOIC packages.

The 1 MB SOJ (DRAM) package is also referred to as 26 position, 20 pin package. This means that it could have a total of 26 pins but has only 20 pins. The leads on each side are split between two groups separated

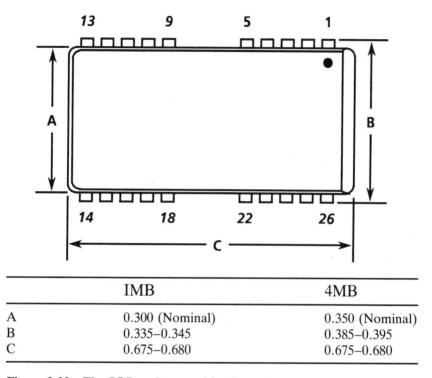

**Figure 3.22 The SOJ package and its dimensions (inches).**

|  | IMB | 4MB |
|---|---|---|
| A | 0.300 (Nominal) | 0.350 (Nominal) |
| B | 0.335–0.345 | 0.385–0.395 |
| C | 0.675–0.680 | 0.675–0.680 |

by a large center gap. The gap provides a space for traces to pass under the package. However, there are some Japanese packages used for static RAMs that do not have center gaps.

As shown in Figure 3.22, the pin count goes from 1 to 5; pins 6, 7, and 8 are skipped; then the pin count resumes at 9. On the right-hand side, pins 19, 20, and 21 are skipped, and the pin count resumes at 22 and ends at 26. The body width of the 1 MB DRAM is 300 mils, and it is referred to as EIA package MO-060.

The SOJ package for 4 MB DRAMs is available in a 26 position, 20 pin package. However, the body width is 350 mils. There is also some speculation that the body size will shrink to 300 mils. If both packages become available, the narrow package will be more real estate efficient, but the wider package may have lower thermal resistance.

## 3.4.5 Fine Pitch Packages

From both the real estate efficiency and package manufacturing stand-points, packages having more than 84 pins become impractical with 50 mil

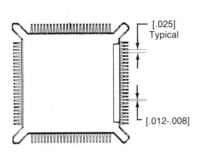
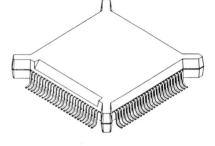

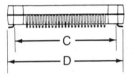

| TERMINAL COUNT | C | D |
|---|---|---|
| 84 | .710–.750 | .775–.785 |
| 100 | .810–.850 | .875.–885 |
| 132 | 1.010–1.050 | 1.075–1.085 |
| 164 | 1.210–1.250 | 1.275–1.285 |
| 196 | 1.410–1.450 | 1.475–1.485 |
| 244 | 1.580–1.620 | 1.645–1.655 |

**Figure 3.23   The U.S. version of a fine pitch package with corner bumpers (dimensions in inches). (Intel Corporation.)** NOTE: **D is maximum terminal dimension and not maximum body dimension. Only dimension C and D are needed for land patter design (Chapter 6).**

lead centers. When the package becomes large, maintaining lead coplanarity is difficult. Therefore the industry is moving toward finer pitch packages for pin counts of 84 and up. As shown in Figure 3.13, with the trend toward the usage of larger pin count packages, the lead pitch must decrease to ensure real estate efficiency. Also, when the lead counts increase as will be the case with the increased usage of application specific integrated circuit (ASIC) devices, finer pitch devices will become very common.

The current standard adopted by all the VLSI manufacturers in the United States is the 25 mil center, gull wing package known as the minipack (Figure 3.23). Even though the J-lead PLCC has been accepted as the industry standard, the fine pitch packages had to be of the gull wing type. As lead thickness and width decrease, J-lead packages are harder to manufacture.

The most exotic feature of the U.S. version of the fine pitch package is its corner bumpers, which extend about 2 mils beyond the leads to protect them during handling, testing, and shipping. For this reason this package type is usually called a "bumper" pack. Board real estate is not wasted by the bumper because the lands extend at least 10 mils beyond the leads to form solder fillets. The fine pitch packages are available in tube, tray, or tape and reel.

The Electronics Industry Association of Japan (EIAJ) has fine pitch packages known as quad flat packs (QFPs) with various body sizes, thicknesses, and lead pitches covering pin counts of 44 through 160. All these packages have one major disadvantage. They are all suceptible to lead damage and distortion of lead planarity during shipping, handling, and placement. The corner leads are especially prone to damage.

Loss of lead planarity in a five-pitch package may not be of concern if the hot bar soldering process is used instead of reflow (vapor phase or IR) soldering.

The Japanese packages require packaging in waffle packs (if there are no bumpers, packages need to be handled in waffle packs to prevent lead damage), and a placement machine has problems in handling waffle packs. For example, special waffle-pack handlers are required and a pick-and-place machine has a limited waffle-pack handling capacity. Most pick-and-place machines will accommodate a maximum of six waffle packs, thus their capacity is limited to a maximum of six different part numbers. If a board uses more than six fine-pitch part numbers, it will require additional passes on the placement machine. This is not desirable in a manufacturing environment. The American version of fine-pitch devices with corner bumpers do not pose such manufacturing constraints since the bumpered devices can be packaged in tubes, tape, and reel, and waffle packs (if necessary).

Lead damage has been a concern in all gull wing type leads, but the problem is compounded in the fine pitch packages because the leads are thin (about 10 mils) and quite fragile. The Japanese QFP packages are only 80 mils thick. The lower package thickness compounds the thermal problems in Japanese packages.

The bumpered package, being much thicker (140 mils), enhances package thermal conductivity through the use of heat spreaders. As shown in Table 3.3, the bumpered package with a heat spreader coupled to its copper lead frame drops the junction-to-ambient thermal resistance value of a 100 lead package by approximately 60% relative to the QFP [2]. The thinner

**Table 3.3 Comparison of thermal resistance of thinner (Japanese) and thicker (U.S.) fine pitch packages [2].**

| | $\Theta_{ja}$ (°C/W) | | |
|---|---|---|---|
| LEAD COUNT | QFP (JAPAN) | FINE PITCH (U.S.) | FINE PITCH WITH HEAT SPREADER (U.S.) |
| 64/68 | 100 | 60 | — |
| 80/84 | 85 | 47 | 35 |
| 100 | 80 | 44 | 33 |
| 132 | — | 40 | 30 |

QFP packages are good for applications in which height is a constraint and all other devices are of the same height.

A challenge to the users of fine pitch packages will be to accurately place the package on the substrate. Use of an accurate placement machine fitted with vision capability is necessary. The fine pitch packages have thinner leads and require a thinner (12 mils) land pattern design. Other challenges for the users of fine pitch packages include getting consistent and good solder paste printing on a 12 mil wide pad and repairing bridged joints.

## 3.5 MISCELLANEOUS COMPONENTS

There are many other miscellaneous devices that we have not discussed, but their availability in surface mount is critical for having a full Type I surface mount board. Examples of such devices are inductors, trimmers, transformers, filters, potentiometers, thermistors, switches, crystal oscillators, delay lines, sockets, and connectors. All these devices are available from limited vendors, but many are not compatible or desirable for surface mounting. Some of the devices are available in packages that cannot withstand the reflow temperatures. Adverse effects may also be seen when a through-hole mount part such as a DIP is converted into surface mount by just cutting or bending of leads. The parts that cannot withstand the reflow temperatures are really not surface mountable.

Progress in developing standards for some of these components is being made. For example, the surface mount versions of connectors and sockets are slowly becoming available for both through-hole and surface mount components. The IPC 406 committee is developing standards for connectors. The EIA is also developing standards. Unfortunately, there is no general agreement on lead configuration, size, or pin counts.

In general there has not been a rush by users to acquire surface mount connectors and sockets because these devices do not provide any real estate benefit at this time. Also, they generally require some form of mechanical hold-down mechanism to reduce stress on solder joints. There is concern that the solder joints in large connectors may not withstand the mechanical stress. While the industry is struggling to develop standards for surface mount connectors, many users are using custom surface mount connectors to meet their unique needs. This approach adds to the cost and further impedes the development of standards for connectors.

The standards for other components are also evolving slowly. For example, switches are being developed with the same lead configuration and body dimensions as an SOIC package, jumpers (zero ohm resistance) have the same dimensions as 1206 resistors; and crystal oscillators are taking the same lead configuration as PLCCs, although they have only four leads. Even the dimensions of ceramic resistors and capacitors, which are well-settled and widely accepted components, are being considered for change by the EIA committees. Specifically, it is contemplated to have the same mechanical outline for both. Currently, the ceramic capacitors are twice as thick as resistors of the same size (1206, for example).

As mentioned throughout this book, surface mount technology is in a state of transition. As the usage of this emerging technology grows, market forces will eventually settle on the lead and body configurations and sizes of different components.

## 3.6 FUTURE COMPONENTS

As packaging density increases and higher and higher pin count devices are used in surface mount instead of pin grid array (PGA) packages, the pitch or the lead-to-lead center distance is decreasing. Currently 25 and 20 mil center packages are available in lead counts of about 200. The 25 mil center package is the limit as far as the conventional pick-and-place, screen printing, and reflow soldering techniques are concerned. As the lead pitch falls below the 25 mil center, new manufacturing techniques will be necessary. For example, for the 20 mil center package developed by National Semiconductor, the pick-and-place machine and solder application techniques must be modified.

It is simply inevitable that lead pitches will drop much further to achieve more and more package shrinkage. However, these smaller packages will have very fragile leads and will require manufacturing technologies entirely different from those currently being used for surface mounting.

The next stage is to get rid of the packages entirely and directly mount the semiconductor device on the board. This will be achieved by means of the chip-on-board (COB), tape automated bonding (TAB), and flip chip technologies, which are now being used in specialized applications such as smart credit cards and watches. Refer to Chapter 1 (Section 1.7) for additional details.

The IPC is developing a new document on the design and assembly techniques of future packages. Details can be found in IPC-SM-784 (Guidelines for Chip-On-Board Technology Implementation). It is possible that there will be only three types of device in the electronics assemblies of the future: 2 pin passive resistors and capacitors, 28 or 32 pin memory devices, and high pin count (application-specific integrated circuit) (ASIC) devices.

There are some premolded packages that were not discussed in this chapter. For example, the EPIC package (premolded leadless plastic) by British Telecom is very similar in concept to the leadless ceramic chip carrier but is made of plastic. The user must put the die inside the package, wire bond it, and seal it. This is a difficult process for most SMT manufacturing houses to coordinate with the wire bonding process. The British package has not really caught on and is rarely used today, but may see service in the future.

## 3.7   MAJOR ISSUES IN COMPONENTS

Many issues in component packaging need to be resolved. Some are technical, some are related to supply and demand, and others are due to lack of standards. In this section we discuss some of the major component issues such as lead coplanarity, lead configuration, lead or termination finish, packing components for shipping and handling, and standardization. Finally, we will discuss component procurement strategy in light of the issues dealt with in this section.

### 3.7.1   Lead Coplanarity

One of the critical issues in surface mount packages with leads is the coplanarity of leads. "Coplanarity" is defined as lying or acting in the same plane. Mathematically speaking, therefore, coplanarity is the distance of component leads above and below a common plane defined by a plane that passes through the average length of all the leads.

This definition is difficult to implement on the manufacturing floor.

Thus noncoplanarity, a simplified term, is the maximum distance between the lowest and the highest pin when the package rests on a perfectly flat surface (see Figure 3.24). This definition represents a package sitting on a PC board on at least three leads.

What this tells us is that even if one lead out of 68 or 84 leads is out of an acceptable range, many leads, not just that particular lead, may not solder properly. In the worst case, coplanarity will cause open solder joints. Let us discuss the solder-open situation first, and then the solder joint weakness caused by lead coplanarity.

The common problem that the users encounter because of coplanarity is the phenomenon known as solder wicking; that is, the solder paste wicks up the lead, causing open solder joints. It should be kept in mind that poor board or lead solderability or uneven and fast heating during the reflow process also can cause open solder joints. See Chapter 12 for a discussion on wicking.

Even if we can reduce the incidence of wicking by depositing thicker solder paste and exercising good process control, coplanarity may cause poor solder reliability. The farther away from the seating plane a lead rests, the more solder volume resides between the lead and the pad. Since the solder itself is weaker than the intermetallic bond at the base metal interface, failure occurs within the solder volume. Hence the concern that PLCC lead coplanarity has a direct effect on the strength of its solder joints [3]. The more solder present between a lead and pad, the weaker the joint will be.

For these two reasons (solder open and solder joint weakness), the user naturally wants all the lead ends to lie perfectly in the same plane, to avoid manufacturing problems.

The component suppliers, however, have a difficult time in supplying a perfectly planar package. The JEDEC committee established a maximum coplanarity of 0.004 inch. Is this adequate, however, for preventing open solder joints? With solder paste 8 mils thick, leads with 6 mil coplanarity can be soldered with proper process control. But the author of Reference 3 found that solder joint strength shows a sharp decrease beyond 0.004

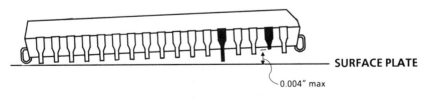

(DIMENSIONS NOT TO SCALE)

**Figure 3.24　Lead coplanarity in PLCC.**

inch coplanarity. It should also be kept in mind that even if the packages are supplied in perfect planarity, 1 to 2 mils can be added during shipping and handling, especially if methods other than tape and reel are used. The issue of tape and reel is discussed in Section 3.7.3.

An 0.004 inch maximum coplanarity is generally accepted by both users and suppliers, and it is also the JEDEC standard. Since the solder paste thickness varies from 0.006 to 0.010 inch, users can live with 0.004 inch coplanarity. In any event, suppliers have a very difficult time meeting tighter requirement, especially for the large packages. Shadow graphs, vision systems, or gauges are generally used for determining coplanarity. However, measuring coplanarity in a production environment is not an easy task for the user. Honaryar, of the University of Lowell (Massachusetts) Center for Productivity Enhancement, has undertaken the development of a pneumatic sensor device for measuring lead coplanarity just before component placement [4].

All these different methods for measuring coplanarity are either slow, sophisticated and expensive, or still under development. Thus the user should insist that the supplier ensure that the 0.004 inch coplanarity requirement is met. It should be noted that the number for acceptable lead coplanarity is not ±0.004 or ±0.002 inch, but 0.004 inch maximum, and it should be given as such in the procurement specification.

## 3.7.2   Lead Configuration

The configuration or the shape of leads is important for reasons having to do with manufacturability and reliability. The two predominant lead shapes that have been used are gull wing leads, which bend down, and out and J-shaped leads. Gull wing leads are used in SOICs and J leads in PLCCs. Now the industry is also using some butt leads by simply cutting DIP leads for surface mounting.

Each lead configuration has its advantages and disadvantages, as summarized in Figure 3.25. Gull wing leads have been used very extensively in Japan, but the J-lead configuration is more common in the United States. In lower pin count devices, the gull wing shape is common in every country because SOIC is more space efficient than a PLCC package.

The major advantages of gull wing leads are their compatibility with trends toward the thin, fine pitch packages of the future. They can be soldered by various of soldering processes including hot bar. Their main disadvantage is their susceptibility to lead damage during shipping and handling especially in fine pitch packages without corner bumpers. (See Section 3.4.5.)

The J-lead configuration is more space efficient than the gull wing

| SHAPE OF TERMINAL / ITEM | GULL-WING  | J-LEAD 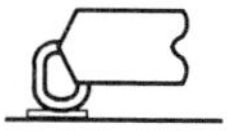 | BUTT-LEAD 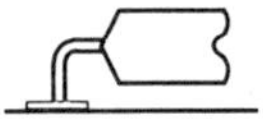 |
|---|---|---|---|
| Compatibility with Future Multi-Pin Package Trend | ◎ | ○ | △ |
| Package Thickness | ◎ | ○ | △ |
| Lead Rigidity | △ | ◎ | ○ |
| Compatibility With Multiple Soldering Methods | ◎ | ○ | ○ |
| Ability To Self-Align During Reflow | ◎ | ○ | △ |
| Accessibility For Inspection After Soldering | ○ | ○ | ○ |
| Ease Of Cleaning | ○ | ◎ | ◎ |
| Real Estate Efficiency | △ | ◎ | ○ |

◎ = EXCELLENT          ○ = GOOD          △ = INFERIOR

**Figure 3.25   Surface mount lead configuration characteristics.**

form and can be soldered by most reflow processes except hot bar. J leads are sturdier than gull wing leads, hence stand up better in shipping and handling. However, the gull wings are a lower profile package than the J-lead configuration. This may be an advantage when all packages must have lower profiles.

It is sometimes wrongly believed that gull wing leads are better for inspection. Actually the inner fillets are difficult to inspect in any lead configuration. It is worth noting that the outer and inner fillets are the major fillets in J-lead and gull wing devices, respectively. Thus inspection, if it is a problem, should be a concern in gull wing devices.

Butt lead, also known as I lead, devices are not commonly used. The advantage claimed for butt leads is the possibility of converting a through-hole component into surface mount by clipping the leads and accomplishing the soldering of all components in one reflow operation. Since all the components are not available in surface-mounted form, this is a convenient way to reach the goal of a full Type I surface mount assembly. However, through-hole packages generally are not meant for reflow soldering. If they are to be reflow soldered, they must be able to withstand the reflow soldering temperatures.

Butt mounting of DIPs does not provide any real estate saving but sometimes allows a one-step soldering process and provides some cost saving. This saving may not always be realized, however. In Chapter 7 (Section 7.4) we discuss situations in which it may or may not be appropriate to use butt leads for surface mounting. But are the butt joints reliable? Dennis Derfiny of Motorola conducted a reliability study on the solder joints formed by butt, J, and gull wing leads [5], using lead forms made of the same lead frame materials mounted on glass epoxy boards. Pull, shear, and thermal cycling tests were conducted for reliability determination.

While the relevance of the pull and shear tests for solder joint reliability may be debated since the controlling failure mechanism is fatigue, the data give relative reliability of different lead configuration. The butt lead joints were found to have 65% less pull and shear strengths than J or gull wings. The test results were corroborated by $-40$ to $+125°C$ thermal cycling tests. It was also found that the butt lead configuration was more sensitive to process-related handling, placement, and soldering when compared to the J or gull wing lead configuration.

There is an interesting trend in Japan on butt leads. Some leading Japanese companies are using butt lead configuration and finding considerably lower defect (bridging and open joints) when compared to gull wing devices.

Such a trend is supported by other researchers who not only prefere butt leads over other types but even suggest that gull wing leads be twisted to form butt [6]. The twisted lead configuration, a variation of the butt lead, will allow more routing space between leads and, by piercing through the solder paste, will prevent paste displacement [6].

At this time, considerable work is going on in the industry on the advantages and disadvantage of butt lead configuration. Given the satisfactory reliability, manufacturability data and the experience in the industry on the gull wing, and "J" lead devices, it will be some time before butt leads become as common as "J" leads, if they ever do.

Another variation of the gull wing was suggested by authors from IBM [7], who provide a via hole in gull wing or J leads. The via is intended to provide an escape path for volatiles and will prevent the internal voids common in reflow solder joints. The via hole will also allow the solder to penetrate through and encapsulate the lead frame, both top and bottom (in addition to side fillets as in gull and J leads); hence the solder joint formed will be stronger.

It should be noted that the concepts suggested in References 6 and 7 are just that. Concepts. They have not been adopted by the industry, and it is not clear whether they will ever be used. However, they do deserve consideration for the obvious benefits pointed out by their proponents.

The reliability of solder joints formed by the most commonly used lead configurations, the J and the gull wing, is considered to be acceptable by the industry in general and is not an issue. But there is a difference in the impact of these configurations in manufacturing. The selection of J on gull wing should not be based only on the pros and cons of these packages, because both are acceptable and each has a long history. The determining factors should be availability from the vendor of choice and real estate considerations.

Butt leads, especially those created by cutting DIP leads, should not be selected until the reliability issues have been resolved. It should be noted that DIPs may be damaged during bending operations for J or gull wing shapes or during cutting operations for butt lead shapes.

### 3.7.3    Standardization

It cannot be overemphasized that standardization, one of the key issues in promoting any new technology, is almost mandatory for surface mount technology because of the need for automation. The lack of surface mount standards has been a major concern for SMT users. Moreover, where standards do exist, they may not be adequate for the manufacturing processes desired.

The problem of standardization is exacerbated because the Japanese, European, and U.S. manufacturers tend to go their own way. In any event, tolerances in existing standards are so poor that they may be of little help if used as procurement specifications. Proliferation of various packages was cited in Section 2.2.2: users have been accepting packages any way they can get them, with or without standards. Use of tape and reel is another example.

Tape and reel packaging is best for accuracy and throughput in component placement. EIA-RS-481 establishes a standard for tape and reel, but it is not without problems. The maximum size of the reel is set at 13 inches. There is no minimum size. Thus a reel can be any size up to 13 inches.

Since active components have generally been available only in the 13 inch reel size, inventory cost may be a real problem for small users and even large users during developmental stages of a product. There should be some effort in EIA to change this standard to allow 7 and 10 inch reels to meet the needs of both small and large users. Refer to Chapter 11 (Section 11.4) for details on the advantages and disadvantages of feeders of different types and the standardization issues in tape and reel. Also refer to Chapter 2 for a better understanding of how standards are set and how one can influence development of standards through the Electronics

Industries Association and the Institute of Interconnecting and Packaging Electronic Circuits.

## 3.8 COMPONENT PROCUREMENT GUIDELINES

Developing in-house component procurement specifications is important for any surface mount component. This is especially true because the technology is in a transition stage. There are certain general guidelines, as discussed below for developing in-house specification.

- One should deal with a limited number of vendors, establishing specifications in cooperation with the preferred vendors. The current trend is to rely on one vendor.
- Only limited number of packages that enjoy widespread support should be selected for use. Limiting the number of part types provides such benefits as increased purchasing leverage, reduced inventory cost, and reduced number of feeders for the pick-and-place equipment.
- The components selected should be qualified for performance and reliability, solderability, component tolerance and compatibility with land pattern design, and compatibility with processes and equipment used in manufacturing.

Within the framework of the general guidelines above, some specific requirements should be incorporated into the procurement specifications.

1. Solder-dipped terminations or leads provide good solderability but may compound coplanarity problem. Hence the tin-lead plated terminations or leads are becoming widely used. Most users want to minimize the intermetallic thickness by asking that terminations or leads not be fused before shipment.

   The coating thickness of solder should be 300 microinches (0.3 mil) and solder should contain 60 to 63% tin. Both types of solder coating are acceptable as long as they meet coplanarity and solderability requirements. See Chapter 10 (Section 10.6) for a discussion of solderability and the impact on this property of plating and dipping.

2. The preferred lead frame material for better thermal conduction is copper, and this metal should be required.

3. The component tolerances should be closely specified. Industry standards and tolerances should be specified where acceptable.

The component tolerances should be measured after solder dipping or fusing.

4.  A nickel barrier underplating of 0.05 mil between the solder termination and the gold or silver adhesion layer should be required to prevent leaching or dissolution of gold or silver during the soldering process. This requirement is necessary primarily for passive components but should be specified for the leads of active or passive devices if precious metal underplating is used.

5.  The parts must be able to withstand at least two cycles in the soldering environments used in manufacturing. Specifying the soldering profile is more relevant, but 60 seconds at 215°C in vapor phase or 2 minutes at 230°C in an infrared oven or at least 10 seconds in molten solder at 260°C should be used. If all these soldering methods are used in manufacturing, the parts should be able to withstand all these temperatures for the specified time without any measurable impact on performance or reliability.

6.  The parts must be resistant to solvents at the temperatures (generally 40°C) used for cleaning, including exposure to ultrasonic cleaning if necessary. The requirement of 1 minute at a frequency of 40 kHz and a power of 100 W/ft$^2$ is sufficient [1]. Ultrasonic cleaning for active devices is generally not recommended for fear of breaking the internal wire bonds.

7.  Component marking should be required for active devices. For passive devices marking is not necessary, especially if there is a cost premium and if tape and reel packaging is acceptable. Tape and reel should be specified as the packaging format whenever feasible, as discussed earlier. The embossed plastic reels are preferable to paper or punched reels because the former are less prone to delamination and placement problems. Parts that are sensitive to electrostatic discharge must be supplied on conductive reels.

8.  Surface mount standards, even when they exist, are not supposed to be used as procurement specifications. The general intent of most standards is to accommodate the needs of all member companies. This tends to make a given standard too loose. Therefore the industry standards should be used as a guideline, not as a substitute for in-house procurement specifications.

## 3.9   SUMMARY

Surface mount components are no different from through-hole components as far as the electrical function is concerned. Because they are smaller, however, the SMCs provide better electrical performance. Not all components are available in surface mount at this time; hence the full benefits of surface mounting are not available, and we are essentially limited to mix-and-match surface mount assemblies.

While only a few types of conventional DIP packages meet all the packaging requirements, the world of surface mount packages is vastly more complex. The package types and the package and lead configurations available are numerous. In addition, the requirements of surface mount components are far more demanding. SMCs must withstand the higher soldering temperatures and must be selected, placed, and soldered more carefully to achieve acceptable manufacturing yield.

There are scores of component types available for some electrical requirements, causing a serious problem of component proliferation. There are good standards for some components, whereas for others standards are inadequate or nonexistent. Some components are available at a discount, and others carry a premium.

Surface mount technology is in a transition phase, the electronics industry is making progress every day in resolving the economic, technical, and standardization issues associated with surface mount components. In the meantime, one must develop in-house procurement specification and not wait for the standards to firm up before taking advantage of surface mount technology.

## REFERENCES

1.  ANSI/IPC-SM-782. 1987 Surface Mount Land Patterns (Configuration and Design Rules). IPC, Lincolnwood, IL.

2.  Knuduson, E. "Surface mount packages for VLSI device." *Proceedings of the SMART III Conference* January 1987. Technical paper SMT III–33. IPC, Lincolnwood, IL.

3.  Smith, W. D. "The effect of lead coplanarity on PLCC solder joint strength." *Surface Mount Technology,* June 1986, pp. 13–17.

4.  Honaryar, B. "Progress report on a pneumatic inspection device for coplanarity sensing of surface mounting PLCC." *Proceedings of the SMART III Conference,* Technical paper III-19, IPC, Lincolnwood, IL, January 1987.

5.  Derfiny, D., and Dody, G "On optimizing lead form." *NEPCON Proceedings,* 1987, pp. 251–256.

6.  Buckley, D. "SMC packaging causes problems. "*Electronic Production,* April 1986, pp. 32–33.

7.  Kang, S. K. and Moskowitz, P. A. *IBM Technical Disclosure Bulletin,* Vol. 29, No. 4, September 1986, pp. 1612.

# Chapter 4
# Substrates for Surface Mounting

## 4.0   INTRODUCTION

The substrate, also referred to as the packaging and interconnecting structure, plays a crucial role in ensuring the electrical, thermal, and mechanical reliability of electronic assemblies. Before choosing from among the many types of substrate that are available for military and commercial applications, however, it is necessary to determine the properties that will be required. Then a substrate material that meets all the requirements in a cost-effective manner can be selected.

The substrate properties that will be required depend on the type of application (commercial or military) and the type of packages (ceramic or plastic). Ceramic packages (generally leadless ceramic chip carriers: LCCCs) are used in military applications to provide the needed hermeticity. In addition, as mentioned previously, if there is any difference between the coefficients of thermal expansion of component packages and substrates, the stress induced by CTE mismatch can strain the solder joints and cause fatigue failure after repeated thermomechanical cycles.

There are essentially three approaches to overcoming the stress on solder joints when LCCCs are used on the substrate: (a) developing substrates with matching CTE to prevent the generation of stress on solder joints, (b) developing substrates with compliant top layers that can absorb the stress, and (c) adding leads to the surface mount packages to take up the stress.

Each method has its advantages and disadvantages. Approach (a), the use of a substrate with a compatible CTE, is the most common, but these exotic and expensive substrate materials require development. Moreover, even when a substrate has compatible CTE values, failures in the via holes are encountered. To prevent via hole cracking, either the board thickness is reduced or the via hole diameter is increased. A modification of approach (a) is to use a substrate material with a lower modulus of elasticity in the vertical direction to prevent via hole cracking.

Approach (b), the use of a substrate with a compliant top layer, has been tried but has not really caught on. Solder joint cracking may be prevented, but the reliability problems are shifted to surface traces and vias.

Approach (c), the use of leaded packages, allows the selection of relatively inexpensive FR-4 glass epoxy substrates if they meet the thermal and electrical requirements. The paper-based substrates used in many consumer applications are even less expensive than glass epoxy materials. However, even the decision to use leaded ceramic packages does not completely eliminate solder joint cracking problems if the lead material and lead configurations are not very compliant. The use of leaded plastic packages on a glass epoxy substrate is the sure way to prevent reliablity problems. But this approach is limited mostly to the commercial sector and is not adequate for military applications. There are no easy answers in substrate selection.

In this chapter we discuss some commonly used substrates for approaches outlined above. However, we concentrate on the glass epoxy substrate not only because it is most widely used but also because the final process steps in the fabrication of glass epoxy boards are applicable to most other substrates. But first let us discuss some basic concepts. The notions of glass transition temperature and coefficient of thermal expansion are useful no matter what substrate is used.

## 4.1 GLASS TRANSITION TEMPERATURE ($T_g$)

Except for the ceramic substrate, almost all substrate laminates contain polymers. Polymers, unlike metals, undergo major structural changes at certain temperatures. The temperature at which the physical structure of the laminate changes from hard and brittle, or glasslike, to soft and rubbery (rubberlike) is called the glass transition temperature $T_g$. The laminate materials go through a structural change above $T_g$ and lose a considerable amount of their mechanical strength.

The glass transition temperature is the characteristic property of most polymers, hence is unique to a particular polymer material. Next to CTE, $T_g$ is the most important property for the laminate and is critical in the selection of a laminate material.

Stress-deformation curves for an epoxy-glass laminate illustrate the criticality of $T_g$. As shown in Figure 4.1, at a temperature below $T_g$, to attain a specific level of deformation in the laminate $d_0$, a stress $F_0$, must be applied. If the same stress is applied above $T_g$, a much larger defor-

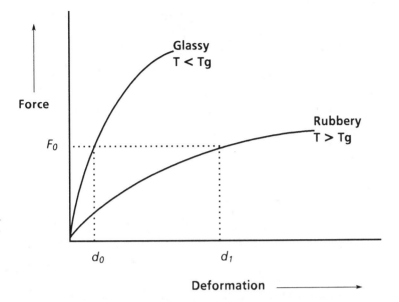

**Figure 4.1 The deformation in laminate materials at temperatures above and below their glass transition temperature $T_g$.**

mation, $d_1$, will result. This larger deformation will generate considerable warpage and will degrade the reliability of the printed board. Figure 4.1 shows the effect of force on deformation at temperatures above and below $T_g$. However, like most materials, the laminate also expands with an increase in temperature even if no stresses are being applied. Such expansion increases exponentially above $T_g$ and linearly below $T_g$.

The value of $T_g$ plays an important part in temperature-related expansion. Let us compare the expansions in two commonly used laminate materials: FR-4, which has a $T_g$ of about 120°C, and polyimide, which has a $T_g$ of about 230°C. For example, as shown in Figure 4.2, the unconstrained thermal expansion of epoxy glass is about 0.019 inch per inch at 150°C. However, for polyimide it is minuscule in comparison—about 0.001 inch per inch. [1] The reason for this disparity is that at 150°C, the $T_g$ for the epoxy glass laminate has been exceeded and the exponential expansion rise has already set in, whereas for polyimide, the temperature is still much below the material's $T_g$, hence laminate expansion is still in the relatively low linear range.

**Temperature in °F**

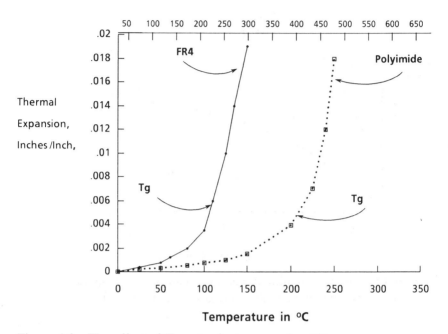

Figure 4.2   The effect of $T_g$ on laminate expansion [1].

## 4.2   X,Y, AND Z COEFFICIENTS OF THERMAL EXPANSION

Above the glass transition temperature, the laminate materials—that is, resins, not the glass material in the laminate—expand greatly. The thermal expansion values shown in Figure 4.2 are for the free and unconstrained expansion of the resin ingredient of the laminate. When the laminate is used in a multilayer printed circuit board, however, the other elements of the multilayer board exert constraining effects on the laminate.

The constraining effects are not identical in all dimensions. This anisotropy in constraining effects is the cause of many major problems in the manufacturing process. In the X and Y directions, laminate thermal expansion is constrained by the glass fibers in the laminate, the copper traces and pads on all the layers, and the barrel plating around the plated through hole (Figure 4.3). However, the constraining effect is very small in the Z direction, only the plated through-hole (PTH) barrel having some effect near the regions around the barrel.

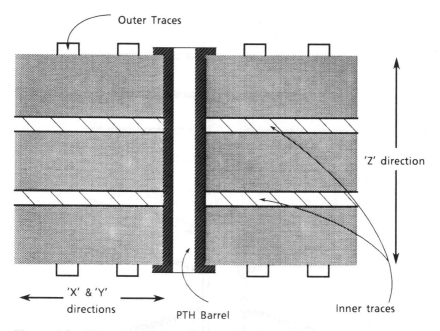

**Figure 4.3  The constraining effects on laminate in a multilayer board.**

A look at thermal expansion for an epoxy glass laminate based multilayer board will illustrate this point better. Below $T_g$ for the X and Y directions, the coefficient of thermal expansion is between 12 and 16 microinches (0.000,016 inch) per inch per degree Celsius (ppm/°C). For the Z direction, the expansion coefficient is 100 to 200 ppm/°C. This is an order of magnitude greater than the CTE for the X and Y directions, since there is no fiber weave to provide restraint.

The Z axis thermal expansion in and around a multilayer PTH board was empirically measured by researchers at IBM [2]. They found that the laminate (50 mils away from the barrel of a plated through hole) and the barrel increased in length at quite similar rates. However, they found a sudden, sharp increase in the expansion of both laminate and barrel. The rate of expansion was still linear, but the laminate expansion rate was higher than that for the barrel. Hence, as temperature increases beyond 110°C, the disparity in the Z axis expansion between the laminate and barrel becomes larger. The effect of this phenomenon on a plated through hole is illustrated in Figure 4.4.

As shown in Figure 4.4, the free expansion rate of the plated through-hole barrel (copper and solder) is lower than that of the surrounding lam-

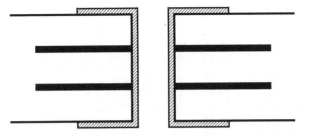

**Room temperature - no strain**

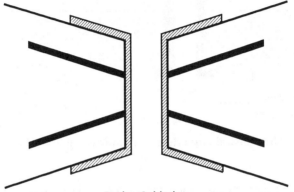

**During Soldering**

**Figure 4.4   Different rates of expansion in materials used in a multilayer board can cause deformation in internal layers, as shown in the bottom sketch.**

inate. This implies that the barrel is being pulled along by the laminate because the laminate expands faster than the barrel. This strain condition creates tensile stresses in the barrel, and as the temperature increases, the tensile stresses will increase. If and when these stresses exceed the fracture strength of the barrel plating, the plating will fracture, thereby causing board failure. The higher rate of laminate expansion causes a dramatic increase in the stresses within the plated through-hole barrel, inner traces, and pads. These stresses can result in cracks in the barrel, traces, or pads or even in the laminate, especially if the substrate is operated above its $T_g$. Due to uneven heating, this problem is further compounded during hand soldering than in wave or reflow soldering.

## 4.3 SELECTION OF SUBSTRATE MATERIAL

As indicated in Table 4.1, all the numerous substrate materials available on the market today for military and commercial applications have their advantages and disadvantages [3]. During fabrication, assembly, and service, these materials may be subjected to high temperatures, high humidities, corrosive chemicals, and mechanical stress. They are expected not only to survive these severe conditions but also to perform electrically. Therefore, it is imperative that designers and manufacturing personnel have a good understanding of substrate materials and their properties. This knowledge will enable designers to predict the behavior of materials under fabrication, assembly, and service conditions. In any event, in case of a failure, the follow-up analysis requires a knowledge of material properties.

Each of these materials has its own particular characteristics and properties. Consequently, the materials behave differently under varying conditions of temperature, humidity, stress, and so on. For the assembly to function properly, the electrical and mechanical requirements of each of the following materials used in substrates must be met under their characteristic operating conditions and environments.

1. The laminate, which forms the basic structural backbone and determines the mechanical and electrical characteristics of the substrate.
2. The solder mask or conformal coating, which is applied on the outer layers of the board.
3. The metal platings in the plated through holes and via holes, as well as the PTH pads and surface mount lands, which provide electrical interconnections.
4. The inner layer and outer layer copper traces, which form the pathways for the signals to flow from one device to another.
5. The legend ink on top of the solder mask, which serves for component identification.

Among the important properties that must be considered in selecting substrate materials are cost, performance, and reliability. The principal substrate properties that should be taken into account in the selection process are summarized in Table 4.2 [3]. The specific properties of various substrates commonly used in the industry are shown in Table 4.3 [3].

**Table 4-1 Advantages and disadvantages of various types of substrates [3].**

| TYPE | MAJOR ADVANTAGES | MAJOR DISADVANTAGES | COMMENTS |
|---|---|---|---|
| **ORGANIC BASE SUB-STRATE**<br>Epoxy fiberglass | Substrate size, weight; re-workable; dielectric prop-erties; conventional board processing | Thermal conductivity, X, Y, and Z axis CTE. | Because of its high X-Y plane CTE, it should be limited to environments and applications with small changes in tempera-ture and/or small pack-ages. |
| Polyimide fiberglass | Same as epoxy fiberglass plus high temperature Z axis CTE; substrate size; weight; reworkable; di-electric properties | Thermal conductivity; X and Y axis CTE; moisture absorp-tion | Same as epoxy fiberglass |
| Epoxy aramid fiber | Same as epoxy fiberglass; X-Y axis CTE; substrate size; lightest weight; re-workable; dielectric prop-erties | Thermal conductivity; X and Y axis CTE; resin microcrack-ing; Z axis CTE; water ab-sorption | Volume fraction of fiber can be controlled to tailor X-Y CTE. Resin selection critical to reducing resin microcracks. |

| Material | | | |
| --- | --- | --- | --- |
| Polyimide aramid fiber | Same as epoxy aramid fiber; Z axis CTE; substrate size; weight; reworkable; dielectric properties | Thermal conductivity; X and Y axis CTE; resin microcracking; water absorption | Same as epoxy aramid fiber. |
| Polyimide quartz (fused silica) | Same as polyimide aramid fiber; Z axis CTE; substrate size; weight; reworkable; dielectric properties | Thermal conductivity; X and Y axis CTE; Z axis CTE; drilling; availability; cost; low resin content required | Volume fraction of fiber can be controlled to tailor X-Y CTE. Drill wearout higher than with fiberglass. |
| Fiberglass/aramid composite fiber | Same as polyimide aramid fiber; no surface microcracks; Z axis CTE; substrate size; weight; reworkable; dielectric properties | Thermal conductivity; X and Y axis CTE; water absorption; process solution entrapment | Resin microcracks are confined to internal layers and cannot damage external circuitry. |
| Fiberglass/Teflon laminates | Dielectric constant; high temperature | Same as epoxy fiberglass; low temperature stability; thermal conductivity; X and Y axis CTE | Suitable for high speed logic applications. Same as epoxy fiberglass. |
| Flexible dielectric | Lightweight; minimal concern to CTE; configuration flexibility | Size | Rigid-flexible boards offer tradeoff compromises. |

**Table 4-1** *(Continued)*

| TYPE | MAJOR ADVANTAGES | MAJOR DISADVANTAGES | COMMENTS |
|---|---|---|---|
| Thermoplastic | 3-D configurations; low high-volume cost | High injection molding setup costs | Relatively new for these applications. |
| **NONORGANIC BASE** Alumina (ceramic) | CTE; thermal conductivity; conventional thick film or thin film processing; integrated resistors | Substrate size; rework limitations; weight; cost; brittle; dielectric constant | Most widely used for hybrid circuit technology. |
| **SUPPORTING PLANE** Printed board bonded to plane support (metal or nonmetal) | Substrate size; reworkability; dielectric properties; conventional board processing; X-Y axis CTE; stiffness; shielding; cooling | Weight | The thickness/CTE of the metal core can be varied along with the board thickness, to tailor the overall CTE of the composite. |
| Sequential processed board with supporting plane core | Same as board bonded to supporting plane | Weight | Same as board bonded to supporting plane. |

| | | | |
|---|---|---|---|
| Discrete wire | High speed interconnections; good thermal and electrical features | Licensed process; requires special equipment | Same as board bonded to low expansion metla support plane. |
| **CONSTRAINING CORE**<br>Porcelainized copper clad invar | Same as Alumina | Reworkability; compatible thick film materials | Thick film materials are still under development. |
| Printed board bonded with constraining metal core | Same as board bonded to supporting plane | Weight; internal layer registration; delamination; via hole cracking, Z axis CTE | Same as board bonded to supporting plane. |
| Printed board bonded to low expansion graphite fiber core | Same as board bonded to low expansion metal cores; stiffness, thermal conductivity; low weight | Cost; microcracking; Z axis CTE | The thickness of the graphite and board can be varied to tailor the overall CTE of the composite. |
| Compliant layer structures | Substrate size; dielectric properties; X-Y axis, CTE | Z axis CTE; thermal conductivity | Compliant layer absorbs difference in CTE between ceramic package and substrate. |

**Table 4-2   Substrate selection criteria [3].**

| DESIGN PARAMETERS | MATERIAL PROPERTIES | | | | | | | | |
|---|---|---|---|---|---|---|---|---|---|
| | TRANSITION TEMPERATURE | COEFFICIENT OF THERMAL EXPANSION | THERMAL CONDUCTIVITY | TENSILE MODULUS | FLEXURAL MODULUS | DIELECTRIC CONSTANT | VOLUME RESISTIVITY | SURFACE RESISTIVITY | MOISTURE ABSORPTION |
| Temperature & Power Cycling | X | X | X | X | | | | | |
| Vibration | | | | X | X | | | | |
| Mechanical Shock | | | | X | X | | | | |
| Temperature & Humidity | X | X | | | | X | X | X | X |
| Power Density | X | | X | | | | | | |
| Chip Carrier Size | | X | | X | | | | | |
| Circuit Density | | | | | | X | X | X | |
| Circuit Speed | | | | | | X | X | X | |

**Table 4-3  Substrate properties [3].**

|  | MATERIAL PROPERTIES | | | | | | | |
|---|---|---|---|---|---|---|---|---|
| MATERIAL | GLASS TRANSITION TEMPERATURE | XY COEFFICIENT OF THERMAL EXPANSION | THERMAL CONDUCTIVITY | XY TENSILE MODULUS | DIELECTRIC CONSTANT | VOLUME RESISTIVITY | SURFACE RESISTIVITY | MOISTURE ABSORPTION |
| *UNIT OF MEASURE* | *(°C)* | *(PPM/°C) (note 2)* | *(W/M°C)* | *(PSI × 10^{-6})* | *(At 1 MHz)* | *(Ohms/cm)* | *(Ohms)* | *(Percent)* |
| Epoxy Fiberglass | 125 | 13–18 | 0.16 | 2.5 | 4.8 | $10^{12}$ | $10^{13}$ | 0.10 |
| Polyimide Fiberglass | 250 | 12–16 | 0.35 | 2.8 | 4.8 | $10^{14}$ | $10^{13}$ | 0.35 |
| Epoxy Aramid Fiber | 125 | 6–8 | 0.12 | 4.4 | 3.9 | $10^{16}$ | $10^{16}$ | 0.85 |
| Polyimide Aramid Fiber | 250 | 3–7 | 0.15 | 4.0 | 3.6 | $10^{12}$ | $10^{12}$ | 1.50 |
| Polyimide Quartz | 250 | 6–8 | 0.30 |  | 4.0 | $10^{9}$ | $10^{8}$ | 0.50 |
| Fiberglass/Teflon | 75 | 20 | 0.26 | 0.2 | 2.3 | $10^{10}$ | $10^{11}$ | 1.10 |
| Thermoplastic Resin | 190 | 25–30 |  | 3–4 | $10^{17}$ | $10^{13}$ | NA |  |
| Alumina-Beryllia | NA | 5–7 21.0 | 44.0 | 8.0 | $10^{14}$ | $10^{13}$ |  |  |
| Aluminum (6061 T-6) | NA | 23.6 | 200 | 10 | NA | $10^{6}$ | NA | NA |

**Table 4-3** (*Continued*)

| MATERIAL | MATERIAL PROPERTIES | | | | | | | |
| --- | --- | --- | --- | --- | --- | --- | --- | --- |
| | GLASS TRANSITION TEMPERATURE | XY COEFFICIENT OF THERMAL EXPANSION | THERMAL CONDUCTIVITY | XY TENSILE MODULUS | DIELECTRIC CONSTANT | VOLUME RESISTIVITY | SURFACE RESISTIVITY | MOISTURE ABSORPTION |
| Copper (CDA101) | NA | 17.3 | 400 | 17 | NA | $10^6$ | | |
| Copper-Clad Invar | NA Note 1 | 3–6 | 150XY/ 20Z | 17–22 | NA | $10^6$ | | NA |

*Notes*: 1. The X and Y expansion is controlled by the core material and only the Z axis is free to expand unrestrained. Where the Tg will be the same as the resin system used.

2. Figures are below glass transition temperature, are dependent on method of measurement and percentage of resin content. NA—Not applicable.

## 4.3.1 CTE Compatibility Considerations in Substrate Selection

Although many properties must be considered in the selection of a substrate material (see Table 4.2), surface mounting has made the coefficient of thermal expansion requirement almost primary as a selection criterion, especially for military applications that use leadless ceramic packages. The reason for this is very simple. As illustrated in Figure 4.5, if the CTEs of the substrate and the packages are different, they will expand and contract at different rates during power and thermal cycling. This difference in rates of expansion caused by CTE will stress the joint, and in extreme cases the solder joint will be fractured.

As stated earlier, the three main approaches used to prevent the cracking of solder joint when using leadless ceramic packages are as follows: using substrates with matching CTEs, using substrates with a compliant top layer, and using leaded ceramic chip carriers.

Most of the effort in the United States has focused on the first approach—that is, the use of substrates with compatible CTEs to relieve stress on the solder joint. Examples of substrates selected for this approach are ceramic (Section 4.4), porcelainized steel (Section 4.4.1), and constraining core substrates such as copper-invar copper and graphite expoxy (Section 4.5). This approach has not been entirely satisfactory: it is very expensive, and the solder joint reliability results have been less than desirable.

Even when the solder joint cracking problem is solved by using substrates with compatible X-Y CTE values, failures have been shifted to the plated through and via holes. The via failures were initially attributed to an increased CTE in the Z axis direction caused by the constraints of the X-Y axis. However, the test data reported by Gray at the Printed Circuits World Conference 1987 [4] show that via hole reliability is a fundamental problem involving the plated through-hole aspect ratio (board thickness divided by drilled hole diameter). The large diameter plated through holes in conventional through-hole technology did not fail because the aspect ratio was less than 2.5. For the small via holes used in SMT, thin boards will be needed to maintain high PTH reliability. There are some practical limits for thin boards, because they also tend to be about eight layers because of increased interconnection density. Also, the via hole size may be increased to maintain a low aspect ratio (<3). Larger via holes, which take up routing area, still may not match the Z direction CTE of a copper-plated hole.

Another potential solution may be to use blind and buried vias, illustrated in Chapter 5 (Figure 5.2). Since the depth of blind or buried vias is

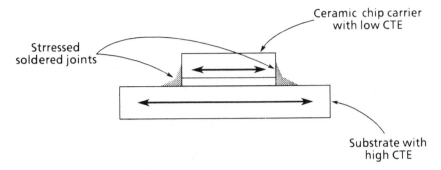

**Figure 4.5   CTE mismatch between substrate and component package can stress the solder joint.**

small, smaller via hole sizes can be tolerated without increasing the aspect ratio, hence without compromising via hole reliability. However, the technology for blind and buried vias is very immature at this time. Very few vendors have production capability, which tends to increase significantly the cost of substrates with such a configuration.

A possible modification of the compatible CTE approach is to use a substrate material having a lower modulus of elasticity. This does not require an increase in via hole size to maintain a low aspect ratio, hence good via hole reliability. Since the modulus of the substrate material is low, it is soft and cannot rupture via holes even if there is a great deal of Z direction expansion of the substrate material. An example of a substrate material with low modulus of elasticity is the Rogers product R02800. This material has a vertical CTE of only 24, but its modulus is only one-tenth that of epoxy. This property would make cracking unlikely even with very small via holes. However, the industry does not have much experience with this approach, which has not been tried and is rarely used. There are not enough data on the material.

The second approach, which has been tried, is to use a compliant substrate top layer, thus transferring solder joint stress to the "buttery" top layers. However cracking of the traces and via holes remains a chief reliability concern. The compliant layer substrate is discussed in Section 4.6.

The third approach is to use leaded instead of leadless ceramic packages on substrates with unmatched CTEs, such that the stress on the solder joint is taken up by the compliant leads. There are many problems with this approach also. For example, the commonly used Kovar or alloy 42 leads are not very compliant and do not completely eliminate solder joint cracking problems unless a substrate with a compatible CTE is used. In addition, the leaded ceramic packages do not address the need for intimate contact between the substrate and the package, to ensure effective cooling.

For military applications, other approaches that should be considered are design related. For example, the solder joint stress may be reduced by reducing the thermal stresses between substrate and the packages. If the package/substrate combination does not change temperature much, failures related to CTE mismatch will be reduced. Thermal cooling aids such as thermal vias, heat sinks, and immersion cooling have been used in this connection. These issues are further discussed in Chapter 5 (Section 5.6: Thermal Considerations) and Section 5.8: (Solder Joint Reliability Considerations).

The safest way to prevent reliability problems without increasing cost is to use plastic leaded packages on glass epoxy substrates. This approach is almost universally used. Any stress generated by minor differences in CTE values is easily accommodated by compliant copper leads in the PLCCs. The main problem associated with this approach is that it is limited mainly to commercial applications. But the commercial use of glass epoxy boards accounts for the majority of the PCB firms' business, and these boards are the most widely used in the industry. Fabrication of glass epoxy substrates is discussed in detail in Section 4.7.

## 4.3.2    Process Considerations in Substrate Selection

Not all substrates can withstand all manufacturing environments equally well. Thus when selecting a substrate, the process to which the material will be subjected must be considered. The surface mount assembly processes can have an adverse impact on the substrate material if extra care in processing is not exercised with respect to the application of thermal and mechanical stresses, the removal of corrosive contaminants from the surface of the board, and so on.

Wave soldering, reflow soldering, repair/rework, and burn-in are assembly processes during which heat, or mechanical stresses, or both are applied to the entire board or to a localized region. Aqueous cleaning and Freon cleaning are processes during which flux and other contaminants are removed from the board. Most assembly processes are carried out at high temperatures.

Printed circuit boards are subjected to high temperature processes that include baking to drive off absorbed moisture; preheating before soldering, reflow soldering, or wave soldering; and repair/rework of boards and components. During wave soldering, the board reaches an even higher temperature than during baking or reflow soldering, but for a shorter time. The major side effect of these processes is to generate thermomechanical strains in the board as a result of the different materials expanding at different rates as the temperature increases above the ambient. These strain differences cause major stresses at crucial points in the board. The higher the temperature at which these processes take place, the higher the strains and stresses generated in the boards during processing.

Besides the material elongation with temperature, other minor effects are observed at high temperatures. The laminate will creep to a considerable degree if left at high temperatures for an extended time ("Creep" is defined as the elongation of a material with time at elevated temperatures. Unlike thermal expansion, creep is an irreversible diffusion-controlled phenomenon.)

Shrinkage of the laminate will also occur as volatiles (i.e., moisture) are evolved. If the rate of moisture evolution is high, the resin in the laminate will shrink quickly and cause recession from the plated through-hole walls. Hence, the rate of temperature increase should be restricted. A low rate of temperature rise also diminishes thermal shock in the metal platings. As previously discussed, the laminate around a plated through hole was observed to expand more than the hole barrel and to increase in temperature. This disparity in elongation increased above the glass transition temperature.

Repair and rework generally require the application of heat (with a soldering iron or hot air) to a localized region on the surface of the board. The localized region is usually a pad and/or a component lead. Localized heating is even worse than heating the entire board to a particular temperature because it generates strains and stresses in one particular region. Such stresses are typically higher than those incurred when the entire assembly is heated to the same temperature.

Consider the case of a rework operation that may require the application of a 800°F iron to a hole pad to remove a component lead. The metal platings in the hole barrel and elsewhere are much better conductors of heat than the laminate. When the top of the soldering iron is applied to the pad, the hole barrel will become heated to a temperature higher than that of the surrounding laminate. Close to the barrel, the laminate will expand more than the barrel. But, when compared to the laminate expansion about 50 mils away from the barrel, the expansion of the barrel will be more than the expansion of the laminate.

Thus, the repair/rework operation generates more stress on the board than any other assembly process; therefore it must be undertaken with the utmost care. Otherwise, deformation of internal traces, as shown previously in Figure 4.4, will result. The time of repair/rework or the time of applying a soldering iron or hot air to the localized region should be kept to a minimum. Operator skill and minimum pressure on the pad are also very important. See Chapter 14, Section 14.6.1 for necessary precautions during repair and rework.

## 4.4 CERAMIC SUBSTRATES

The general appeal of ceramic substrates is that they have the same CTE as the ceramic package and do not pose any problems resulting from CTE mismatch. Ceramic substrates are widely used in hybrid applications, where small size is not a major concern. There are some other differences between the substrates used in hybrids and in surface mounting, however. For example, hybrids use bare or unpackaged ICs.

The commonly used compositions are pure (99%) alumina, 96% alumina, and beryllia. The high purity alumina, which is more expensive, is used only where higher strength is necessary. When high thermal dissipation is required, beryllia is used. The higher thermal conductivity of this metallic oxide which is electrically insulative, is very useful for high power devices; however, beryllia dust does pose a health hazard.

Both thin film and thick film processes are used for building ceramic substrates. The thin film process is a subtractive method of defining conductors and resistors on a ceramic substrate. In thin film technology a resistive material such as tantalum nitride or nichrome is first deposited in vacuum; then a metallic surface (generally gold) is plated over it. Then, as in the subtractive process used in printed circuit boards, the gold is removed from the areas in which resistors are desired.

Gold wire bonding between IC pads and the gold pattern on a hybrid substrate poses no problem. In surface mounting applications, packaged ICs are used and are soldered to the surface. The gold conductors on the substrate are leached by the material used for soldering, and this may embrittle the solder joints. The industry has achieved some success in using copper instead of gold to prevent the embrittlement problem and to also reduce cost. Also, the high dielectric constant of ceramic materials prevents their use in very high speed circuitry.

The exact values of resistors are defined by laser trimming. The thin film process is expensive, but very fine line widths (0.002 inch) are easily achieved. This somewhat compensates for the failure of thin film technology to accommodate more than a single layer of conductors. The thin film process does allow precise resistance on the substrates, but this is not a real advantage in surface mounting for two reasons. First, very precise resistance values are not necessary in the digital circuits commonly used today. In any event, discrete surface mount resistors with tight tolerances of 1% are available for surface mounting, if necessary.

In the thick film process, line widths of 0.004 inch are possible, but generally 0.01 inch widths are produced. The thick film process has two general variations. In the conventional process the various conductive, resistive, and insulative layers are screen printed, dried to remove volatiles, and then fired. The firing of the paste is done at a much lower temperature (900°C) than the firing of the ceramic substrate, which is done at 1500°C. As many layers of paste as desired (up to 20 layers) can be sequentially deposited and fired to achieve the specified interconnect.

A variation of the conventional thick film process is the cofired process in which "green" layers of ceramic are printed with conductive, resistive, and insulative inks. Each individual ceramic sheet is dried, whereupon all the layers are stacked and fired simultaneously. This is somewhat similar to the lamination process used in printed circuit boards except that printed board lamination is accomplished at a much lower temperature than is used for cofiring.

The sequential printing and firing of various layers of paste produces undulations or peaks and valleys of resistors and conductors in the conventional thick film process. The cofired process, on the other hand, gives a very flat surface, which is much more desirable for surface mounting. It is not possible to deposit, fire, and trim resistors in the cofired process, however, as can be done in conventional thick film work.

Generally circuits are formed in "multipacks" (several identical circuit boards in panel form), and the individual boards are cut off by scribing or laser. Ceramic substrates are limited by smaller size, higher dielectric constant, and high cost. Commonly used sizes 2 in.$^2$, but sizes up to 7 in.$^2$ are possible at an increase in cost.

## 4.4.1   Porcelainized Steel Substrates

Certain disadvantages of ceramic substrates, such as their smaller size and higher dielectric constant, can be overcome by using porcelainized steel instead of plain ceramic. For example, a 14 inch by 24 inch substrate size is possible with porcelainized steel. A well-known application of porcelainized steel substrates in mass production is in the Flashbar for Polaroid SX-70 cameras [5].

Like any other substrate, the porcelainized substrate has its advantages and disadvantages. For example, the CTE of a porcelainized substrate is very high (13.3 ppm/°C), rendering the material unsuitable for leadless chip carriers [5]. Where lower CTE is desired, a substrate of porcelainized copper-clad invar should be used instead of porcelainized steel. Both these substrates have lower dielectric constants than ceramic and can be used in relatively faster circuitry.

The procedure for screening the conductive, insulative, and resistive layers on porcelainized steel is the same as that used in conventional thick film technology, as discussed earlier. The firing of the paste is done at lower temperature (about 650°C), however, because of the lower melting point of steel.

## 4.5 CONSTRAINING CORE SUBSTRATES

The constraining core approach for matching the CTEs of a component package and a substrate has served successfully in producing reliable sur-

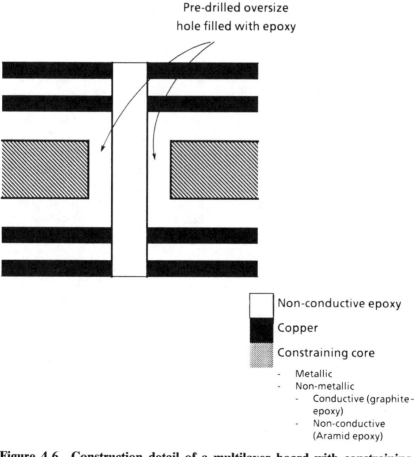

**Figure 4.6 Construction detail of a multilayer board with constraining core.**

face mount assemblies using leadless chip carriers. The basic idea is to produce a "sandwich" substrate in which the core has a very low CTE.

The constraining core can be either functional or nonfunctional electrically, and it can be a metal or a nonmetal. Even when electrically nonfunctional, the core acts as a heat sink and supporting plane. When the core is electrically functional, it also serves as the power and ground planes. The core is predrilled and filled with compatible nonconductive resin before lamination. The outer layers are processed as in conventional printed circuit glass epoxy board fabrication. Figure 4.6 presents construction details of a substrate with a constraining core.

Metal cores of various types are used to restrain the substrate from its normal expansion. Examples are alloy 42, copper-clad molybdenum, Kovar, porcelainized invar, and copper-invar-copper. The copper-invar-copper type of substrate is more commonly used.

### 4.5.1 Copper-Invar-Copper Constraining Core Substrate

A metal core material pioneered by Texas Instruments that is gaining very wide acceptance in constraining core applications is a three-layer conductive metal composite of copper, invar, and copper generally referred to as Cu-In-Cu or CIC substrate [6]. Invar (an abbreviated form of "invariant") is an alloy composed of 36% nickel and 64% iron; it has practically no CTE. When Invar is clad with copper, the CTE of the composite can be tailored, depending on the thicknesses of the invar and copper laminates, as discussed later on.

Copper-invar-copper consists of pure (99.95%) copper bonded to both sides of an invar layer. Invar, the nickel-iron alloy contains trace amounts of minor impurities. Reference 7 has established the requirements of a CIC metal foil. To form a substrate like the one shown in Figure 4.6, the copper-invar-copper (sandwich) is predrilled and then laminated with outer layers as in conventional multilayer epoxy glass board construction. The outer layers are regular copper-clad epoxy laminates and are processed in the same fashion as the usual glass epoxy substrates discussed in later sections.

By varying the thickness and the composition of the Cu-In-Cu plane, a substrate having desired CTE value can be tailor made. Since copper has a higher CTE than invar, to lower the CTE of the composite laminate core, the thickness of each copper layer is reduced. This also increases the tensile strength of the core. However, as in any other metal, the greater tensile strength is acquired at the expense of ductility.

Since the component packages run hotter than the substrate during power cycling, the CTE of the substrate should be slightly higher than the CTE of the package. Hence, depending on the wattage of the package, the CTE of the substrate should be 1 to 2 PPM higher than that of the package. In the beginning, the Cu-In-Cu sandwich contained 20% copper, 60% invar, and 20% copper, where the numbers refer to the percentage thickness of each metal in the core. The 20-60-20 metal core gives a CTE of about 5.2 ppm/°C (9.4 ppm/°F) [7].

A Cu-In-Cu core of 12.5-75-12.5 will give a CTE of 4.6 ppm/°C (8.2 ppm/°F) [7], slightly higher than that of the ceramic package. To achieve this in the total thickness of a 6 mil core, the top and bottom copper layers must be less than 1 mil thick and the invar layer about 5 mils thick.

Excellent solder joint reliability results were found with an 8-84-8 Cu-In-Cu core. [8]. Even localized etching can be performed to tailor the CTE of the substrate to both connectors and ceramic chip carrier. The CTE value of the commonly used 8-84-8 core is about 3.9 ppm/°C (7 ppm/°F).

Both problems and successes have been reported in connection with Cu-In-Cu substrates. For example, leadless ceramic chip carriers ranging in size from 20 to 84 input/output connections have been tested to 1500 temperature cycles from −55 to +125°C without failure [8]. The problems that generally have been seen are delamination of the core from the glass epoxy outer layers and cracking of the via holes. A new red oxide copper treatment process has been found to improve adhesion of the Cu-In-Cu layers, hence decreasing the potential for delamination.

The cracking of the plated through holes is generally encountered when epoxy (a nonconductive material) must be used to fill in around the holes to prevent the shorting of circuitry from the conductive cores. In thermal cycling, the larger Z axis expansion of the adhesive material causes failure in the plated through holes.

The problem of via hole cracking can be prevented by substituting the clearance hole with a clearance "ring" in the voltage and ground planes. In other words, the size of the predrilled hole is reduced to provide a minimum amount of adhesive fill. Figure 4.6 shows an oversize predrilled hole filled with epoxy. The resulting "nonfunctional" pad reduces the fill volume and reinforces the plated through hole. Plated through holes using this approach have passed 3800 temperature cycles from −55 to +125°C without failure [8]. One must very carefully determine the minimum size of the via holes even in glass epoxy boards. Cracking of via holes is discussed in detail in Section 4.10.

## 4.5.2 Graphite Epoxy Constraining Core Substrates

As mentioned earlier, the constraining core can be either a metal or a nonmetal. One core material that has been used at Boeing with the company's patented process is graphite [9]. Graphite has a very low CTE and can be used to produce a composite substrate with glass epoxy that can match the CTE of ceramic. For example, the graphite filament core has a CTE of 0 to 1.6 ppm/°F [9] and the glass epoxy substrate has a CTE of 7.5 ppm/°F. The composite laminate can be tailored, as in CIC substrates, to produce the desired CTE.

A graphite epoxy substrate is constructed essentially in the same fashion as the copper-invar-copper substrate shown previously (Figure 4.6). First the constraining core, consisting of graphite filament reinforced with thermosetting resin, is prepared. Depending on thermal requirements, the resin can be either epoxy or polyimide. Since graphite, like metal, is electrically conductive, larger via holes must be predrilled and filled wth nonconductive adhesive during lamination.

The graphite filament can be replaced with nonconductive fibers such as quartz or aramid. Otherwise, nonconductive adhesive must be used to prevent electrical shorting. The adhesive does pose the problem of large Z axis expansion. Thus as in CIC substrates, a clearance ring instead of a clearance hole should be used; this will reduce Z axis expansion by minimizing the amount of adhesive.

The outer layers of a graphite epoxy substrate are processed like conventional glass epoxy or glass polyimide substrates. Then the constraining graphite core and the outer layers are laminated, drilled, and processed as in conventional substrates. The internal core and the outer layers are bonded with adhesive.

The graphite epoxy substrate generally suffers from short cracks that separate the epoxy from the graphite laminate. This problem, called microcracking, tends to cause a shift in the CTE values. For this reason, the electronics companies that used graphite epoxy originally have switched to Cu-In-Cu. Nevertheless, the graphite epoxy substrate does have some inherent advantages over the CIC substrate. For example, the graphite core, unlike a metal core, is a lightweight material and is used in many structural aerospace applications where weight is a major consideration.

## 4.6 COMPLIANT LAYER SUBSTRATES

The most comon approach to eliminating the CTE differential problems encountered with ceramic packages and glass epoxy substrates is to use ceramic packages and the copper-invar-copper substrates just discussed. This is an expensive option, however, and substrates with a compliant layer offer an intermediate solution at a minor increase in cost. Because the stresses created by the CTE differential are taken up by the

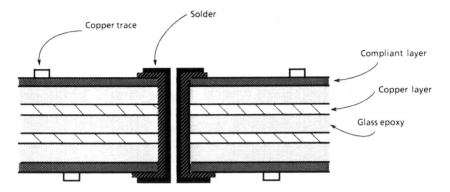

**Figure 4.7 Compliant layer substrate construction.**

compliant or "buttery" top surface, it is possible to greatly reduce the solder joint reliability problem.

Compliant layer substrates are fabricated by the same processes used for the conventional glass epoxy substrates but with a minor difference, illustrated in Figure 4.7. The "compliance" comes from nonreinforced epoxy layers (i.e., layers having no glass fiber), added to one or both sides of the copper layers that surround the epoxy. Epoxy layers on both sides are preferred, to provide balance and to prevent board warpage.

The layers are generally 0.01 inch thick on each side. This dimension is achieved by laminating various sheets of nonreinforced laminates. Generally each sheet is about 0.002 inch thick. The nonreinforced laminate can either be epoxy or polyimide, depending on the base substrate.

The final thickness of the top epoxy layer determines the amount of stress on the solder joint. The higher the thickness, the lower the stress on the solder joint. For example, one researcher has found stresses of 100,000 and 5160 psi/joint in compliant layer thicknesses of 0.001 and 0.007 inch, respectively [10]. Naturally, the lower the stress on the solder joint, the more thermal cycling it can withstand.

Satisfactory reliability data on compliant layer substrates are available now. For example, 1200 failure-free thermal cycles ($-55$ to $+125°C$, 15 minute soak, $<5$ minute transition time between chambers), and 2000 power cycles on a 0.006 inch thick compliant substrate containing 20 to 52 pin leadless ceramic packages have been reported in military applications [10].

There are problems with compliant layer substrates, too, however. For example, the top epoxy laminate layer (nonreinforced laminate material) is susceptible to dissolution by some of the chemicals used in fabrication and assembly. The epoxy resin dissolves in methylene chloride, but alcohol, Freon, and other solvents used in manufacturing do not cause problems.

There is also some concern about Z axis expansion which, depending on the thickness of the top layers, can be significantly higher than it is in a reinforced laminate. In addition to causing cracking in the plated through holes, the "buttery" layer may be responsible for problems such as pad lifting during repair operations.

Also, air bubbles between the laminates may occur during fabrication, causing board delamination later. Hence, the process should be closely controlled, and operators must be instructed to watch out for air bubbles. The ductility of the copper plating on the top surface also must be controlled, to prevent possible cracking in the plating.

Either because of the foregoing problems or for other reasons, compliant layer substrates have not gained much acceptance in the industry despite their significantly lower cost. The least expensive and most widely used substrate is the glass epoxy substrate, which we discuss next in great detail.

## 4.7   GLASS EPOXY SUBSTRATES

When using leaded packages, especially plastic, the conventional glass epoxy substrate is the cheapest option. It is the most commonly used material for single, double, or multilayer printed circuit boards in both commercial and military applications. The substrate is basically a composite material consisting of epoxy (resin) and glass fibers (base material). The base material gives the laminate its structural stability and the resin provides ductility.

A composite structure is advantageous because it combines the favorable properties of its ingredients. For instance, glass fibers are strong but they break easily when bent. Forming a composite of the brittle glass fibers and a tough resin imparts better ductility to the composite because the glass fibers absorb most of the stress during bending.

The glass fibers are made essentially by the same process as any other glass: high temperature melting of sand. The molten glass is turned into glass fibers of different diameters, which are formed into bundles or yarns and assembled in the form of a cloth having the characteristic weave structure that adds strength to the laminate. There are many ways in which the glass fibers are tied together to form the yarns and many ways in which the yarns are woven together.

The epoxy resin dispersed within the weave structure of the glass cloth is the medium that holds together successive layers of the glass cloth weave. Besides the base epoxy, laminate resins contain other chemicals, specially added to impart certain properties. Curing agents promote cross-linking of the epoxy chains, thus hastening the cure. Stabilizing agents prevent the epoxy from decomposing at elevated temperatures during processing, assembly, and use. Flame retardants, such as bromine, make the laminate self-extinguishing, which is very important for safety. Still other additives called adhesion promoters enhance the adhesion of the glass cloth to the epoxy resin. This adhesion should be sufficient to avoid delamination of the glass cloth from the resin under mechanical or thermomechanical stresses.

The traces are made of copper used in the subtractive process. They are the unetched portions of the copper foil from the copper-clad base laminates. The critical characteristic for the traces is peel strength. A low peel strength will lead to trace delamination or pad lifting during assembly, component repair/rework, and handling. For FR-4 laminates, the peel strength should be at least 5 pounds per inch of trace width after a 1 hour conditioning treatment at 125°C.

Most copper foil used in multilayer laminates is electrodeposited copper. The typical purity of this copper is 99.8% and typical tensile strengths are 15,000 and 30,000 psi for 1/2 oz. and 1 oz. foil, respectively.

As we will see later on, various layers of laminates are processed (etched to form the required circuitry), laminated, drilled, and then plated to achieve the final interconnection.

## 4.7.1    Types of Glass Epoxy Substrate

Table 4.4 adapted from Reference 1, page 2–18 summarizes the types of glass epoxy laminate in use today in the printed circuit industry. We discuss now some of the commonly used glass epoxy substrates.

1.    G-10 and G-11. These are glass fiber based laminates with an epoxy resin system; they can be drilled but not punched. G-10 lacks flame retardant; otherwise it is similar to FR-4, the most widely used laminate in the United States today.

2.    FR-2, FR-3, FR-4, FR-5 and FR-6. These laminates are characterized by the presence of a flame retardant (hence the FR designation), which makes the laminates self-extinguishing if they happen to catch fire.

FR-2 is a paper-based material with a phenolic resin system; it can only be punched.

FR-3 is paper based also but has an epoxy-based resin, it can be punched at room temperature only.

FR-4 laminate, the industry favorite, has glass fibers as the base material and epoxy as the resin.

**Table 4.4   Laminate materials used in Printed Circuit boards [1].**

| COMMON DESIG-NATION | RESIN SYSTEM | BASE MATERIAL | DESCRIPTION |
|---|---|---|---|
| XXXP | Phenolic | Paper | Punchable at room temperature. |
| XXXPC | Phenolic | Paper | Punchable at or above room temperature. XXXP and XXXPC are widely used in Japan. |
| G-10[a] | Epoxy | Glass fibers | General purpose material system. |
| G-11 | Epoxy | Glass fibers | Same as G-10, but can be used to higher temperatures. |
| FR-2 | Phenolic | Paper | Same as XXXPC, but has a flame retardant (FR) system that renders it self-extinguishing. |
| FR-3 | Epoxy | Paper | Punchable at room temperature and has flame retardant. |
| FR-4[a] | Epoxy | Glass fibers | Same as G-10, but has a flame retardant. |
| FR-5 | Epoxy | Glass fibers | Same as FR-4, but has better strength and electrical properties at higher temperatures. |
| FR-6 | Polyester | Glass fibers | Designed for low capacitance or high impact resistance; has flame retardant. |
| Polyimide[a] | Polyimide | Glass fibers | Better strength and demonstrated stability to a higher temperature than FR-4. |

**Table 4.4**   (*Continued*)

| COMMON DESIG-NATION | RESIN SYSTEM | BASE MATERIAL | DESCRIPTION |
|---|---|---|---|
| GT or GX[a] | Teflon | Glass fibers | Controlled dielectric laminate. GX has better tolerance of dielectric properties than GT. |

[a]Principal materials for multilayer PCBs.

FR-5 is similar to FR-4 but has better strength and electrical properties at higher temperatures.

FR-6 has a glass fiber base and polyester resin system; it is designed for low capacitance or high impact resistance.

The G-10 and FR-4 laminates are the most common laminates for multilayer boards. These laminates are cheap and easily processible into multilayers by heat and pressure, in addition, they can be drilled wtihout major problems.

Other laminates that are not glass epoxy are commonly used in certain applications. Examples of some of them are given below.

1.    Polyimide. This laminate has a glass fiber base and a polyimide resin system. It has better strength and stability at high temperatures than FR-4. Hence, it finds widespread use in the military.

2.    GT and GX. These laminates have a glass fiber base too, but the resin is Teflon, which imparts to the materials a controlled dielectric property that permits their use when laminate dielectric constant is critical. GX has a better tolerance of dielectric properties than GT and finds use in high frequency applications.

3.    XXXP and XXXPC. These paper-based laminates with a phenolic resin are punchable only; they cannot be drilled. The XXXPC laminate can be punched at or above room temperature, but XXXP can be punched only at room temperature. These laminates are widely used in Japan for consumer products.

G-10, FR-4, polyimide, GT, and GX can be used to make multilayer boards; XXXP and XXXPC are used only for single- and double-sided boards.

**Table 4.5  Highest continuous operating temperature for common printed circuit board materials [1].**

| MATERIAL | HIGHEST CONTINUOUS OPERATING TEMPERATURES | | | |
| | ELECTRICAL FACTORS | | MECHANICAL FACTORS | |
| | °C | °F | °C | °F |
| --- | --- | --- | --- | --- |
| XXXP | 125 | 257 | 125 | 257 |
| XXXPC | 125 | 257 | 125 | 257 |
| G-10 | 130 | 266 | 130 | 284 |
| G-11 | 170 | 338 | 180 | 356 |
| FR-2 | 105 | 221 | 105 | 221 |
| FR-3 | 105 | 221 | 105 | 221 |
| FR-4 | 130 | 266 | 130 | 248 |
| FR-5 | 170 | 338 | 180 | 356 |
| FR-6 | 105 | 221 | 105 | 221 |
| Polyimide | 260 | 500 | 260 | 500 |
| GT | 220 | 428 | 220 | 428 |
| GX | 220 | 428 | 220 | 428 |

## 4.7.2   Operating Temperatures for Glass Epoxy Boards

Each of the laminates has a characteristic highest continuous operating temperature. That is, the laminate will break down and the assembly will not function properly if the board is used above this temperature. The highest continuous operating temperature is determined by either electrical properties or mechanical properties (see Table 4.5 which is adapted from Reference 1, page 2–22). As indicated in Table 4.5, FR-2, FR-3, and FR-6 can be used to only 105°C; XXXP and XXXPC can be used up to 125°C; FR-4 can be used up to 130°C for electrical and mechanical factors. G-11 and FR-5, on the other hand, can be used up to 170°C for electrical factors and 180°C for mechanical factors. Polyimide, which can be used to 260°C, is usually classified as a high temperature laminate. GT and GX can operate up to relatively high temperatures, too: about 220°C.

## 4.7.3    Fabrication of Glass Epoxy Substrates

Depending on the materials, processes, and equipment, there are many ways to fabricate a multilayer board. The major steps in the commonly used processes are listed in Table 4.6 and Figure 4.8. There are two major processes: tin-lead plating and solder mask over bare copper (SMOBC). Steps 1 to 34 (Table 4.6) are the same for both processes; steps 35 to 45 apply to SMOBC only. Table 4.7 compares the final process steps in SMOBC and tin-lead boards. Let us discuss and illustrate some of the process steps in detail.

The final substrate is achieved in three stages. Compounding or mixing the epoxy is called the A stage. In the B stage, very large sheets of the glass fabric are impregnated with the resin and dried to an intermediate cure. In the C stage, copper sheets and the B stage material are pressed together. Handling and storage of the B stage material is very critical in determining its final properties; it must not absorb moisture, and it must be allowed to stabilize before being converted to the C stage (lamination of copper sheet).

A typical board vendor begins the process with a C stage (fully cured) laminate containing glass epoxy and copper. He also uses B stage material (no copper on outer layers) during lay-up for lamination (step 17) to achieve the dielectric separation between copper layers. In steps 2 and 3, the large sheets of C stage material are cut down to size and marked by the fabricator for processing. To improve the dimensional stability, the C stage material is baked (step 4).

After tooling holes have been punched in all layers for accurate alignment, the laminate is ready for developing the image. Dry film resist is laminated on the copper surface; then the area that will eventually have the copper traces is exposed to UV light to receive the artwork image transfer (see Figure 4.9). The dry film that was not exposed to UV light is removed, leaving behind copper covered with dry film and exposed copper.

The exposed copper is etched away in the commonly used subtractive process, with dry film serving as the etch resist. Figure 4.10 shows copper before and after etching. Now the dry film has served its purpose and is removed (Figure 4.11). Then after inspection and testing, the copper is given black oxide treatment to promote adhesion to the resin during lamination (also shown in Figure 4.11).

Now the board is ready for lamination. All the separately processed layers just described are put in the lamination fixture, with "prepreg" as the B stage material is called, between the layers (see Figure 4.12). After

**Table 4.6** **Process steps in fabrication of multi-layer boards.**

| | |
|---|---|
| 1. Raw material | 24. Deburr holes |
| 2. Shear C stage material into panel sizes | 25. Hole cleaning (etch back) |
| 3. Identify material with markings | 26. Plated through-hole sensitization (electroless copper) |
| 4. Extra material cure (bake) [optional] | 27. Scrub for photoresist application |
| 5. Drill or punch registration tooling holes | 28. Apply dry film photoresist (outer layers) |
| 6. Material scrub | 29. Artwork image transfer (outer layers) |
| 7. Apply dry film photoresist (inner layers) | 30. Develop artwork image (outer layers) |
| 8. Artwork image transfer (inner layers) | 31. Electroplate copper |
| 9. Develop artwork image (inner layers) | 32. Electroplate tin-lead |
| 10. Etch inner circuit patterns | 33. Strip dry film photoresist |
| 11. Remove photoresist | 34. Etch copper circuits |
| 12. Inner circuit inspection | 35. Strip tin-lead |
| 13. Inner circuit test | 36. Apply solder mask and legend/cure |
| 14. Scrub for lamination/oxide treatment | 37. Hot air level tin-lead solder/clean |
| 15. Bake | 38. Tape board for edge connector plating |
| 16. Shear B stage and punch registration holes | 39. Electroplate nickel-gold on edge fingers |
| 17. Layup for lamination | 40. Remove tape/clean |
| 18. Lamination | 41. Drill nonplated holes |
| 19. Remove laminate from fixtures | 42. Route boards to final contour and bevel leading edge |
| 20. Laminate trim (shear epoxy flash) | 43. Final test |
| 21. Mark ID number on panels | 44. Final inspection |
| 22. Postlaminate bake | 45. Packaging/shipping |
| 23. Drill plated through-holes | |

pressure and heat have been applied, with proper controls, the excess resin from the B stage material that has oozed out is sheared off.

The laminated board is drilled, and using pressurized air or water, the burrs are removed, and the holes are cleaned with sulfuric acid (see Figure

**Table 4.7   A comparison of solder mask on bare copper (SMOBC) and tin-lead plating process steps. Process steps 1 to 34 (Table 4.6) are the same.**

| SMOBC | TIN-LEAD |
|---|---|
| | 1-34. Same as SMOBC |
| 35. Strip tin-lead | 35. Tape board for edge connector plating |
| 36. Apply solder mask and legend/cure | 36. Plate nickel-gold on edge connector fingers |
| 37. Hot air level tin-lead solder/clean | 37. Remove tape/clean |
| 38. Tape board for edge connector plating | 38. Reflow tin lead |
| 39. Electroplate nickel-gold on edge fingers | 39. Clean boards |
| 40. Remove tape/clean | 40. Apply solder mask and legend |
| 41. Drill nonplated holes | 41. Drill nonplated holes |
| 42. Route boards to final contour and bevel leading edge | 42. Route boards to final contour and bevel leading edge |
| 43. Final test | 43. Final test |
| 44. Final inspection | 44. Final inspection |
| 45. Packaging/shipping | 45. Packaging/shipping |

4.13). Electroless copper is then plated to a thickness of 10 to 20 microinches on the nonconductive glass epoxy.

Now the outer layers are processed in the same fashion as the inner layers. In other words, steps 28, 29, and 30 are a repetition of steps 7, 8, and 9. Then the entire board except for the area covered with dry film receives electrolytical plating (with copper followed by tin-lead). The dry film has served its purpose and is now removed. Next the tin-lead acts as an etch resist while copper is being removed. Then the tin-lead is removed to expose the bare copper underneath, and solder mask is applied over any copper that does not need to see solder later. The areas that will be soldered are not covered with mask. Solder is applied to these areas with the hot air leveling (HAL) process.

After the solder mask application, the legends are applied. The inks used in the legend on the outer layers of a multilayer board are made of a two-part epoxy. That is, the two components are mixed together and

**Inner Layer Process**

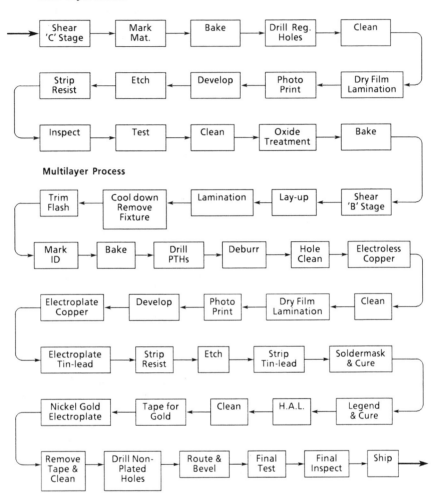

**Figure 4.8  Summary of process steps in the fabrication of glass epoxy board; see text for details.**

screened on to the board, which is then cured. The legend color should be such that the legend is legible on the solder mask. A color that doesn't form a good contrast with the solder mask color should be avoided. Since solder masks are usually green or blue, white, yellow, and black inks are most often chosen for legends.

This completes all the process steps. Figure 4.14 shows a cross section of the final result. It should be noted once again that the fabrication pro-

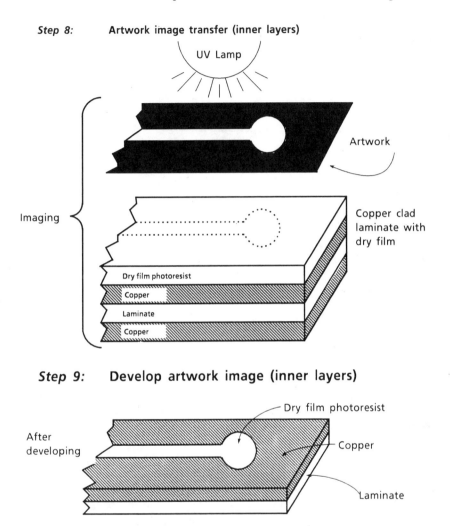

**Figure 4.9  Using dry film photoresist for the development of circuitry (steps 8 and 9, Table 4.6).**

cesses (plating, etching, and lamination) discussed above for glass epoxy boards also apply to boards with compliant layers and constraining core.

## 4.8  PLATING PROCESSES

Plating is a critical process, one of those that determine the quality of the board. There are many different types of metal plating in a multilayer board, including copper, gold, nickel, tin, and solder, and they are found in many locations.

# *Step 10:* Inner layer etch

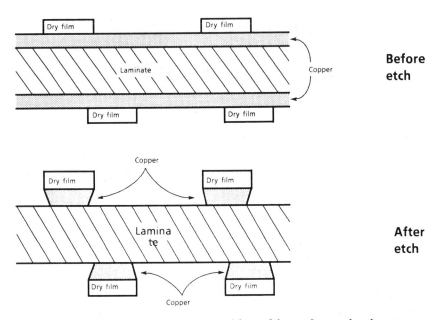

**Figure 4.10 Before and after the etching of inner layer circuit patterns (step 10, Table 4.6).**

The bonds between the various platings differ markedly in strength. For example, the bond between the solder plating and the copper electrolytic plating is a metallurgical bond and is very strong because the two plating layers are held together by the compounds called intermetallics, which are formed between them. On the other hand, the bond between the electroless barrel plating and the inner trace is relatively weak because it is a physical bond of adhesion. Any process-related problem such as the presence of smeared epoxy on the edge of the inner trace prior to electroless copper plating is likely to cause a reduction in the contact area between the trace edge and the plating and consequently to weaken the bond even further.

The basic plating mechanism is similar for all metals. A solution containing the metal ions is the source of the metal, and plating occurs when the positively charged metal ions capture electrons from this specific source, depositing the metal on a substrate in the solution.

## *Step 11:* Inner layer - Remove photoresist

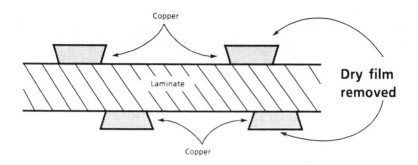

## *Step 14: Black oxide treatment*

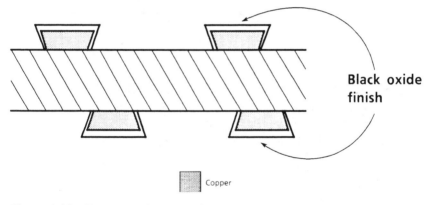

**Figure 4.11   Removal of photoresist and treatment with black oxide (steps 11 and 14, Table 4.6).**

### 4.8.1   Copper Plating

There are two types of copper plating: electroless and electrolytic. The difference lies in the source of the electrons. For electroless copper plating, in which no electrical current is used, the source is a chemical reaction. In electrolytic copper plating, the current that is passed through the plating solution serves as the source of electrons.

The purpose of electroless copper deposition is to make the nonconductive surfaces within the drilled hole conductive so that subsequent plat-

## Step 17: Pinning and Lamination - 8 layer board

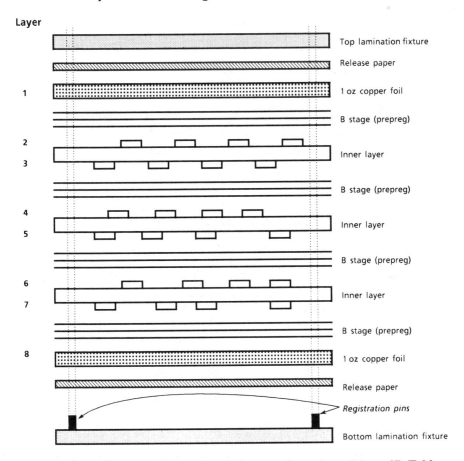

**Figure 4.12** Lay up for lamination of an eight-layer board (step 17, Table 4.6).

ing operations can take place electrolytically. The purposes of electrolytic copper plating are to provide a connection between the layers in a multilayer board and to strengthen the barrel of the plated through hole.

Many plating solutions are used for electrolytic copper deposition. These include copper sulfate, copper fluoroborate, and pyrophosphate solutions. The tensile strength (the maximum stress that the metal plating can withstand under tension) of the plating solutions is about 3 to $5 \times 10^4$ psi. The higher the tensile strength, the stronger the plated through hole. However, the ductility (the extent to which the copper plating can be

*Step 25:* **Desmear and etchback (hole cleaning)**

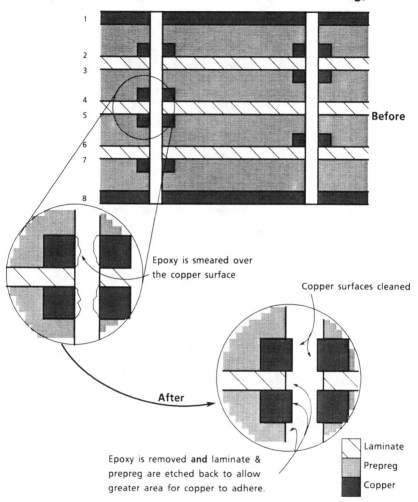

**Figure 4.13  Hole cleaning after drilling (step 25, Table 4.6).**

"stretched" or elongated before fracture) differs markedly. A high value of ductility is preferred to allow "yielding" before failure instead of sudden failure.

Electroless plating offers very low ductility compared with electrolytic platings—a good reason for requiring electrolytic plating for plated through-hole strengthening. The fluoroborate and pyrophosphate plating solutions give a better ductility than the copper sulfate plating bath.

## Step 37:   Hot air leveling (HAL) / Clean

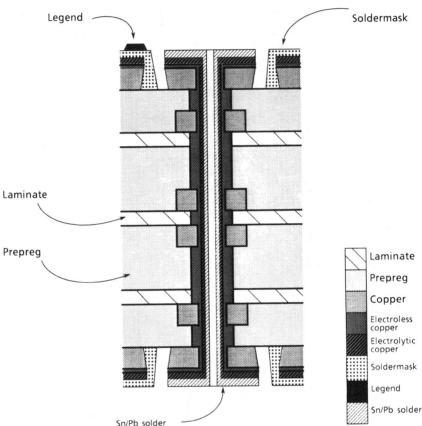

**Figure 4.14   A finished board.**

The residual stress in the platings also differs substantially. It is compressive in electroless copper plating and tensile in the electrolytic form. A compressive residual stress enhances the adhesion of the electroless plating to the nonmetal laminate and to the inner layer trace, comprising another reason for using electroless copper before electrolytic copper.

## 4.8.2   Gold Plating

Gold plating is deposited electrolytically as a contact surface because gold has a low contact resistance and a high resistance to corrosion. Gold is also used as a tarnish-resistant layer because it doesn't oxidize even at elevated temperatures. Gold sometimes is also used as a metal etch resist during board fabrication.

There are four different types of plating solution, which are classified according to their pH as follows: acid, neutral, alkaline cyanide, and alkaline noncyanide. The properties of gold deposits (such as hardness and porosity) depend on the plating solution. Acid gold is the most common plating solution for connector pad plating. Cobalt is used as a brightener in acid solutions.

### 4.8.3 Nickel Plating

Nickel plating, which is usually deposited electrolytically, forms an undercoat to precious metals such as gold to improve wear resistance during use. Nickel also forms an effective barrier layer to intermetallic formation. Intermetallic formation is essential but can be detrimental if thickness is excessive.

### 4.8.4 Lead-Tin Solder Plating

There are two methods by which a lead-tin solder coating is applied: electroplating and solder coating by dipping and hot air leveling. The advantage of HAL over electrodeposition is that HAL solder coatings are dense and have good adhesion (due to intermetallic formation). But the nonuniformity of the coating (across the pad cross section) from HAL can lead to solderability problems, especially in surface mount assemblies. Control of plating thickness is also a problem, especially when thick coatings are required.

Electroplated solder is notoriously porous, but it provides excellent control over the plating thicknesses. Fusing the solder after electroplating gives a solder coating without porosity. Unless sufficient and uniform coating thicknesses can be achieved by hot air leveling, the plating process will continue to enjoy preference because it provides better yield in surface mount assemblies during manufacturing.

### 4.9 SOLDER MASK SELECTION

The solder mask is a polymer. However, unlike the laminate it is not a composite but a homogeneous material. As the name suggests, this material is used to mask off the outer areas of the board where solder is not required to prevent bridging between conductors. In past years not all boards required a solder mask because the traces and pads were spaced quite far apart. The solder bridges between adjacent traces during wave soldering were not critical. But with the advent of fine lines and spaces, the use of a solder mask has become almost mandatory for boards that are going to be wave soldered.

On a full SMT board (Type 1 SMT), where no wave soldering is required, tenting of via holes for testing may be more critical than the bridging of adjacent conductors. Solder mask is also used for the tenting of via holes. Tenting—the application of solder mask to block or plug a via—allows closer spacing between a via and the adjacent conductor lines. Tenting also helps in bed-of-nails testing, where almost all the via holes must be sealed to provide good vacuum for the test fixture.

The two broad categories of soldermasks are permanent and temporary, but there are subdivisions, as shown in Figure 4.15. The focus of this section is on permanent solder masks, but let us briefly mention the temporaries.

Temporary solder masks are washable or peelable. They are used during wave soldering to prevent filling of holes that must be kept open for the installation after soldering of such leaded items as unsealed parts that might not withstand the cleaning or soldering environments. The temporary masks that can be peeled off are also used to mask off gold pads, which must not be soldered.

Washable masks are more convenient than peelable masks because

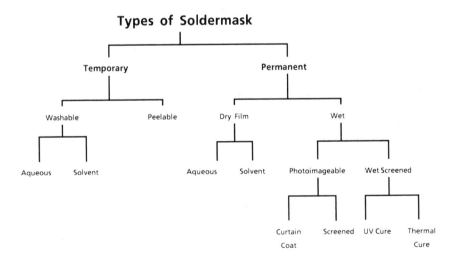

**Figure 4.15  Types of solder mask.**

the former come off in the cleaning operation after wave soldering and do not need an extra step to remove them. The aqueous washable solder masks require an aqueous cleaning system, and the solvent washable masks must be washed with a solvent in a vapor degreaser. Let us now turn to the permanent solder masks.

## 4.9.1    Wet Versus Dry Film Solder Masks

Permanent solder masks come in dry film and wet film. Dry film masks can have an aqueous or a solvent base. In both cases, the mask starts out as a polymer film, which is applied to the board by vacuum lamination.

Wet film solder masks, as the name implies, are liquid or pastelike. They include photoimageable (see Section 4.9.2) and wet screenable solder masks. The latter are differentiated by the method of cure. Some wet screenable solder masks can be cured by UV light and some can be cured thermally in convection or IR ovens.

Each of the solder mask categories has advantages and disadvantages. Wet screenable solder masks have a good vendor base and are the most widely used today. They are inexpensive and highly durable. Being liquid, they flow between conductors and prevent the formation of air pockets. There is no trim waste, and the thickness of the mask can be controlled for each design.

Since the wet film solder masks are screened on (a mechanical process), they are difficult to register and have a tendency to skip over traces, especially on fine line boards. They also tend to bleed onto the pads and surface mount lands during cure. Wet screenable masks are difficult to use on boards with fine lines and spaces (<8 mils).

The wet film solder mask cannot successfully tent via holes. Generally these materials fill the vias partially, which can prevent bridging but is ineffective in sealing holes to prevent vacuum leakage during testing. Partially filled vias trap process chemicals and are difficult to clean.

Dry film solder masks have some advantages over their wet screen counterparts. The former provide very accurate registration, which is critical in preventing solder bridging or bleeding on fine line boards, as well as sharper resolution. Tenting of via holes is also superior with dry film solder masks because they are never in a liquid state and do not drip into the via during vacuum lamination. However some problems arise when trying to laminate a semisolid dry film over an uneven board surface with traces and pads. Any warping and twisting of the board will compound the problem, possibly causing air pockets underneath the dry film near traces.

The curing of a dry film solder mask is extremely critical for achieving a reliable coating. Insufficient cure can cause cleaning problems because of diminished resistance to chemical attack by fluxes or cleaning solutions. Overcure results in a brittle mask, which can crack easily under thermal stress.

Dry film masks are more costly than the wet film variety, and the vendor base is limited. Moreover, the application process for dry film solder masks is very difficult to control. Not many film thicknesses are commercially available, and this can limit flexibility and increase cost. Trim waste also adds to cost.

Most dry film solder masks are not resistant to thermal shock. Cracks develop in the cured masks within 100 cycles of thermal shock cycling from +100 to −40°C. This can be especially a problem in SMOBC boards because of exposed copper traces. Most dry film solder masks are also quite thick, typically 3 to 4 mils.

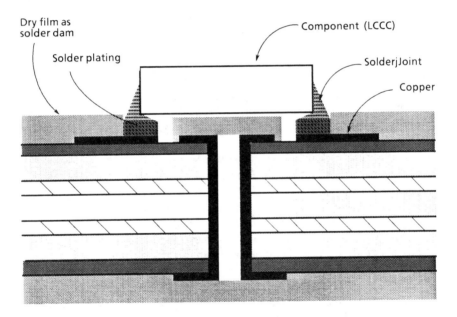

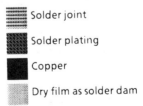

**Figure 4.16** **Use of dry film solder mask for solder dam and tenting of via holes.**

Nonwetting of surface mount boards containing wave soldered chip components sometimes occurs because of the greater thickness. In addition, the thicker mask surrounding the small vias may prevent them from filling with solder during wave soldering (crater effect).

The thicker masks can cause problems in reflow soldering as well. For example, a dry film mask applied between the pads of passive surface mount devices can cause tombstoning (standing up on edge) during reflow soldering because of the rocking effect of the mask. For this reason dry film solder mask should not be used between the pads of chip resistors and capacitors or in assemblies that have components glued to the bottom side for wave soldering. This property of dry film can be desirable in some applications, however. For example, as shown in Figure 4.16, the higher thickness of dry film over copper can be used to provide a solder dam. This can be very helpful for LCCC packages in achieving good solder fillets.

## 4.9.2    Photoimageable Solder Masks

Photoimageable solder masks combine the advantages of dry film and wet solder masks. Dry film is also a photoimageable mask. In this section, however, our discussion is focused on wet film photoimageable masks, which provide accurate registration, are easy to apply, encapsulate the circuit lines totally, have excellent durability, and are cheaper than dry film.[11].

Photoimageable masks can be either screened on or applied by a process called curtain-coating, in which the board is passed at high speed through a curtain or waterfall of solder mask.

The photoimageable mask may have solvent along with photopolymer liquid. If the solvent is added in the mask, the liquid mask is screened on, solvent is dried off in an oven, and then the mask is exposed to UV light by off-contact or on-contact methods. (If no solvents are used, the liquid is 100% reactive to UV light.) The off-contact method requires a collimated light system to minimize diffraction and scatter in liquid. This makes the system very expensive. The on-contact approach needs no collimated UV light source, and the system is relatively cheaper.

As with dry film solder masks, however, the vendor base for photoimageable masks is limited because the processing equipment is very costly. Also the photoimageable mask has not achieved widespread industry acceptance. This will change, because the combined advantages of wet and dry film solder masks are very attractive.

Photoimageable solder masks can tent only very small via holes. Most photoimageable wet film masks will not even tent 0.014 inch via holes because it is difficult to cure polymer in via holes. If tenting is required, dry film is needed because only dry film can tent via holes effectively.

## 4.10 VIA HOLE CRACKING PROBLEMS IN SUBSTRATES

All the substrates discussed in this chapter have at least one problem in common: the cracking of via holes. A via hole is defined as a plated through hole that is used only for interconnection, not for the insertion of leads.

As the packing density on the board has increased, smaller and smaller via holes have become essential, to save on real estate and to compensate for the impact on routing channels of higher desnity. As the via holes get smaller, their reliability becomes questionable. This is because thermo-mechanical fatigue generated by the laminate $T_g$ causes cracking and creep effects. First, let us discuss the explanation for the cracking and then present some practical means to minimize it.

The commonly seen failure modes are barrel plating cracks (Figure 4.17), which are all around the circumference of the hole.

The plating thickness of the barrel determines the stress it can with-stand. As the aspect ratio (board thickness divided by drilled hole diameter) of the hole increases, the plating thickness in the middle of the barrel decreases. This is also described as the dog bone effect in plating, because plating thicknesses are higher near the top and bottom surfaces of the board but very thin in the middle of the hole. For any size hole, laminate expansion at elevated temperatures imposes a tensile force on the barrel plating. This force can be equated to a tensile force acting on the circular edges of a cylinder. However, the barrel plating cracks when stress (not force) exceeds the fatigue fracture stress of the plating. The holes with a large aspect ratio are weak in the middle and generally crack there.

The glass epoxy board is constrained by the glass in X-Y direction. However, there is no constraint in the Z direction. The Z axis CTE of the laminate is driven by resin, its degree of cure, and the ratio of resin to glass. The CTEs of epoxy and polyimide are 60 and 40 ppm/°C, respec-tively, versus 13 ppm/°C, which is the CTE of copper. Thus the basic cause of this stress lies in the different coefficients of thermal expansion in the Z direction between the substrate and the copper barrel.

Reliability problems are encountered when the via hole diameter is reduced to 10 mils. At 18 mil and above, performance is satisfactory. Even 14 mil vias are generally considered reliable as long as the aspect ratio is kept under 3. Note that these are finished diameters after plating. The drilled hole diameter, used for the calculation of aspect ratio, are larger. Via hole failure is also related to poor quality drilling, which creates ragged

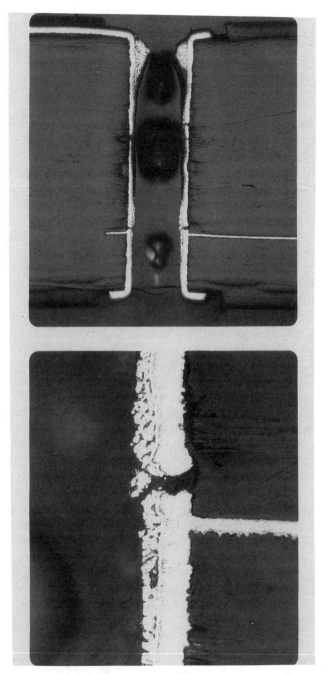

**Figure 4.17    Barrel cracking in a 10 mil via hole at 50X (top figure) and 250X (bottom figure).**

hole surfaces. The uneven surfaces plate nonuniformly and may cause barrel cracking.

How do we minimize the via hole cracking problem? The easiest way to reduce barrel cracking is to use larger holes. This not only increases reliability but also reduces cost. See Figure 5.4 in Chapter 5. But the reason for using smaller via holes is to increase the interconnection density on the board. Thus the trend has been to move toward smaller and smaller holes. The result is that board thickness must be decreased by using fewer layers. Again, reducing layer count is not always feasible, so the designer must make tradeoffs. In addition to aspect ratio, the plating thickness must be uniform, with a 0.001 inch minimum. Achieving a uniform plating thickness is an industry-wide problem, especially with respect to small via holes.

If vias as small as 10 or 14 mils must be used, nickel has been found to improve reliability. It was the experience of at least one company that nickel underplating over electroless copper followed by usual solder/tin plating reduces the via hole cracking problem by a factor of 4 [12]. Nickel is a tougher metal, and a better defense against the stresses imposed during the thermal cycling was reported.

Generally nickel plating is deposited electrolytically to serve as an underplating to precious metals such as gold, to improve wear resistance during use. Nickel also forms an effective barrier layer to intermetallic formation and prevents leaching of precious metals during soldering. Here, the function is simply to strengthen the barrel or via in anticipation of higher stresses.

What is the best way to evaluate via hole quality? An IPC round robin test [13] found that the military thermal cycle test ($-65$ to $+125°C$, 30 minute dwell time at each extreme, 30 minute transition time) is a good way to weed out poor vendors and poor quality vias. An acceptable via hole must pass at least 300 cycles of this test, which allows one to discriminate among via holes of different quality. Good quality holes with uniform and sufficient copper thickness will not fail this test. The round robin test also established that the commonly used practice of microsectioning to determine via hole quality is not reliable.

The same round robin test also found that the commercial thermal cycle (0 to 100°C, 40 minute dwell at low temperature, 23 minute dwell at high temperature, 30 minute transition time) is not a good test. Almost all the 200,000 via holes (ranging in size from 10 to 20 mils in 30, 60, and 90 mil thick boards, built by 16 vendors) passed this test. Only the very poor quality via holes failed. This test can only determine infant mortality. The findings suggest that the aspect ratio (board thickness divided by drilled hole diameter) be maintained around 3.

## 4.11   SUMMARY

Despite the many considerations (such as CTE, cost, dielectric properties, and $T_g$) to be taken into account when designing surface mount boards, the selection of a substrate is basically determined by the types of component to be used. When leadless ceramic chip carriers are mounted on conventional glass epoxy substrates, solder joint cracking is generally seen in about 100 cycles. The cause of the excessive stress is the CTE differential between the ceramic package and the glass epoxy substrate.

There are three different approaches to solder joint cracking problems: using a substrate with a compatible CTE, using a compliant top layer substrate, and replacing leadless ceramic packages with leaded ones. The most widely used substrate—namely, glass epoxy—entails no CTE compatibility problems when used for plastic surface mount packages. This provides the solution for commercial applications only, however.

The most commonly used substrate for military applications is one with a CTE value compatible with that of the ceramic package that has been specified. Each option has its advantages and disadvantages. The designer needs to carefully balance the constraints of cost with reliability and performance needs. In addition, solder masks and via hole size should be selected carefully.

## REFERENCES

1.   Coombs, Cyde F, Printed Circuit Handbook, McGraw Hill, 2nd edition, 1979, pages 2–18, 2–22, & 23–5.

2.   Lee, L. C., et al. "Micromechanics of multilayer printed circuit board. *"IBM Journal of Research and Development,* Vol. 28, No. 6, November 1984.

3.   IPC-SM-782. Surface Mount Land Patterns: Configurations and Design Rules. March 1987, IPC, Lincolnwood, IL.

4.   Gray, F., Paper WCIV-38 presented at the Printed Circuit World Convention, Tokyo, 1987. Available from IPC, Lincolnwood, IL.

5.   Hughes, E. W., and Beckman, E. C. Porcelain Enameled Metal Substrates for Surface Mount Components. Publication of Ferr-ECA Electronics Company, 3130 West 22nd Street, P.O. Box 8305, Erie, PA 16505.

6.   Foster Gray. "Substrates for chip carrier interconnections," In

*Surface Mount Technology,* International Society for Hybrid Microelectronics Technical Monograph Series 6984-002, 1984, Reston, VA, pp. 57–85.

7. IPC-CF-152. Metallic Foil Specification for Copper-Invar-Copper for Printed Wiring and Other Related Applications. Proposal, August 1987, IPC, Lincolnwood, IL.

8. Hanson R. R., and Hauser, J. L. "New board overcomes TCE problem." *EP&P,* November 1986, pp. 48–51.

9. Jensen, W. M. U.S. patent number 4,318,954, March 9, 1982.

10. Chen, C. H., and Verville, J. M. "Direct soldering of ceramic chip carrier solves radio production problems." *Electronics,* February 1984, pp. 15–17.

11. Denkler, J. D. "The speed of liquid." *Circuits Manufacturing,* May 1986, pp. 21–24.

12. Manufacturing Technology Review of VHSIC Program by Martin Marietta Corp., August, 1987. Wright Patterson, Air Force Materials Lab, Dayton, Ohio.

13. IPC-TR-579 Round Robin Evaluation for Small Diameter Plated Through Holes, September 1988, Available from IPC, Lincolnwood, IL.

# Chapter 5
# Surface Mount Design Considerations

## 5.0   INTRODUCTION

The design engineer is responsible for designing a product that meets certain functional, performance, and environmental requirements. The functions moreover, must fit within the form factor constraints for the product, which may be either cost driven or performance driven. This is not to imply that performance-driven products do not have to be cost effective. It means only that performance is the overriding issue. The designer must also ensure that the product will meet thermal and reliability requirements and can be designed and built in a timely manner to succeed in a given market window.

These requirements can be met by various packaging options: through hole, surface mount technology (SMT), application-specific or custom packages, or a combination of all these. Which packaging option will be chosen and why? Should the packages be ceramic or plastic? What kind of substrate should be used? What are the pros and cons for these various choices?

Once the packaging option has been selected, the product must be designed for manufacturability to ensure that the cost goals are met. To successfully accomplish these goals, the designer must be aware of the manufacturing processes and the equipment needed for building the product.

Before embarking on a particular course, the designer must take into consideration form, fit, function, cost, reliability, and time to market. This is a very broad undertaking indeed. In this chapter we explore the pertinent issues, with the objective of helping the designer to choose the appropriate packaging option.

## 5.1   SYSTEM DESIGN CONSIDERATIONS

Systems analysis starts with determination of product needs in the marketplace. Therefore the designer must look at the proposed product from the systems point of view. Generally the design engineer is concerned only with the device or the board or the system. This is certainly a mistake, since customers do not buy a device or a board; rather, they buy function or product. As Davidow explains it, great devices or boards do not sell in the marketplace. Products do [1].

Since the definition of the product is the responsibility of the marketing department, the definition of the factors involved (form, fit, function, manufacturing cost, selling price, distribution channels, service, etc.) should start with the marketing department.

Many board designs are canceled after extensive development costs have been incurred, sometimes before and sometimes after the product has been introduced in the marketplace. There are as many reasons for cancellation as there are products. The cause in any given instance could be poor design or manufacturing, or inadequate marketing research. If a product is canceled during the prototype stage, losses may be relatively small. In extreme cases, however, a product is canceled after a complete SMT line has been set up just for that product, and the financial position of the company may be badly damaged. What it boils down to is this: failure to consider all aspects of product design can have serious adverse impacts on a business.

Volumes have been written on market analysis and research and strategies for attacking a particular market segment. It is not the objective of this book to discuss marketing strategies but simply to point out that no project should be initiated until the product (not the device, the board, or the box) has been defined. Product definition should be done by the product team, consisting of representatives from marketing, design, manufacturing, quality assurance, reliability, and other affected functional organizations.

After the product has been defined, the system requirements must be established. For example, what is the environment in which the product will function? Is it an office, home, or industrial environment? Is the system intended, instead, for a high reliability military or critical life support environment? Answers to these questions will help determine whether the packages, either through-hole mount or surface mount, should be hermetically sealed (ceramic) or plastic. Selection of components, in turn, will influence selection of substrate material.

The form, fit, and function of the product must be defined next. This will further narrow the selection of packaging options because it will help determine the most cost-effective way to meet the functional requirements. Form, fit, and function constitute one of the most important considerations in determining whether surface mount technology or conventional through-hole technology is required. Section 5.2 offers further discussion of form, fit, and function.

If in the evaluation of form, fit, and function it is determined that surface mounting is needed, various questions must be answered before a particular type of SMT assembly is selected. The real estate requirement is not the only consideration, as is generally thought, although it is the primary one.

The final decision for surface mounting packaging options should be made after cost and such technical issues as thermal management, interconnects, reliability, and design for manufacturability have also been considered. In the sections that follow, we discuss in some detail these items, because the designer must carefully consider them before committing to SMT.

## 5.2  FORM, FIT, AND FUNCTION

To make a product that is successful in the marketplace start with the idea of form, fit, and function. Once these variables have been defined, packaging technologies can be considered.

What do these often-used terms—form, fit, and function—mean? The *form* of a product (complete system or individual board) relates to the physical size, shape, and weight of the item. Another term for form is "factor constraints." Such constraints may be imposed by the various bus architectures (MULTIBUS I, MULTIBUS II, PC BUS, etc.) Different architectures have different slot spacings, which will establish the upper and lower height limits for component protrusions. The size of the board will determine the need for SMT, and the top and bottom height constraints will determine the component and assembly process requirements.

The *fit* of the product refers to its relationship with other functions or products within the system. Fit will determine the need for a given type of connector or input/output (I/O) function.

*Function* refers to a product's basic mission in life. What must it do? Once we have established the functional requirements, we see that they can be met by devices of different types, each with different physical, thermal, cost, and availability characteristics.

## 5.3  REAL ESTATE CONSIDERATIONS

When considering the use of surface mount technology, the most important factor is real estate. This is not to imply that other factors are not

**Table 5.1   Typical real estate savings in different types of SMT assemblies.**

| TYPE OF SMT ASSEMBLY | FUNCTIONAL DENSITY INCREASE |
|---|---|
| Type III SMT | 1.05–1.10 X |
| Type II SMT | 1.20–1.40 X |
| Type I SMT | 2.00–4.6 X |

important. They are. But as a rule, SMT is considered only when a product does not meet the functional requirements within the specified form factors. Once it has been determined that SMT is necessary, the designer must consider the three types of SMT assembly: I, II, and III, which we met in Chapter 1.

To review briefly, Type I SMT refers to assemblies containing only surface mount components; in Type III, components are glued to the bottom side and then wave soldered. Type II SMT, in terms of both process and components, is simply a mixture of the other two types, I and III.

Different types of SMT assembly provide different degrees of real estate savings. Table 5.1 gives a rough estimate of real estate savings for the three types. The mixture of surface mount and through-hole mount components primarily determines the real estate savings, which also will vary for logic and memory boards.

Actual real estate savings will depend on component selection, inter-package spacing rules for design, and the skill of the board designer in taking advantage of every nook and cranny (packing efficiency) on the board. There is generally some packing inefficiency, since the available space is not optimally utilized because of proximity requirements (for some components to be near others), component orientation requirements, form factor restrictions, and most important of all, varying sizes and shapes of different components on the board.

Final real estate requirements can be determined only after actual board layout, but the designer needs a rough idea before beginning the layout. An estimate is possible by multiplying the area required by different devices (size of the land pattern plus interpackage spacing) by their total number on the board and adding a packing inefficiency factor. Typically, the packing inefficiencies vary from 10% for a memory board to 30% for a logic board. For logic board, a minimum of 20% packing inefficiency should be allowed. Irregular sizes of different components cause some area to be wasted. An approximate estimate of real estate requirement can be represented by the following formula:

$$A = [(a_1 + I_1)n_1 + (a_2 + I_2)n_2 + (a_3 + I_3)n_3 + \cdots]K + M$$

where  $A$ = total real estate area requirement

$a_1, a_2, \ldots$ = land pattern areas of different components (see Chapter 6 for actual values; Chapter 6 does not show the land pattern requirement for through-hole components)

$I_1, I_2, \ldots$ = interpackage spacing requirments (see Chapter 7 for actual values, including interpackage spacing between surface mount and through-hole components)

$n_1, n_2, \ldots$ = total number of each component

$K$ = packing inefficiency constant: 1.10 for memory board and 1.30 for logic boards

$M$ = reserved area required for miscellaneous purposes such as clearance for edge card guide, automated test equipment, (ATE), card ejector brackets and front panel etc. Using the above formulas, Table 5.2 gives an example of real estate area that should be allowed for some selected through hole and surface mount components. The table shows the land pattern length and width and the inter-package spacing requirements for the package. The last column shows the area which takes into account both the land pattern and inter-package spacing. A similar table can be generated for all the packages to be used on a given board. The required area (A) can be obtained by multiplying the area for the packages (last column in Table 5.2) by the number of each packages on the board. To take into account the packing inefficiency, the total area must be multiplied by $K$ (1.1 for memory, 1.2 or 1.3 for minimum and maximum limits respectively for logic boards). Now add to this number, the reserved area. As mentioned earlier but worth repeating, the final number is only an estimate for the real estate requirement. Such an estimate is only intended to determine if all the components will fit on one side of the board. The final determination can be made only after placement in CAD by using the land pattern and interpackage spacing requirements discussed in Chapters 6 and 7 respectively.

**Table 5.2**  Land pattern/package area and interpackage spacing requirements for some selected surface mount and through hole components.

| PACKAGE/PIN COUNT | WIDTH | LENGTH | W-SPACE | L-SPACE | AREA |
|---|---|---|---|---|---|
| SOIC8 | 0.292 | 0.197 | 0.050 | 0.050 | 0.084 |
| SOIC14 | 0.292 | 0.344 | 0.050 | 0.050 | 0.135 |
| SOIC16 | 0.292 | 0.394 | 0.050 | 0.050 | 0.152 |
| R-PACK14 | 0.352 | 0.325 | 0.050 | 0.050 | 0.151 |
| R-PACK16 | 0.352 | 0.390 | 0.050 | 0.050 | 0.177 |
| SOICL16 | 0.452 | 0.413 | 0.050 | 0.050 | 0.232 |
| SOICL20 | 0.452 | 0.512 | 0.050 | 0.050 | 0.282 |
| SOICL24 | 0.452 | 0.614 | 0.050 | 0.050 | 0.333 |
| SOICL28 | 0.452 | 0.713 | 0.050 | 0.050 | 0.383 |
| SOJ26 | 0.382 | 0.680 | 0.060 | 0.050 | 0.323 |
| PLCC16 | 0.370 | 0.370 | 0.050 | 0.050 | 0.176 |
| PLCC20 | 0.420 | 0.420 | 0.050 | 0.050 | 0.221 |
| PLCC24 | 0.470 | 0.470 | 0.050 | 0.050 | 0.270 |
| PLCC28 | 0.420 | 0.620 | 0.050 | 0.050 | 0.315 |
| PLCC32 | 0.520 | 0.620 | 0.050 | 0.050 | 0.382 |
| PLCC44 | 0.720 | 0.720 | 0.050 | 0.050 | 0.593 |
| PLCC68 | 1.020 | 1.020 | 0.050 | 0.050 | 1.145 |
| PLCC84 | 1.220 | 1.220 | 0.050 | 0.050 | 1.613 |
| PQFP84 | 0.810 | 0.810 | 0.100 | 0.100 | 0.828 |
| PQFP100 | 0.905 | 0.905 | 0.100 | 0.100 | 1.010 |
| PQFP132 | 1.105 | 1.105 | 0.100 | 0.100 | 1.452 |

| | | | | | |
|---|---|---|---|---|---|
| 0805(RES/CAP) | 0.050 | 0.150 | 0.050 | 0.050 | 0.020 |
| 1206(RES/CAP) | 0.060 | 0.210 | 0.050 | 0.050 | 0.029 |
| 1210(RES/CAP) | 0.100 | 0.190 | 0.050 | 0.050 | 0.036 |
| DIP16 | 0.300 | 0.800 | 0.100 | 0.100 | 0.360 |
| DIP18 | 0.300 | 0.900 | 0.100 | 0.100 | 0.400 |
| DIP20 | 0.300 | 1.000 | 0.100 | 0.100 | 0.440 |
| DIP22 | 0.300 | 1.150 | 0.100 | 0.100 | 0.500 |
| DIP24 | 0.300 | 1.250 | 0.100 | 0.100 | 0.540 |
| DIP32 | 0.600 | 1.650 | 0.100 | 0.100 | 1.225 |
| DIP40 | 0.600 | 2.050 | 0.100 | 0.100 | 1.505 |
| PGA68 | 1.165 | 1.165 | 0.100 | 0.100 | 1.600 |
| PGA96 | 1.165 | 1.165 | 0.100 | 0.100 | 1.600 |
| PGA132 | 1.450 | 1.450 | 0.100 | 0.100 | 2.403 |
| PGA148 | 1.560 | 1.560 | 0.100 | 0.100 | 2.756 |
| PGASOCKET68 | 1.075 | 1.075 | 0.100 | 0.100 | 1.381 |
| PGASOCKET132 | 1.470 | 1.570 | 0.100 | 0.100 | 2.622 |
| PGASOCKET149 | 1.570 | 1.570 | 0.100 | 0.100 | 2.789 |
| DIPSOCKET22 | 0.500 | 1.150 | 0.100 | 0.100 | 0.750 |
| DIPSOCKET32 | 0.700 | 1.600 | 0.100 | 0.100 | 1.360 |
| DIPSOCKET40 | 0.700 | 2.000 | 0.100 | 0.100 | 1.680 |
| STAKEPIN | 0.070 | 0.070 | 0.050 | 0.050 | 0.014 |
| OSCILLATOR | 0.520 | 0.835 | 0.100 | 0.100 | 0.580 |
| CRYSTAL | 0.435 | 0.550 | 0.100 | 0.100 | 0.348 |
| AXIALCAP2.2MF | 0.020 | 0.447 | 0.100 | 0.100 | 0.066 |
| AXIACAP22MF | 0.180 | 0.420 | 0.100 | 0.100 | 0.146 |
| SIP6 | 0.200 | 0.600 | 0.000 | 0.000 | 0.120 |
| SIP8 | 0.200 | 0.800 | 0.000 | 0.000 | 0.160 |
| SIP10 | 0.200 | 1.000 | 0.000 | 0.000 | 0.200 |

## 5.4 MANUFACTURING CONSIDERATIONS

Manufacturing, especially with new technologies, is a specialized discipline. The designer is not expected to understand the manufacturing processes thoroughly, but there must be a manufacturing engineer on the team. It is not in the best interest of the company to design a product without considering manufacturability. A design that cannot be built or tested in a cost-effective manner is a very poor design indeed, any theoretical excellence notwithstanding.

Most designers become very concerned about "wasting" space and tend to pack boards as tightly as possible. This is short-sighted. Look at the IBM System/2 PC boards—there is plenty of room between packages for easy inspection, repair, and test. Sufficient space should be set aside for routing and interpackage spacing requirements to ensure manufacturability. Chapters 6 and Chapter 7 discuss in detail such design-for-manufacturability areas as land pattern, cleaning, repair, testability, inspection.

In addition to easy manufacturability, the designer must consider the processes to be used. As discussed in Chapter 7, different soldering processes have different rules. Process selection is also tied to component selection, and the effects on space saving and cost must not be ignored.

Many through-hole mount components are being used as surface mount components by modifying the leads as gull wing or butt leads. Naturally these modifications do not save space; rather, they are done to save a process step in hand or wave soldering. As a general rule, however, through-hole mount components are NOT expected to withstand the higher temperatures of reflow soldering. Before selecting items for use as surface mount components, it should be verified that they will not be damaged at the soldering temperatures contemplated.

Surface mount active components are generally designed for reflow soldering and should not be wave soldered. This means that they must all fit on the primary side of the board. Some surface mount components may not withstand the reflow temperature, either. This fragility may be encountered if an industry standard does not exist for a particular device, as is not uncommon with an evolving technology. For example, we had a surface mount delay line that had its internal connections soldered with eutectic solder with a melting point of 183°C. When this component was reflow soldered at the usual reflow temperature of 220°C, the internal connections melted and the solder came oozing out of the package.

Space and cost are also functions of the various possible combinations of components to go on the primary or the secondary side of the board. If all components could be had in surface mount, this would not be a problem. Since all the components are not available in surface mount, however, very creative combinations are needed to fit all the components on the board.

If the form factor must not change and no functions can be taken off, and the board must have active surface mount devices on top and bottom along with through-hole devices on the top, there are two options. Neither of them is really desirable because of adverse impact on cost or on product reliability. Refer to Chapter 7 (Sections 7.4: Soldering Considerations) for details on this complex situation.

Finally, the designer must consider what manufacturing capability exists and what processes are preferred by the manufacturing organization. If the capability for a Type II SMT exists, all options including through-hole assembly are available. If the capability for only Type III or Type I processes exist, the manufacturing options are limited. Even when there is full SMT capability, only the most appropriate technology should be used, because a Type I or Type III assembly may be more cost effective than a Type II assembly.

The selection of a packaging option should also be based on the volume requirement. For example, if product volume is very high, many functions can be combined in an application-specific integrated circuit. An ASIC package, depending on the real estate requirement, can be used in addition to or in lieu of SMT. For example, an ASIC package may make it possible to use Type III assembly instead of Type II.

The designer should not commit to SMT yet. The cost impact of the packaging option, discussed in the next section, remains to be dealt with.

## 5.5    COST CONSIDERATIONS

The switch to SMT is made for various reasons, but cost must not be neglected. When comparing the cost of a through-hole mount assembly (THMA) to that of a surface mount assembly (SMA), a basic distinction should be made between cost per assembly and cost per function. If surface mounting is used because all the functions do not fit on a given form factor, the cost per function is a more appropriate index than the cost per assembly.

The cost per assembly of a densely packed SMA is going to be higher because of the higher component count and finer lines and spaces on the board. But the cost per function will be less because it would take two THMAs to provide the functional equivalent. It will be more expensive to procure, handle, assemble, and test two THMAs than one SMA. There are additional savings if the need for connectors or cabling is reduced because there are fewer assemblies.

There are three major contributors to the cost of an assembly: the

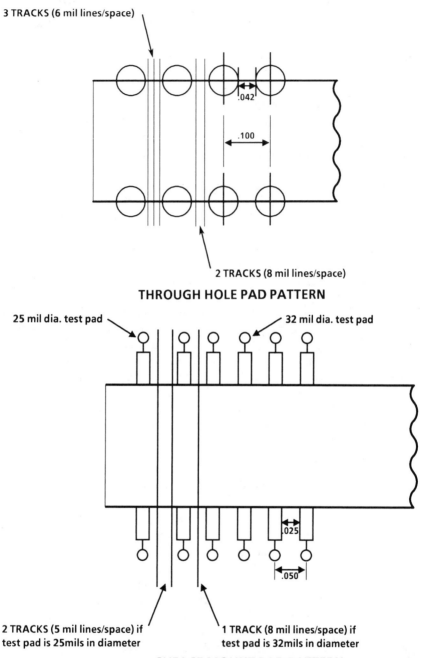

3 TRACKS (6 mil lines/space)

.042

.100

2 TRACKS (8 mil lines/space)

**THROUGH HOLE PAD PATTERN**

25 mil dia. test pad

32 mil dia. test pad

.025

.050

2 TRACKS (5 mil lines/space) if
test pad is 25mils in diameter

1 TRACK (8 mil lines/space) if
test pad is 32mils in diameter

**SUFACE MOUNT PAD PATTERN**

**Figure 5.1   Need for finer lines and spaces in surface mount boards. Note 3 six mil lines (3 track) in through hole board (top figure) versus only 2 five mil lines (2 track) in surface mount boards. This is due to change in lead/pad pitch from 100 mils in through hole board to 50 mils in surface mount boards.**

printed circuit board (PCB), the components, and the assembly. These are discussed in turn.

## 5.5.1    Printed Circuit Board Cost

The requirements for surface mount PCBs are not significantly different from those for conventional through-hole assemblies. However, the 50 mil center surface mount components reduce the gap between the adjacent pads to only 25 mils. Generally 0.050 mil center devices require a pad width of 0.025 inch. Since even micro vias generally require pads larger than 0.025 inch, less than 0.025 inch is left between the pads. This tiny gap demands that finer features (trace width and space) be used for routing traces between pads and vias. Refer to Figure 5.1, which compares the line width and space requirements for through hole (top figure) at 100 mil center with 58 mil (typical) pad sizes and surface mount (bottom figure) at 50 mil center with 25 mil (typical) devices.

As shown in Figure 5.1, the line width can be 8 mils for routing two traces in through hole and only one trace in surface mount pad patterns. For routing two traces in surface mount, the line width must drop to 5 mils. If larger than 0.025 inch test pads are used for test, only one trace (6 to 8 mils) can be routed. For ATE, 0.032 inch test pad vias generally are used to provide a larger test target for the test probes (see Chapter 7). As is clear from Figure 5.1, SMT boards will require of finer lines on the PCBs.

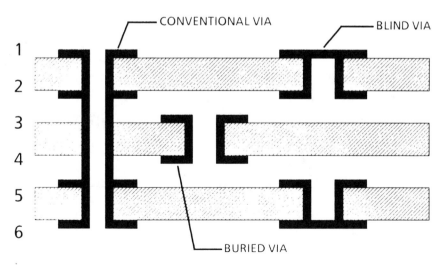

**Figure 5.2   Blind and buried vias.**

If surface mount active logic devices are mounted on both sides of the board, use of blind and buried via technology will become necessary to achieve the greatest density. The vendor base for such dense, double-sided, fine line boards with blind and/or buried vias is limited at this time. Also, there is a heavy cost premium associated with blind and/or buried via technology.

Figure 5.2 illustrates blind and buried vias. Blind vias extend from an outer layer to one or more inner layers. These vias can trap contaminants, affecting reliability. Buried vias, which connect inner layers, may also have an adverse effect on reliability. Blind and buried vias, although they increase cost, do reduce layer count. If this technology matures, therefore, cost increases due to blind/buried vias may be offset by a cost decrease due to reduced layer count.

The cost of the board is primarily driven by layer count and board trace width and space, as shown in Figure 5.3. As the line width decreases, the yield decreases due to opens and neckdowns. This combination in-

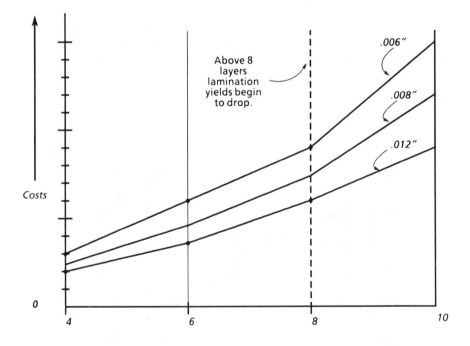

Number of Layers (.062″ thick board)

**Figure 5.3  Effect of layer count and trace width/space on PCB cost.**

creases cost. The cost of the board also increases with increase in layer count, rising significantly if the layer count goes beyond 8 layers for a given trace space and width. In such a case it is cost effective to switch over to finer lines and spaces. As the vendor base matures, significant cost differences between 6/6 and 8/8 trace/space board technologies may disappear.

The PCB cost for surface mount board, as compared with through-hole board, will be lower if the component count is the same in both cases. In other words, if the functionality of the boards is the same, the packaging density on the surface mount board will allow enough space between packages for routing purposes to permit the layer count to be reduced. Reduction of layer count from 8 to 6 layers or from 6 to 4 layers can generally result in PCB cost savings of about 25%. Looser component packaging is uncommon.

Generally the real estate freed up by smaller surface mount components is used to provide additional functions. Hence the component packaging generally becomes tighter in a surface mount board. Tighter

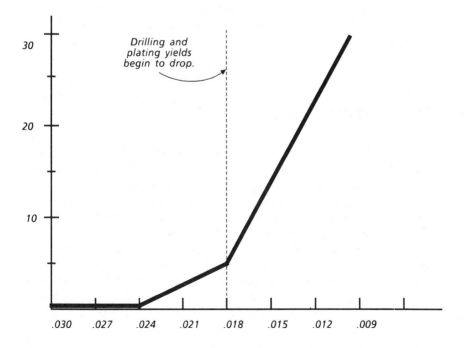

Finished hole diameter in inches

**Figure 5.4   Effect of finished via hole size on PCB cost.**

packaging density requires boards with finer lines and spaces, thus increasing board cost. If, however, increased packaging density, makes it possible to replace two boards with one, the overall PCB cost will be lower.

Surface mounting also increases the need for a smaller via hole size. A decrease here saves real estate, but again, the reliability of the hole suffers as a result of increased aspect ratio (see Chapter 4, Seciton 4.10: Via Hole Cracking). In any event, the vendor base for small vias is limited. As shown in Figure 5.4, there is generally a cost premium for via holes of 18 mils and less.

In general (either loose or tight packaging), the switch to SMT requires finer lines and spaces. Any board can be designed with wider lines and spaces, but the process usually is manual and takes more time in the CAD cycle. This may not be acceptable because of schedule constraints. Also, staggered interconnection or test vias must be used to allow routing with wider traces. Staggered vias require extra real estate, hence take up some of the real estate benefits of SMT. Thus there is seldom a need to use traces wider than 8 mils because there is essentially no savings in PCB cost.

The cost of the PCB also depends on the type of substrate used. With plastic packages, the commonly used FR-4 glass epoxy boards are necessary. For ceramic packages, substrates with a compatible coefficient of thermal expansion are necessary. See Chapter 4 for details on substrates.

## 5.5.2    Component Cost

Component cost constitutes the major portion of the total cost of assembly, but it is very difficult to generalize the cost impact of surface mount components. Some components are cheaper, others are more expensive in surface mount, and others are equivalent in cost. For example, most passive components such as chip resistors and capacitors are cheaper. There is some premium on tantalum capacitors.

Some of the active devices may be more expensive, bearing a premium of up to 15%. This is slowly changing, however. Indeed, for many other components, depending on the supplier, there may not be any cost difference between surface mount and through-hole mount devices. For special or custom components, the cost depends heavily on the volume and on the business relationship between the supplier and the user.

In general, surface mount components may be harder to purchase on a quick-turn around basis. Thus the schedule should allow a longer lead time for their procurement. This situation is likely to change with time.

## 5.5.3    Assembly Cost

Assembly cost includes component placement, soldering, cleaning, repair, and test costs. Soldering, cleaning, and test costs are essentially the same for both through-hole and surface mount assemblies. The cleaning and soldering operations for through hole and surface mount take roughly the same amount of time. Since the same equipment is used for testing, the test cost is the same, although the cost of the surface mount test fixture is generally higher.

The repair cost is lower for multileaded surface mount devices because they are easier to remove. In addition, the thermal damage during repair is almost nonexistent in surface mount devices, since the use of hot air eliminates the thermal damage caused by the pressure and high temperature of soldering iron (see Chapter 14). It is difficult to quantify savings here because the amount of repair required is based on the process control, the adequacy of the design for manufacturability, and the quality of incoming materials and components.

For all practical purposes, among the various assembly processes that determine product cost, placement is the most significant. In general, the pick-and-place time for surface mounting is considerably less than for the insertion of the through-hole mount devices. The basic reason for higher insertion time and cost for through-hole components is that the insertion operation requires setup and operation of three or four machines such as a component sequencer, an axial inserter, a radial inserter, and a DIP inserter.

On the other hand, all surface mount components can be placed by a single placement machine. This can reduce setup time considerably, as

**Table 5.3    Assembly cost comparisons for through hole and surface mount technology.**

| TYPE OF ASSEMBLY | SAVING IN PLACEMENT COST[a] |
| --- | --- |
| Through-hole assembly | (Used as standard) 0 |
| Type III SMT | 0 |
| Type II SMT | 19% |
| Type I SMT | 30% |

[a]Does not include any savings in storage, shipping, handling, and repair cost, which are expected to be lower. The test cost is assumed to be the same for both through-hole and surface mount assemblies.

well as queue time, handling time, and work in progress. In some applications, if versatility is not as important as speed, more than one dedicated piece of surface mount equipment for different component types can be used for increased speed. Thus, since placement time depends on the speed of pick-and-place equipment and on whether the manufacturing line is set for versatile or dedicated placement machines, it is difficult to generalize savings in this area.

Dedicated placement equipment is much faster than versatile equipment. The savings also depend on the feeder input capacity of the machine and on whether components are fed by tube or by tape and reel feeders. No matter what system is used, about 25% savings can be expected in placement time. This is shown in Table 5.3 for versatile, hence relatively slow, placement equipment.

## 5.6   THERMAL CONSIDERATIONS

With increased levels of integration on silicon chips (i.e., more transistors per unit area of silicon), thermal management remains a concern even with through-hole devices. Surface mount technology only compounds the problem. The surface mount packages, being smaller, yet containing the same die used in the roomier DIP packages, have higher thermal resistance. In addition, because of increased packing density provided by smaller packages, the power density per unit area of the substrate increases.

Thermal management is necessary not just for the sake of reducing the internal temperature of the electronic equipment. Rather, it is necessary to cool the devices on the board so that they can operate more reliably and fail less frequently. One does not have to radically change the basic thermal management approach (e.g., from forced air cooling to liquid cooling) just because of SMT. Instead, thermal management is undertaken mainly to achieve reliability at both package and system levels.

Thermal management is a complex subject, and here we simply highlight the thermal problems that are compounded by SMT. With the advent of SMT, such problems have become more critical for both suppliers and users of the packages. Since, moreover, the problems exist at both the package and the assembly levels, suppliers and users must take certain steps to minimize these difficulties.

The existing thermal problem that is exacerbated by SMT involves an increase in junction temperature ($T_j$), which can adversely affect the long-term operating life of the device. Junction temperature is a function of three items: power dissipation, junction-to-ambient thermal resistance, and ambient temperature. It can be calculated by the following equation:

$$T_j = (\Theta_{ja} \times P_d) + T_a$$

where $T_j$ = junction temperature (°C)

$\Theta_{ja}$ = thermal resistance, junction to ambient (°C/W)

$P_d$ = power dissipation at $T_j$(W)

$T_a$ = ambient temperature (°C)

The ability of the package to conduct heat from junction to ambient is expressed in terms of thermal resistance ($\Theta_{ja}$). It represents the total thermal resistance from junction to ambient. This property generally is separated into two parts: thermal resistance from junction to case ($\Theta_{jc}$) and thermal resistance from case to ambient ($\Theta_{ca}$).

Several factors control the junction-to-ambient thermal resistance. By regulating those factors, the junction temperature can be controlled and the device kept cooler when powered. The suppliers of surface mount devices are reducing the thermal resistance by changing the lead frames from the Kovar (54% Fe, 29% Ni, and 17% Co) and alloy 42—(Cu 58% and Ni 42%) alloys to copper itself, which is a better thermal conductor. Thus by changing the lead frame material, package suppliers have been able to reduce the thermal resistance of surface mount packages: the packages can get hotter, but the heat is conducted away more efficiently by the copper leads.

Another change component suppliers are making consists of providing heat spreaders inside the package, to conduct heat away much more rapidly. Changes in the die attach materials, use of larger die pads, and specification of better molding compounds also help in this endeavor.

Despite the changes just described, surface mount packages have a junction-to-ambient thermal resistance somewhat higher than that of DIPs (Table 5.4). The problem is more critical for smaller and thinner SOIC packages than for PLCCs.

Users must also take appropriate steps to ensure that even the use of packages with greater thermal resistance does not increase the junction temperature to unacceptable levels, compromising reliability.

**Table 5.4   Thermal resistance ($\Theta_{ja}$) of DIP and surface mount packages.**

| TYPE OF PACKAGE | LEAD FRAME MATERIAL | THERMAL RESISTANCE, $\Theta_{ja}$ (°C/W) |
| --- | --- | --- |
| DIP 20 pin | Kovar | 68 |
| SOIC 20 pin | Copper | 97 |
| PLCC 20 pin | Copper | 72 |

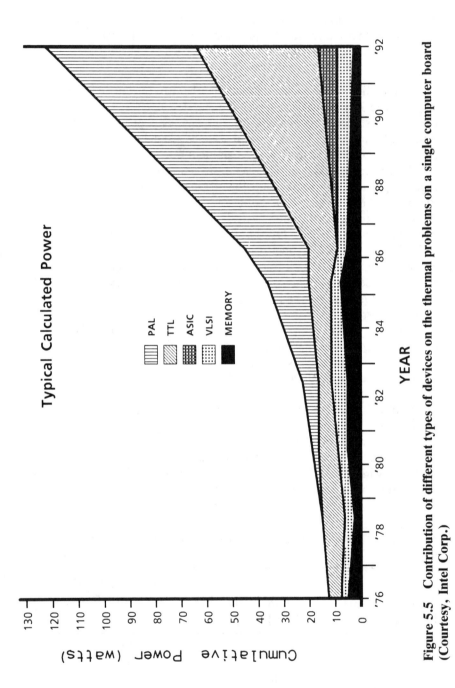

**Figure 5.5   Contribution of different types of devices on the thermal problems on a single computer board (Courtesy, Intel Corp.)**

The very first thing the user must do to analyze a board thermal problem is to determine the types of device that are the major contributors of heat on the board. For example, as shown in Figure 5.5, programmable logic arrays (PLA) devices, which account for a significant poriton of the power on the board, are the culprit in most single-board computers. The second greatest contributors are the transistor-transistor logic (TTL) devices. The situation shown in Figure 5.5 may not be applicable to every case, but similar charts can be generated for other situations to determine the significant contributors to thermal problems.

When feasible, 10 to 15 PLA or TTL devices can be converted into a single ASIC gate array package. This will not only reduce the power on the board but will provide plenty of room on the board for other packages. However, it is essential to avoid the temptation to fill up the space savings created by the ASIC device with additional functions. That would bring us back to where we were in the first place.

The user must also develop a design rule for interpackage spacing to minimize thermal problems. As discussed in Chapter 7 (Section 7.6), interpackage spacing rules are required for repair, cleaning, inspection, and testing. Interpackage spacing rules should also take thermal cooling into consideration and, when developing thermal design rules, the user can establish requirements for placing hotter devices near cooler devices. In addition, the minimum acceptable interpackage spacing between components can be increased to maintain a maximum wattage per unit area of the board.

The thermal design rules should also establish the maximum power rating ($P_d$) for each type of package (SOIC, PLCC, etc.) The $P_d$ can be derived from the equation given earlier. A maximum acceptable junction temperature (generally about 100 or 115°C) should be established to ensure the desired reliability. The next important variable is the ambient temperature, which depends on the market in which the product is designed to operate.

The final variable, junction-to-ambient thermal resistance, is supplied by the vendor. By plugging in the values for these parameters, an acceptable power rating for various packages can be established. Table 5.5 can be used in developing the maximum acceptable power rating for various surface mount packages for different values of $T_{ja}$, $T_j$, and $T_a$.

In addition, the user can take the following steps (a) use conductive adhesive between the package and the substrate, (b) provide thermal pads for the packages such as LCCCs and mount them flush with the substrate, (c) connect devices with thermal vias and wider traces with the ground planes, and (d) use heat sinks on top of the package to conduct heat.

Another approach is to select different substrates with better heat dissipation characteristics. With ceramic packages, for example, the me-

**Table 5.5 Maximum power rating (Pd) for surface mount packages for junction temperatures of 115 and 130°C and an ambient temperature of 70°C. (M. Kastner, and P. Melville, Ed.** *Signetics SMD Thermal Considerations.* **Signetics Publication 98-9800-010, 1986.)**

| PACKAGE | $\Theta_{ja}$ (°C/W)[a] | POWER (W) FOR JUNCTION TEMPERATURE | |
|---|---|---|---|
| | | UNDER 115°C | UNDER 135°C |
| SO 14 | 110–130 | 0.340 | — |
| SO 16 | 110–120 | 0.375 | — |
| SOL 16 | 90–110 | 0.450 | — |
| SO 20 | 80–90 | 0.500 | 0.720 |
| SO 24 | 70–80 | 0.560 | 0.810 |
| SO 28 | 70 | 0.640 | 0.930 |
| PLCC 20 | 70–80 | 0.560 | 0.810 |
| PLCC 28 | 59–73 | 0.620 | 0.890 |
| PLCC 44 | 44–53 | 0.850 | — |
| PLCC 52 | 38–42 | 1.070 | — |
| PLCC 68 | 41–47 | 0.960 | — |
| PLCC 84 | 32–38 | 1.180 | — |

[a]These typical thermal resistance values are based on copper lead frames with components mounted in sockets. Actual thermal resistance will vary with die size, component mounting, and molding compound.

tallic constraining core used for matching the package and substrate coefficients of thermal expansion also helps in conducting heat.

Changing the airflow rate from, say, 200 to 400 linear feet per minute (lfpm), derating the equipment for a lower ambient temperature, or changing the cooling medium entirely (from air to liquid cooling) are some of the more drastic steps that users can take. These steps are generally not warranted because of SMT alone.

## 5.7 PACKAGE RELIABILITY CONSIDERATIONS

Package reliability and solder joint reliability are different in the surface mount and through-hole technologies. Solder joint reliability (discussed in Section 5.8) has been an issue for ceramic packages in military applications because of CTE mismatch between package and substrate. Although the plastic packages mounted on commonly used glass epoxy substrates do not have solder joint reliability problems, the packages themselves may be susceptible to cracking.

When moisture-saturated surface mount packages are subjected to reflow soldering (vapor phase, IR, hot plate) conditions, the moisture expands inside the package and may crack the plastic body. If deleterious contaminants such as flux can seep in through such a crack, they may cause bond pad corrosion, seriously degrading the long-term reliability of the package.

It should be emphasized here that the impact of cracking on the reliability of plastic packages is not fully understood. Corrosion failures have been seen only under extreme test condiltions. Even under experimental conditions, no cracking is observed unless the package is saturated with moisture above a certain threshold limit. No failures have been reported in the field. Steiner and Suhl [2] have concluded that the problem is industry wide. [2].

Cracks are difficult to visually inspect because they occur on the bottoms of the packages. The cracking problem is more widespread in larger PLCCs (68 pin and greater). However, some cracking has been reported in smaller packages such as DRAMs and larger SOICs. Various long- and short-term solutions are being proposed by the industry for this relatively new problem. For example, the Package Cracking Task Force of the Surface Mount Council chaired by the author has published a white paper [3] summarizing the status of the problem with recommendations for the industry. As mentioned in Chapter 2, the Council is only a management body chartered by the industry to coordinate the efforts of the Government and the industry to promote SMT. Since the Council itself does not write standards, it has recommended that the industry organizations address this issue through development of standards. As a result the IPC has formed a committee also chaired by the author to study the problem and will publish standards for use by vendors and users. Other industry organizations may also follow suit. Since development of a standard is a time consuming process because of long and necessary process of reaching consensus by the industry, publication of a final standard takes time. However, the proposed standard on package cracking currently under development and coordination is available from the IPC. Let us discuss factors that control package cracking and then focus on short- and long-term solutions.

## 5.7.1 Package Cracking Mechanism

The plastic DIP packages have never been thought to be moisture resistant, but no cracking has been reported in them. Why, then, since the industry is using the same molding compound in PLCCs as in DIPs, have the plastic DIPs not been found to be susceptible to cracking? In wave soldering, only the leads undergo the high soldering temperatures. The package body rarely is heated beyond 150°C. However, in reflow soldering (vapor phase or infrared), both the packages and the leads are subjected to temperatures of 215 to 240°C. If such a package is saturated with moisture, the moisture will expand during reflow and may crack the package.

It has been pointed out that no cracking is seen if there is no die or a very small die in the package. Thus we can conclude that along with moisture, die size plays an important role. The thickness of the plastic body is also important: a thicker package is less susceptible to cracking, but the package will remain in a stressed condition because the moisture attempts to escape.

Steiner and Suhl [2] have determined that the threshold temperature for cracking is 180 to 200°C. They also found that the phenomenon is not related to the rate of heating or cooling. This means that as far as package cracking is concerned, no reflow process (vapor phase or infrared) is better than another. However, this problem will not occur if the soldering process (e.g., laser or hot bar soldering) does not subject the packages to higher soldering temperatures.

The specific mechanism of package cracking is depicted in Figure 5.6 [4]. The top diagram shows the package before moisture absorption, including the lead frame, the plastic encapsulant, and the plastic thickness below the lead frame. The package absorbs an amount of moisture that depends on the thickness of the plastic and the storage conditions. When the package is soldered by the reflow process, the entrapped moisture expands (Figure 5.6: middle diagram). If the adhesive strength of the plastic bond with the die is exceeded, a delamination void results, such that the package cracks, usually at the edges (Figure 5.6: bottom), and the pressure dome collapses because of the escape of moisture. The crack generally emanates from the corner of the die paddle (lead frame directly below the die), bottom diagram in Figure 5.6. Side cracking has also been observed, however. See Chapter 10, Figure 10.8 for a cross-sectional photograph of die sitting on the die pad.

Cracks initiate from points of high stress at the edge of the lead frame die attach pad and propagate along the path of least resistance—downward through the thinnest portion of the plastic molding or radially out along the lead frame interface. This phenomenon was reported first by Oki Electric on flat packs but has recently been observed in small outline devices containing large static or dynamic random access memories and mostly high pin count (68 pin) PLCCs. Some cracking has also been observed in lower pin count PLCCs as well. Die size is more critical than package size or pin count.

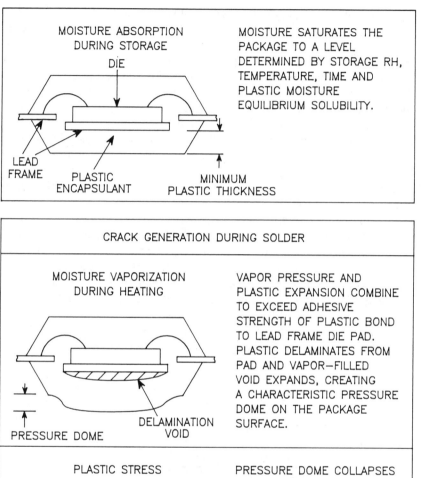

**Figure 5.6    Mechanism of cracking in large plastic surface mount package [4].**

Fukuzawa et al. [5] of Oki Electric observed cracks that developed in plastic surface mount packages (56 pin, plastic quad flat pack) after being subjected to temperatures from 215 to 270°C during the attachment process. They theorized that the package temperature experienced during reflow was far greater than the glass transition temperature (165°C) for the silica-filled novolac epoxy, resulting in localized delamination between the die paddle and the plastic. Vapor pressure from absorbed moisture in the package, in combination with CTE mismatch between the die pad and the surrounding molding compound, caused cracks to develop in the thinner bottom part of the package. Further evidence of this rupture was the dome that appeared on the underside of the package (Figure 5.6: bottom). Water condensation in the gap between the chip and the plastic led to aluminum pad corrosion.

Similarly, Steiner and Suhl [2] also reported that vapor pressure from absorbed moisture during reflow soldering causes package cracking. The packages were subjected to a temperature of 85°C and 85% relative humidity (RH) for 168 hours. Ito et al. [6] confirmed that moisture absorbed by the molding compounds coupled with the temperature rise experienced by the packages during reflow will promote delamination between the molding compound and the die pad interface.

Figure 5.7 shows the moisture absorption levels in a 68 pin PLCC at 85°C and three relative humidity conditions. After the preconditioning at

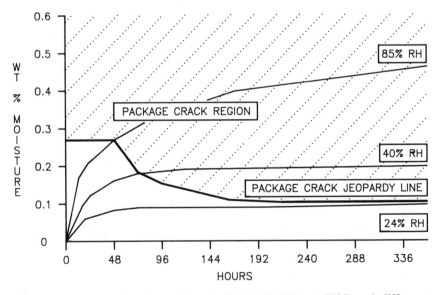

**Figure 5.7   Moisture absorption in 68 pin PLCCs at 85°C and different relative humidities [4].**

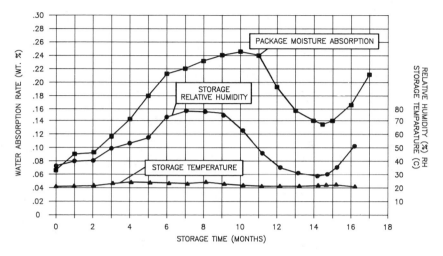

**Figure 5.8    Effect of relative humidity on moisture absorption in PLCC at 25°C storage over a period of 18 months [4].**

various times, Alger et al. [4] subjected the packages to vapor phase soldering. Packages with moisture content levels above the "package crack jeopardy line" (Figure 5.7) may crack when subjected to reflow soldering environments.

Figure 5.7 shows moisture absorption in accelerated experimental conditions. Is this level of moisture absorption likely to be attained in normal storage conditions over time? The answer to this question, which is very critical in establishing whether the problem is real or simply a theoretical possibility, is contained in Figure 5.8 [4]. A Japanese Company (Shinko) collected the moisture absorption data shown in Figure 5.8 at 25°C for over 18 months. As the humidity in the environment changed, so did the moisture absorption level. Thus we see that moisture levels can exceed the threshold limit to cause package cracking. The threshold limit is 0.1% (weight percent) of moisture, as shown in Figure 5.7.

It should be pointed out that total moisture content alone is not the determining factor for package cracking. As the packages are saturated with moisture, a moisture gradient occurs, and the location of the moisture is very important. For example, moisture absorption in the outer regions of the package is not as critical as absorption toward the center. Other important factors include reflow soldering temperature, lead frame design and surface preparation, strength of the plastic, die size, and plastic thickness. The die size and the thickness of the molding compound are particularly significant.

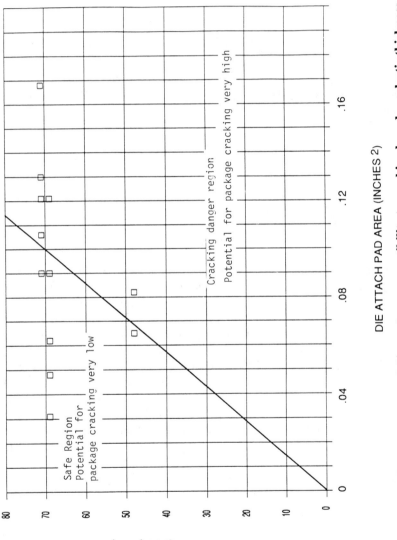

**Figure 5.9 Guideline for determining package susceptibility to cracking based on plastic thickness and die attach pad area [4].**

Because of greater CTE mismatch between the metallic lead frame paddle and the plastic molding compound, cracking severity increases with die size. A rule of thumb is that die sizes greater than 4 mm (160 mils) on each side should be considered to be susceptible to cracking. Based on die size and plastic body thickness (bottom side of package), Figure 5.9 [4], which can be used as guideline in determining a package's susceptibility to cracking, shows safe and unsafe regions. All packages that fall into unsafe reigons should be declared susceptible to cracking. When using Figure 5.9, note that there will be some exceptions. The best way to correctly characterize a package is to subject it to the reliability tests discussed in Section 5.7.3.

The larger die sizes also generate more heat, hence require heat spreaders for heat dissipation. Although heat spreaders effectively reduce the thickness of the molding compound, they compound the original problem by increasing the risk of cracking because the plastic body is made relatively thinner.

## 5.7.2    Solutions to Package Cracking

The industry is struggling with short- and long-term solutions to the package cracking problem. It has been found, for example, that increasing the thickness of the molding compound around and below the die can prevent cracking. Similarly, it is now known that package manufacturing variables that promote better lead frame adhesion, such as rougher internal leads, perforated lead frames, and anchor holes, can minimize cracking by allowing the plastic to adhere more tightly to the lead frame. Use of new molding compounds that are more impervious to moisture penetration will provide the true long-term solution. However, it is unknown when or indeed whether such a material can be developed. If success is achieved someday, the use of expensive ceramic packages will drastically diminish, along with the solder joint cracking problems associated with them.

Another potential solution is to prevent the absorption of moisture at the interface of the lead frame and the molding compound. If moisture vapor does not collect at this interface, cracking will not occur. Suhl [7] reported excellent results from applying hexamethyl disilazane (HMDS) vapor to lead frame and die just before the molding operation. HMDS has low flash point (13°C) and is hygroscopic (absorbs moisture) and should be handled accordingly. It is used to promote adhesion of silicon wafers to novolac photoresist, hence is a familiar chemical to the semiconductor industry.

Suhl [7] treated one group of samples with HMDS and used the other half as a control. The samples in both groups were considered susceptible

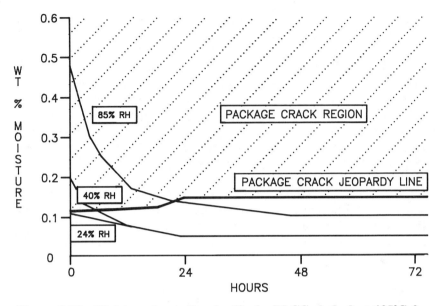

**Figure 5.10   Moisture desorption in 68 pin PLCCs baked at 125°C for different times [4].**

to cracking. All the samples were saturated with moisture and subjected to vapor phase reflow conditions. Suhl found that the packages treated wth HMDS showed significantly lower cracking (5%) than the untreated standard packages (100% cracking) that served as controls.

Approaches based on use of a better molding compound or treatment with HMDS are still in the experimental stages, but baking is evolving as the short-term solution. Figure 5.10, which shows moisture desorption for packages saturated with moisture at various relative humidities [4], suggests that the absorbed moisture can be baked out. The "package crack jeopardy line" in Figure 5.10, develops when the packages are baked at 125°C for various times.

To prevent cracking, packages should be baked to a moisture content below 0.1% (by weight). In addition, baked packages must be shipped in desiccants to prevent moisture absorption during shipping and storage at the user's facility.

A construction of a shipping container with desiccant is shown in Figure 5.11. Although not hermetic, desiccant containers can be stored for up to 6 months before rebaking [4]. If the shipping containers are damaged and absorb moisture or if the shelf life exceeds 6 months, the packages must be baked before soldering. Two baking options can be used to provide

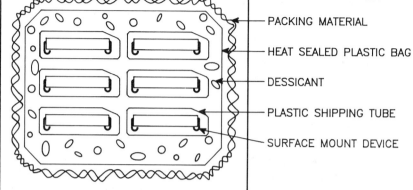

SHIPPING CONTAINER

PACKING MATERIAL

HEAT SEALED PLASTIC BAG

DESSICANT

PLASTIC SHIPPING TUBE

SURFACE MOUNT DEVICE

**Figure 5.11 Moisture-sensitive packing to minimize moisture absorption during shipping and storage [4].**

flexibility in manufacturing. These are the high and low temperature baking cycles. The process conditions for the high temperature bake are:

Temperature: 125 (±5)°C
Duration: 24 (+1-0) hours
Chamber relative humidity: <50%

The high temperature baking process is not suitable for plastic tubes or tape and reel because they cannot withstand the baking temperature and the packages must be removed from the tubes or tapes and reels before being baked. They can be placed either in a metallic tube or in waffle packs, which can withstand the baking temperature.

Each package that was not received baked and bagged from the vendor should be subjected to a maximum of TWO high temperature bake cycles before being soldered to the board. Packages received in a baked and bagged condition should be subjected to only ONE high temperature bake cycle before being soldered. This is mainly because repeated baking may cause solderability problems.

If repeated baking is to be used, any adverse impact on component solderability should be determined. Baking of packages at 125°C for 24 hours (one baking cycle) will add about 15 to 20 microinches to the copper-tin intermetallic thickness and should not pose solderability problems. The tin-lead is more than 200 microinches thick.

Low temperature baking, which provides much more flexibility for a manufacturing environment, may be especially helpful for packages that come in plastic tube or on tape and reel. Low temperature baking takes longer, however. The process conditions for the low temperature bake are:

Temperature: 40 ( +5-0)°C
Duration: 192 hours minimum (8 days)
Chamber relative humidity: <5%

The low temperature baking process is suitable for plastic tubes as well as tapes and reels. Packages can be subjected to as many low temperature bake cycles as required.

Whether baking is done at low or high temperature, packages that are not used within a certain period after baking generally must be rebaked. One possible way to avoid rebaking is to store parts in controlled environments (cabinets or rooms). Desiccant bags containing parts from vendors do not require storage under controlled conditions unless the bag seals have been broken or the humidity indicator cards inside the bags indicate exposure to moisture above acceptable levels. Table 5.6 lists storage conditions and maximum times before soldering.

If the humidity indicator cards in a vendor's bags indicate exposure to moisture above acceptable levels (if a bag seal is broken), or if the exposure time of in-house baked packages exceeds the time given in Table 5.6, the packages must be baked according to one of the alternatives discussed earlier. The storage conditions and maximum time before reflow soldering given in Table 5.6 also apply to packages that have been baked in-house.

Storage conditions apply only to packages that have been baked first. Simply storing a part in a low temperature, low humidity environment will

**Table 5.6  Maximum exposure times at various storage conditions before re-baking for reflow soldering to prevent package cracking.**

| STORAGE CONDITIONS | | |
| --- | --- | --- |
| RH (%) | TEMPERATURE (°C) | EXPOSURE TIME (HOURS) |
| 5 | 40 | Unlimited |
| 60 | 30 | 200 |
| 70 | 30 | 120 |
| 85 | 30 | 90 |

not desorb moisture from the lead frame/molding compound interface. Thus any package that has absorbed moisture above its threshold limit must be first baked. A controlled storage environment is not a substitute for baking. It should be only used to eliminate the need for rebaking.

## 5.7.3    Reliability Tests for Package Cracking

Cracking is difficult to spot during routine manufacturing operations such as inspection and electrical testing. To prevent potential package

**Table 5.7    Preconditioning test sequence to determine susceptibility of a plastic package to cracking during reflow soldering.**

| STEP | TEST CONDITIONS |
|------|-----------------|
| 1. | 20 temperature cycles per Condition B of Military Specification 883, (Optional, necessary only on suspect parts) |
| 2. | 48 hours at 125°C (bake out to expel any moisture) |
| 3. | 85°C at 85% RH exposure (unbiased 168 hours), except for crack-sensitive products for class I, II, and III parts respectively. |
| 4. | Weight gain determination (optional, to be followed only if step 1 is used) |
| 5. | Reflow one cycle at 220°C for 60 seconds |
| 6. | Solvent clean |
| 7. | Electrically test for cracking. |
| 8. | Hot air exposure (4 cycles at 200°C) (nearest neighbor rework) |
| 9. | Flux exposure |
| 10. | Deionized water wash |
| 11. | Hot air exposure (2 cycles at 200°C) (component removal/ rework) for 60 seconds each. |
| 12. | 85°C and 85% RH exposure (unbiased 168 hours) |
| 13. | Reliability stress testing |
| 14. | Biased humidity stress |
| 15. | Temperature cycling |
| 16. | Examine for cracking visually or by non-destructive testing. If a crack permeates more than 1/2 way through the plastic body (see Figure 5.6), the part is considered susceptible to cracking. |

cracking problems, moisture-sensitive packages should be qualified for use. To simulate a worst case, packages should be saturated with moisture in the required temperature and humidity environment and then subjected to process conditions. Alger et al. [4] have proposed a precondition sequence to simulate the surface mount process for determining a plastic package's susceptibility to cracking. The precondition sequence shown in Table 5.7 is adapted from Reference 4. All components would be subjected to this test sequence before reliability stress testing. The test conditions suggested in Table 5.7 are being considered by the industry as a standard test for susceptibility of a plastic package to cracking during reflow soldering.

Packages that have been subjected to the preconditioning test flow sequence of Table 5.7, should be cross sectioned and examined under a microscope. A package is considered to be susceptible to cracking if it shows visible cracks inside the package. Any such package must be baked before reflow soldering.

Cracks are difficult to find, but they commonly occur along the die edges or in the center of the die, then proceeding toward the leads. Complete cracking of packages on the bottom is not very common. Industry is also investigating cheaper and not destructive tests such as ultrasonic and die penetrant tests to locate package cracks.

## 5.8  SOLDER JOINT RELIABILITY CONSIDERATIONS

In through-hole mount assemblies, solder joint reliability has not been an issue because the plated through hole provides added mechanical strength to the joint. However, solder joint reliability causes concern in surface mounting because the surface mount lands provide both mechanical and electrical connections. Are surface mount solder joints less reliable because of the absence of plated through holes? Solder joint reliability is of concern only in military applications.

In commercial applications using plastic leaded packages (J or gull wing) or small ceramic chip components on a glass epoxy board, solder joint reliability is not critical. Since difference in CTE between the plastic packages and the glass epoxy boards is very minor, the leads of the packages take up the small expansion mismatch created by thermomechanical cycling.

Ceramic chip resistors and capacitors are also widely used on glass epoxy boards. But because of the small size of these components, the stresses they generate do not exceed the mechanical strength of the solder

**Table 5.8  Thermal cycling results on solder joint reliability for Type I, II and III SMT. See Table 5.9 for test conditions. (Intel Corporation.)**

|  | TYPE III ASSEMBLY | TYPE I ASSEMBLY | TYPE II ASSEMBLY |
|---|---|---|---|
| Random vibration | Pass | Pass | Pass |
| Mechanical shock | N/A | Pass | Pass |
| Humidity storage | Pass | Pass | Pass |
| Temperature cycle (operating) | N/A | Pass | Pass |
| Temperature cycle (nonoperating) | Pass | Pass | Pass |
| Life test | Pass | Pass | Pass |
| Total solder joint cycles | 210,000 | 580,000 | 1.3 million |

joint; hence there is no reliability concern. When using properly designed land patterns (see Chapter 6), the surface mount solder joints (plastic packages and ceramic chip components mounted on glass epoxy boards) pass all the environmental reliability tests, as shown in Table 5.8. The tests were conducted on functional assemblies using the test parameters given later in Table 5.9 (Section 5.8.1).

As the size of the ceramic package increases, the stresses can increase substantially. This is generally the case in ceramic capacitors larger than 0.250 inch and in high pin count leadless ceramic packages. For example, it is well known that because of mismatch in CTE values, LCCCs mounted on glass epoxy substrates cause solder joint cracks in fewer than 200 thermal cycles. There are three commonly used approaches to solder joint cracking problems related to CTE mismatch: using substrates with compatible or matched CTE, using substrates with top compliant layers that can absorb the stress, and using leaded ceramic packages.

The compliant top layer approach is rarely used because the reliability problems may be shifted to the traces or the via holes. Using substrates with compatible CTEs is the way most commonly chosen, but even this approach has its problems. For example, cracking may be eliminated in the solder joints, but there are failures in the plated through holes. Even if the solder joints do not crack as a result of compatible CTE values, the cracking in the plated through holes has the same effect: unreliable assembly.

To prevent via hole cracking, either the board thickness is reduced or the via hole diameter is increased. Use of a substrate material with a lower modulus of elasticity has been tried to prevent via hole cracking. Since the lower modulus material is soft, it cannot also damage vias even under stress. See Chapter 4 (Section 4.3: Selection of Substrate Material), and sections dealing with substrates.

In the third approach, using leaded instead of leadless ceramic packages on substrates with unmatched CTEs, the stress on the solder joint can be taken up by the compliant leads. Use of leaded packages not only reduces the overall cost but almost assures the reliability of the solder joints. The leads can be either gull wing or J type. However, the use of leaded ceramic package has not been a very successful approach.

The leaded ceramic packages typically have leads made from Kovar or alloy 42 instead of copper, to ensure that the lead frame material matches the ceramic body. These leads are not ductile like copper leads. They may, however, be adequate for smaller chip carrier sizes that do not generate excessive stress on the solder joints. In addition to having a ductile lead material, ceramic packages require proper lead design. For example, gull wing leads provide better compliancy than J leads. And even S-shaped leads are being proposed for best results. However, S-shaped leads are neither widely available nor commonly used. To use leaded ceramic packages on inexpensive glass epoxy substrates calls for additional work in the areas of lead frame material selection and lead configuration design, to ensure compatibility with mass soldering.

The concept of using leaded ceramic packages for military application is not new. As a matter of fact, originally the military wanted to use leaded packages but abandoned the idea for various reasons. The leadless ceramic packages were selected because they were easier to assemble. In addition, they can be placed closer together (giving a higher density assembly), they have superior thermal performance, and they are cheaper than leaded packages. As discussed earlier, the use of leadless packages requires ceramic substrates or exotic substrates. This adds significantly to the cost of the assembly. In the final analysis, the overall increase in assembly cost far exceeds any savings from the leadless packages. Now the trend is shifting back to leaded packages for the same reasons—reliability and cost concerns.

Achieving adequate solder joint reliability at reasonable cost for military applications is no easy matter. The leadless packages give the most problems. The U.S. Department of Defense has put LCCCs on a reliability suspect list even though the devices are mounted on substrates with matched or tailored CTE values. This is by no means to imply that leadless packages do not have any application. They do, especially where it is critical to have the shortest possible signal path as well as efficient cooling. The use of the leadless packages can and should be reduced whenever possible, however. Leadless packages should be selected only when leaded ceramic or plastic packages are not feasible at all. Moreover, when using leadless packages, via hole and solder joint reliability should be extensively evaluated before a substrate material is chosen.

## 5.8.1   Solder Joint Reliability Tests

What are the relevant solder joint reliability test methods? What constitutes a failure? How should the failures be monitored? The answers are not as obvious as they might appear. This is basically because there are no hard data to correlate the accelerated reliability test results with what is known of field failures or intended service life. Among the various types of test methods available, which one should be chosen? The answer will depend on the purpose of the test. A test method that closely resembles the field application will give the most representative results.

The electronics industry uses many test methods for ascertaining solder joint reliability, but the most common are powered functional cycling, thermal cycling, and mechanical cycling. Table 5.9 shows an example of test parameters for functional, thermal, and mechanical cycling tests used on Type I, II, and III surface mount assemblies for commercial/industrial applications.

For any application, the power functional cycling is most desirable; it

**Table 5.9   Environmental test conditions to which surface mount assemblies were subjected to. See Table 5.8 for test results. (Intel Corporation.)**

RANDOM VIBRATION:
(0.01 $g^2$/Hz at 10 Hz sloping to 0.02 $g^2$/Hz from 20 Hz to 1 kHz)
MECHANICAL SHOCK:
(50 $g$ sine for 11 ms)
HUMIDITY STORAGE:
(5 days at 55°C, 95% RH)
TEMPERATURE CYCLING (OPERATIONAL)
($-15$ to $+70$°C for 5 days)
TEMPERATURE CYCLING (NONOPERATIONAL)
($-40$ to $+125$°C; 2500 cycles)
LIFE TEST:
(3000 hours at 60°C)

is a recommended test regardless of whether any other tests are conducted. Engelmaier has shown that powered functional and thermal cycling are more realistic tests to induce solder joint fatigue [8]. Functional cycling tests should be conducted in an environment that closely simulates the field environment.

The other commonly used test, thermal cycling, is appropriate for comparison purposes and for assemblies with large CTE mismatch, such as ceramic chip carriers or chip components on FR-4. For thermal cycling to produce useful results in a reasonable time frame, a CTE mismatch of at least 3 ppm/°C must exist between the component and the substrate.

The dwell temperatures and times and the ramp rates are the most critical parameters in functional and thermal cycling. It is important to note that dwell times at extreme temperatures are significantly shorter in accelerated testing than in the field, and thus the stress relaxation is less complete. Therefore the number of cycles to failure in the field is always less than the number of cycles to failure in an accelerated test. This discrepancy is explained by the stress-strain curves of Engelmaier [9] in (Figures 5.12 and 5.13, which illustrate different cyclic stress-strain behaviors of metals. Figure 5.12 shows a hysteresis loop without the stress relaxation characteristic of fatigue cycling of copper or even solder with zero dwell times. It is more typical of engineering metals, such as copper, aluminum, or steel, that do not creep or relieve stress at room temperature. Figure 5.13, on the other hand, shows the hysteresis loop for a solder joint with cyclic dwells long enough for complete stress relaxation, as would occur in field use.

The area in the cyclic hysteresis loop is a measure of the fatigue damage per cycle. Somce cyclic fatigue damage is cumulative with each fatigue cycle and exhausts the available fracture toughness, the larger the hysteresis loop, the lower the number of cycles to failure. This means that the difference in the hysteresis loops generated in accelerated testing and in the field needs to be accounted for when establishing product reliability. It also means that to achieve high reliability, the component/substrate system needs to be designed to produce a small hysteresis loop. This can best be accomplished by a combination of component and substrate CTE tailoring (taking into account the tempearture difference between component and substrate in a functional environment) and attachment (lead, solder, and substrate) compliancy.

Thermal cycles of various types with different dwell temperatures and times are used to test solder joint reliability. There is absolutely no consensus in the industry about the upper and lower temperatures and the ramp rate. In addition to thermal cycle (one-chamber) tests, thermal shock (two-chamber) tests with rapid change in temperature extremes are used. These conditions rarely are encountered by a product in service, however,

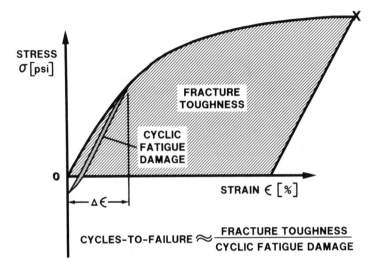

**CYCLES-TO-FAILURE** $\approx \dfrac{\text{FRACTURE TOUGHNESS}}{\text{CYCLIC FATIGUE DAMAGE}}$

**Figure 5.12** **Illustration of cyclic stress-strain behavior of solder; stress relaxation in the hysteresis loop is not shown [9].**

so unless field conditions warrant, thermal shock test results may be misleading [10].

For industry-wide round robin tests of surface mount attachment reliability, thermal cycling from −55 to +125°C for 500 cycles for military applications and 0 to +100°C for 1000 cycles for commercial applications has been proposed [11]. The dwell time is 30 minutes at each temperature extreme and the transition time between temperature extremes is 30 minutes, as shown in Figure 5.14.

The third commonly used method of determining solder joint reliability, mechanical cycling, is generally preferred because it is a highly accelerated test as compared with thermal cycling. For example, 20,000 mechanical cycles can be performed in the same amount of time (84 days) needed for only 1000 thermal cycles. Since both mechanical and functional cycles subject the solder joints to shear stress, the failure mechanism in either case is by low cycle fatigue (as long as the joint is not overstressed).

Mechanical cycling generally calls for 40 to 80 inch radius alternate convex-concave bending of the circuit board. A frequency of 240 cycles per day until failure is commonly used with mechanical testing [10]. Equipment for conducting mechanical cycling tests is commercially available [12]. The tester was originally designed for the Compliant Lead Task Force of the Institute of Electrical and Electronics Engineers (IEEE). One of the problems with both mechanical and thermal cycles has been that it takes

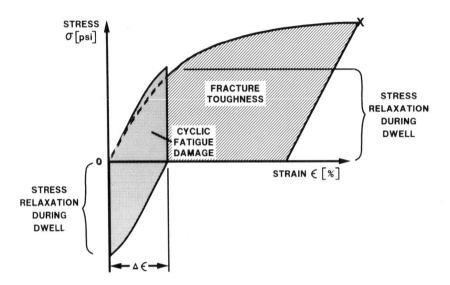

**Figure 5.13 Illustration of cyclic stress-strain behavior of solder; stress relaxation is shown in the hysteresis loop, which indicates loss of ductility and premature fatigue failure [9].**

around three months to see any failure. For mechanical cycling, only thin boards (40 mil thick) were used because the tester mentioned in reference 12 could not accommodate thicker (62 mils) boards. (There is no thickness limit for thermal cycling, but it still takes about 1000 cycles or 80 days to see the first failure). A thinner board being very flexible, took 20,000–30,000 cycles (84–126 days) to show the first failure. Now, with some modification in this tester, 62 mil thick boards can be used and failure is seen within 8 days (under 2000 cycles). The existing testers can be modified to accommodate thicker boards. Now IPC has adopted the round robin test plan for use on 62 mil thick board [11]. Also, caution should be exercised when using the tester mentioned in reference 12. Variations of as much as 15% in the convex and concave bending from the set values may be seen. Such a wide variation, if allowed to occur, will give questionable test result.

No matter which method (thermal or mechanical cycling) is used, the failure criteria should be the same. When evaluating solder joint reliability, daisy-chained packages should be *continuously monitored*. The majority of test monitors in the industry today, however, are ordinary data loggers that do not continuously monitor solder joints. Depending on the number of channels available and the number of resistance loops being monitored,

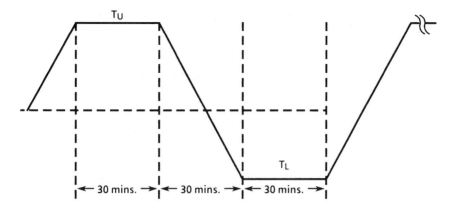

**Figure 5.14 Recommended thermal cycle for commercial applications on FR-4 substrate [11].**

every loop is monitored every 5 to 6 seconds. Since most (not all) solder joint failures occur during the transient phase (between the upper and lower test limits), the solder joint will be open for only a few microsceconds. A multiplexer data logger could easily miss the first failure indication, causing the final results to be off by 10 to 30%. In other words, failures might not be seen until some time after they really happened. This is true of failures caused by shear stress, as is the case for solder joints.

When the solder joint failure mode is due to tensile stress instead of shear stress, the resistance loops do not open and close as quickly. Thus a multiplexed data recording instrument can easily catch the first open. For example, when monitoring plated through holes in a printed circuit board, failures are likely to be spotted early because the copper barrels fail due to tensile stress and are open for a longer period.

For failure, discontinuities lasting at least one microsecond need to be observed and recorded. A discontinuity, that is, loop resistance of 1000 ohms or more, should be confirmed by 10 such failures [11]. Equipment and software are commercially available to continuously monitor resistance, hence solder joint failures [13]. However, the number of suppliers for such equipment is very limited.

Failure also may be defined by the occurrence of a change in resistance of some percent (50-200%), but this approach is less exacting. Visual observation can also be used, but it is highly unreliable and labor intensive. However, valid confirmation of a totally cracked joint can be achieved by way of visual examination.

Having completed the reliability tests, can we predict the field performance of the product accurately? No. But the accelerated tests results can be used to compare the results of several designs and to help us select superior land pattern designs and better package/substrate combinations.

## 5.9   INTERCONNECT CONSIDERATIONS

As mentioned earlier, surface mounting has essentially forced the industry into using fine line boards. As trace spacings become smaller and circuit speeds become higher, the potential for circuit malfunction increases as a result of cross-talk or coupling energies from adjacent traces. With the use of SMT and faster devices, therefore, the design must reflect consideration of cross coupling effects [14].

Present indications are that active lines (lines with impedance to ground of 100 ohms or less) can be spaced at 6/6 or even 5/5 without significant cross coupling. However, signal to ground plane spacing should be within 8 mils, and adjacent traces should run parallel for less than 3 inches.

With surface mounting, the integrity of the power and ground planes is increased because the number of component mounting holes is reduced. This enables the designer to route key signal traces on these planes without damaging their effectiveness as reference planes for the signal traces on adjacent layers. Prime candidates for these traces are high energy clock lines and high impedance lines, which require greater decoupling from the other signal traces.

When working on surface mount boards, the designer must develop rules and guidelines for acceptable trace width and spacing, maximum allowable lengths of parallel circuit lines, and the optimum dielectric thickness for the substrate material. Typical guidelines for the use of these traces on ground and power planes are shown in Figure 5.15 and are summarized as follows:

1.   Maximum trace length of 1.5 inches.
2.   All traces routed parallel with minimum spacing of 200 mils.
3.   Adjacent signal plane lines routed perpendicular to traces on ground and power planes.

There is some good news in SMT interconnect design. For example, surface mount decoupling capacitors are more effective than through-hole mounted capacitors because the former have less lead inductance to the ground and power planes. A rule of thumb would indicate that half as many capacitors are necessary in surface mount design, versus through-hole board design. Also, because lead lengths are short, there is less inductance loss when using surface mount components.

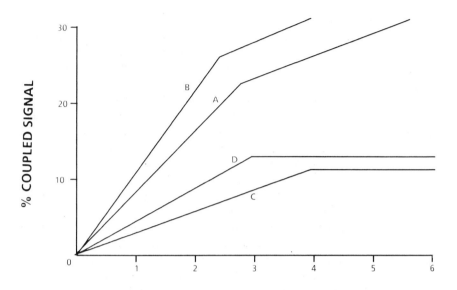

**COUPLED LENGTH - INCHES**

**Figure 5.15 Guideline for maximum allowable parallel active lines free of cross-talk. A and B: traces on external layers; D and C: traces on internal layers; B and D: lightly loaded; A and C: heavily loaded. (Courtesy of Intel Corporation.)**

Propagation delay is a function of the total lead length (outside lead length and inside wire bond length) of the package. As we move from the through hole technology to SMT, considerable improvement in propagation delay can be expected because of drastic reductions of total lead length in SMT. The propagation delay gets even smaller in packages such as chip-on-board or tape automated bonding (TAB) since they have smaller total lead lengths than "conventional" SMT packages (See Chapter 1, Section 1.7 for discussion on COB and TAB). Let us clarify this point by an example.

The propagation delay is defined as the time required for travel of the electrical signal through wire bond (package body) and outside lead (air). To calculate delay time, speed of signal is divided by the **square root** of dielectric constant. The speed of travel for electrical signal is $3 \times 10^{10}$ cm/second or 12 inches/nano second (same as the speed of light) when the dielectric constant of the medium is 1 (air). In other words, the propagation delay will be 1 nano second for a 12 inch long lead length in air (12 divided by the square root of 1). Since the signal travels in package body (wire bond) and air (outside package lead), the propagation delay is a function of the effective dielectric constant of these two mediums. To keep our calculations simple, let us assume a dielectric constant of 9 for the medium (for ceramic package and air combination) the propagation delay will be 1 nano second for every 4 inch of lead length (12 divided by the square root of 9).

Let us refer back to Chapter 1, Figure 1.14 which shows the propagation delay of 0.3 nano seconds for a 64 pin DIP package. This means that the authors in Reference 1, Chapter 1 have assumed the total lead length of 1.2 (4X0.3) inches in a DIP package (The corner leads could have this lead length). By comparison the propagation delay is shown to be 0.05 nano second in a 64 pin LCCC. This means that the total lead length of .2 inch (0.05X4) in 64 pin LCCC. The propagation delay in TAB is further reduced in half (0.025 nano second).

The propagation delay for plastic package would be even lower than these number just discussed because of lower dielectric constant of plastic material. If we carry this analysis further, the propagation delay in a flip chip (not shown in Figure 1.14) will be practically non-existent since a flip chip has no lead and is directly mounted on the substrate.

## 5.10   CAD LAYOUT CONSIDERATIONS

Most computer-aided design (CAD) systems were designed for through-hole boards. Various requirements imposed by SMT call for additional capability in the CAD software. It is safe to generalize that more work is needed by the CAD system suppliers to completely meet the CAD requirements of SMT boards. The designer must take this into account and should be prepared for the impact of the CAD tool on the product schedule as well as on cost.

There are many types of CAD system available for SMT applications, and all have their pros and cons [15]. In particular, the product designer must be aware of the implications for PCB features due to SMT. For example, changing to SMT design generally also means using boards with narrow trace width and space (refer to Figure 5.1, above). It may take longer to lay out a board using SMT, and an increase in time here may have a bearing on the time to market for the product, especially if engineering changes are reqiured after prototyping, as is generally the case.

Certain design requirements are different for SMT boards, and the CAD system must address them successfully. For example, in through-hole boards, the location of via holes is not an issue. But in surface mounting, vias are required for interconnection, and testing, and specific guidelines for their location must be developed. SMT also requires specific guidelines for via placement, since errors in this respect can have a negative impact on real estate, testability, and manufacturability.

To preserve as much space as possible for routing traces, the number of vias must be minimized. This means that the CAD system must be able to route from the surface pads. Most CAD systems, however, look for familiar through-hole vias for autorouting, hence must be manually routed, which takes more time in CAD and also increases the potential for errors.

Typically, SMT may require about 20% additional time in the CAD cycle because surface mount calls for such postprocessing steps as generating a separate layer of solder paste artwork and generating test maps for assembly testing. Additional time may also be required if the CAD software does not allow complete autorouting and design rule checking.

Although most CAD systems allow placement of components on both sides, very few allow for "flip-flop screens," which permit the CAD designer to view the opposite side of the board [16]. Also, placement of components on both sides, especially in logic boards, may require use of blind and buried vias. Not all CAD systems have this capability.

Some CAD systems are limited in the number of pad sizes they can accommodate in their library. This may necessitate a compromise on the optimum pad size for best reliability and manufacturability results.

## 5.11 SUMMARY

There are many design issues that must be settled before a particular packaging course is selected. The designer must consider market needs, time to market, real estate, and cost per function. This is not enough, however. The product must also satisfy thermal and reliability requirements. As the packaging density increases, moreover, thermal problems are compounded, with a potential adverse impact on overall product reliability.

Solder joint reliability for surface mount is a source of concern because of CTE mismatch between ceramic packages and FR-4 glass epoxy substrates. This problem, which is confined to the military applications, is being addressed by using expensive substrates with matched CTE. Success, however, has been less than complete.

Because the plastic packages used in commercial applications have compliant leads, they do not experience problems related to CTE mismatch. The large plastic packages may be prone to cracking at reflow soldering temperatures, however; this is an industry-wide problem. Long- and short-term solutions are still evolving, but baking before reflow offers one answer.

The increase in package density has necessitated the use of fine lines at closer spacings. This can increase cross-talk between the lines, especially if they carry a high speed signal. The product design also is influenced by the type of CAD system that is available. The cost or the schedule or both may be affected by the type of CAD system used for SMT board design.

## REFERENCES

1. Davidow, W. H. *Marketing High Technology*. New York: Free Press, 1986.

2. Steiner, T. O., and Suhl, D. "Investigation of large PLCC package cracking during surface mount exposure. *IEEE Transactions on Component, Hybrid, and Manufacturing Technology,* Vol. CHMT-10, No 2, June 1987, pp. 209-216.

3. Prasad, Ray, Intel; Spalik, Joe, IBM; Fehr, Gerald, LSI Logic, and Nixon, David, Aerospace Corporation. Moisture induced plastic IC package cracking, Surface Mount Council white paper, May 1988, Available from IPC.

4. Alger, C., Huffman, W. A., Gordon, S., Prough, S., Sandkuhle, R., and Yee, K. "Moisture affectss on susceptibility to package cracking in plastic surface mount components." *IPC Technical Review,* February 1988, pp. 21-28.

5. Fukuzawa, I., et al. "Moisture resistance degradation of plastic LSIs by reflow soldering." *Proceedings of the 23rd Reliability Physics Symposium,* 1985, pp. 192-197.

6. Ito, S., et al. "Special properties of molding compounds for small outline (SO) packaged devices." *Proceedings of the 36th Electronic Components Conference,* 1986, p. 360.

7. Suhl, D. "The mechanism of plastic package cracking in SMT and two solutions." IPC-TP-683 Technical paper presented at the IPC spring meeting, Hollywood, FL, April 1988.

8. Engelmaier, W. "Functional cycles and surface mount attachment reliability." IN *Surface Mount Technology,* International Society for Hybrid Microelectronics, Reston, VA, Technical Monograph Series 6984-002, 1984, pp. 87-114.

9. Engelmaier, W., "Test method considerations for SMT solder joint reliability. *Proceedings of IEPS,* October 1984, pp. 360-369.

10. Engelmaier, W. "Is present-day accelerated cycling adequate for surface mount attachment reliability evaluation?" Technical pa-

per IPC-TP-653, presented at the IPC fall meeting, Chicago, October 1987.

11.  IPC Round Robin Test Program for Evaluation of Surface Mount Land Patterns Defined. In IPC-SM-782, Lincolnwood, IL, August, 1988.

12.  IEEE Compliant Lead Task Force Flex Tester, Penn Tech Services, Inc., Southampton, PA.

13.  Solder Joint Monitor. STD Series, Models 32, 128, and 256, Anatek, Wakefield, MA.

14.  DeFalco, J. A., "Reflection and crosstalk in logic circuit interconnections. "*IEEE Spectrum* 1970, pp. 44-50.

15.  VLSI System Design Staff. "Survey of circuit board CAD systems." *VLSI System Design,* March 1986, pp. 62-77.

16.  Collet, R. "Board layout strikes a new chord." *Electronic System Design Magazine,* September 1987, pp. 31-37.

# Chapter 6
# Surface Mount Land Pattern Design

## 6.0 INTRODUCTION

Surface mount land patterns, also referred to as footprints or pads, define the sites at which components are to be soldered to a printed circuit board. The design of land patterns is critical because it not only determines solder joint strength, hence reliability, but it also influences the areas of solder defects, cleanability, testability, and repair/rework. In other words, the very producibility of surface mount assemblies depends on land pattern design. However, the producibility of SMA is not determined by pad design alone. Materials, processes, components, and board solderability also play very important roles. These issues are covered elsewhere in this book.

Poor component tolerances and the lack of standardization of surface mount packages have compounded the problem of standardizing the land pattern design. Numerous package types are being offered by the industry, and the variations in a given package type can be numerous, as well. More important, the tolerance on components varies significantly, causing real problems for land pattern design and adding to the manufacturing problems for SMT users.

Consider the component tolerance found in "standard" 1206 resistors and capacitors. These items are anything but standard: the tolerance of the most important dimensions—the termination width—varies from a minimum of 10 mils to a maximum of 30 mils (nominal dimension is 20 mil). With a tolerance variation of this magnitude, the design of land patterns is challenging indeed.

The technical societies, such as the Institute of Interconnecting and Packaging Electronic Circuits, the Electronics Industries Association, and the Surface Mount Council, are doing everything they can to improve the current situation as far as component standardization and tolerances are concerned. For example, the IPC chartered the Surface Mount Land Pattern Committee (IPC-SM-782) chaired by the author to standardize on the land patterns for surface mount components [1].

Through the efforts of the Surface Mount Council, now there is a significant amount of coordination between the EIA and the IPC to ensure compatibility between component outlines (established by the EIA) and their land patterns (established by the IPC). As chairman of IPC-SM-782, I had the privilege of working with many individuals from EIA and IPC member companies in developing land pattern designs for surface mount components. For example, any new component outline considered by the EIA parts committee is first reviewed by the IPC land pattern committee before a final decision is made [2]. The IPC-SM-782 land pattern guidelines offer a good example of the industry's effort to promote standardization in surface mount technology.

Standards for miscellaneous components such as crystals, oscillators, connectors, sockets, switches, and delay lines are still evolving. Since complete standards for these components do not exist, their land patterns were not included in this chapter. However, we discuss the basic concepts of land pattern design for different package types, including the formulas that serve as the basis for the land patterns. This information can also be used for developing land patterns for newer components as they become available or when the dimensions of the existing components change.

The formulas were developed after extensive testing for reliability and manufacturability of assemblies soldered by reflow and wave soldering (reflow soldering of active components and reflow and wave soldering of passive components). The test boards were tested for reliability by thermal and mechanical cycling for solder joint failures. Some of the results of the test program have been published [3].

To simplify the land pattern design guidelines discussed in this chapter, let us divide surface mount components into categories. For each category, we first discuss the basic concept and then the formulas for land pattern design. Finally, the specific land pattern dimensions are given.

## 6.1   GENERAL CONSIDERATIONS FOR LAND PATTERN DESIGN

Surface mount technology is still maturing. New packages are being introduced constantly, and their dimensions are being standardized. In such a situation, it is important to keep in mind certain design considerations for land pattern design. For example, it is not advisable to use industry standards such as JEDEC/EIA standards on component outline as procurement specifications. Thus users must develop in-house specifications and qualify a limited number of vendors that meet those specifications. (Limiting the number of vendors reduces the tolerances the land pattern design will have to support.)

The land pattern design is also process dependent; that is, the same design may give different defects depending on the process variables. For example, for solder paste thicknesses, degree of solder joint reliability, and number of manufacturing defects will be different. Additional impact should be expected depending on the type of solder mask and the reflow processes. Not surprisingly, therefore, we find some variations in land pattern dimensions used by various companies. Yet most of these different land pattern dimensions may be correct, since companies do not use the same process variables. In addition, all companies have different expectations of manufacturability and reliability. Thus it is very important that the land pattern design accommodate reasonable tolerances in component packages; also, it should take into account the processes and equipment used in manufacturing. Hence it is recommended that the basic concepts discussed in this chapter or in any other publication serve only as guidelines.

The user must develop an in-house land pattern design that covers materials, processes, and reliability requirements. Final land patterns should be validated for manufacturability and reliability before being implemented on products. The IPC-SM-782 round robin test plan [4] can be used as a guideline to validate any new land pattern design for reliability and manufacturability. This test plan was developed by the IPC land pattern committee to validate the reliability and manufacturability of land patterns established in IPC-SM-782. [1]

## 6.2 LAND PATTERNS FOR PASSIVE COMPONENTS

Passive components are soldered by wave, reflow (vapor phase and infrared), and other processes, each with its own unique thermal profile characteristics and requirements. For example, in wave soldering there is an unlimited supply of solder from the solder pot. In reflow soldering, however, the supply of solder is limited and can be controlled to a limited degree by regulating the metal content and the thickness of the solder paste. Also since components in wave soldering are held in place by adhesive, solder defects such as part movement and tombstoning (component standing on its end) do not occur and need not be addressed in designing. Even in reflow soldering, the incidence of part movement in vapor phase and infrared soldering is different because of the dissimilar thermal profile characteristics of these two processes. See Chapter 12.

Given the process and thermal profile differences among soldering methods, should the land pattern designs for each be different? From the standpoint of optimizing the land pattern design, it is advantageous to have different land patterns for different soldering process. But from a systems standpoint, having different land patterns for different processes poses some problems. For example, some CAD systems limit the total number of land pattern sizes that can be supported.

Even if the CAD capability is not a concern, it would be very confusing for a CAD designer to have to maintain different sets of land patterns for different soldering processes for the same part in the CAD library. In many cases this approach could be very restrictive if the company uses different soldering processes for various applications because no one process meets all requirements.

The primary reason for having different land pattern for reflow and wave soldering is the susceptibility of components to moving or standing up during soldering. During wave soldering, since the components are glued, part movement is not a concern. Thus a land pattern design that is optimum for reflow soldering will work also for wave soldering. Obviously, then, it is desirable for the land pattern design for components to be transparent to both wave and reflow processes. It should be kept in mind that this issue of common land pattern is applicable only to discrete devices, such as resistors, capacitors, and SOTs. Active devices such as PLCCs and SOICs generally are not wave soldered; hence for them this issue does not arise.

Whether using the same or different land pattern designs for different soldering processes, the land patterns must be validated for reliability and manufacturability before being used on products. In this connection, some additional do's and don'ts about the land pattern design of passive components should be kept in mind. For example, component misalignment can be caused both by improper land pattern design and by excessive component misplacement. Components that have poor land pattern design but are accurately placed can misalign after reflow soldering. Nor do components self-align when they have good land pattern design but are misplaced.

There are many causes of tombstoning. But they all have one thing in common—uneven downward forces on each side of the pad. It is the downward force generated by the surface tension of solder on the pad that keeps the chip in place. This uneven surface tension force can be generated by uneven solderability of component termination on each end, uneven termination width (poor tolerance, a common industry problem), uneven solder deposit during screen printing, misplacement of component, rapid rate of heating (in vapor phase soldering), and shadowing of a chip component by a large neighboring component (in IR soldering). These factors make it possible for the solder joint on one end to reflow slightly before the other end causing unequal surface tension force. Depending on the degree of force differential, either tombstoning or minor part movement or something in-between will occur. The user should be aware of these causes in taking appropriate corrective action.

In addition to noting the above commonly known causes and solutions mentioned above, Giordano and Khoe suggest that the vendors should supply chip components with wider bottom termination width and maintain tight tolerances at least on the bottom side to allow a greater margin of manufacturing flexibility, eliminating expensive rework [5].

To prevent misplacement and tombstoning, the component should be placed accurately so that both terminations fall symmetrically on solder paste. Whether components self-align during reflow depends on the degree of misplacement. Minor misplacement in passive devices and smaller active devices does get corrected, but one cannot count on it.

The incidence of misalignment and tombstoning is also minimized by the elimination of solder mask, especially a dry film solder mask, between pads, and proper preheat before reflow. This is discussed in Chapter 7. In addition to symmetric placement of terminations, the gap between pads plays a very important role. The formulas for the land patterns of resistors and capacitors discussed later on take this factor into account.

## 6.2.1 Land Pattern Design for Rectangular Passive Components

Rectangular passive components such as ceramic capacitors and resistors are the most widely used, but some companies select a land width narrower than the component width and others use lands wider than the components for the same soldering process. To determine the effect of land geometry on reliability and manufacturability (solder defects), a study was conducted by the author. Two different test boards with various land pattern configurations were designed. One board was used for wave soldering and the other for vapor phase soldering.

An analysis of the test results showed that the solder joint failures were dependent on land lengths, not land widths [3]. Therefore, although the width of the land and the width of the component should be roughly equal, a minor variation in land width is not important. However, failure in solder joints was noticeable for land lengths of less than 50 mils for 1206 resistors. This study also found that resistors are more susceptible to part movement and tombstoning if larger land geometries are used.

To accommodate these findings, I developed formulas for resistor and capacitor land patterns using minimum and maximum component dimensions instead of nominal dimensions. *The nominal dimension is meaningless because of poor tolerances in components.* When using minimum and maximum dimensions, a distinction is made between resistors and capacitors because resistors are about half the thickness of capacitors. If this distinction is not made, the land patterns for resistors and capacitors of a given component size will be the same, causing part movement in resistors because of their smaller thickness, hence lower mass. This is why the lands for resistors and capacitors, for a given component size, are of different lengths. However the widths of the land and of the gaps between the lands are the same for both resistors and capacitors, as shown in Figure 6.1, and as summarized below.

*Land patterns for resistors and capacitors*

Pad width

$$A = W_{max} - K$$

Pad length

$$B \text{ (Resistor)} = H_{max} + T_{min} + K$$

$$B \text{ (Capacitor)} = H_{max} + T_{min} - K$$

Gap between lands

$$G = L_{max} - 2T_{max} - K$$

*where* $L$ = length, $W$ = width, $T$ = solderable termination width, $H$ = height, and $K$ = constant (recommended to be 0.01 inch). Note that the formulas for A & G are the same for both resistors and capacitors, but they are different for B (pad length).

The land dimensions of commonly used ceramic resistors and capacitors shown in Figure 6.2 are based on these formulas.

## 6.2.2    Land Pattern Design for Tantalum Capacitors

For most passive components, the solderable termination width and the component width are equal, as are the component height and the solderable termination height. However, molded plastic tantalum capacitors are an exception. For these components, the solderable termination width and height, depending on the package style and vendor, may be less than the component body width and height. Hence although the general concept is the same as in Figure 6.1, the exact formulas shown in this figure cannot be used for tantalum capacitors.

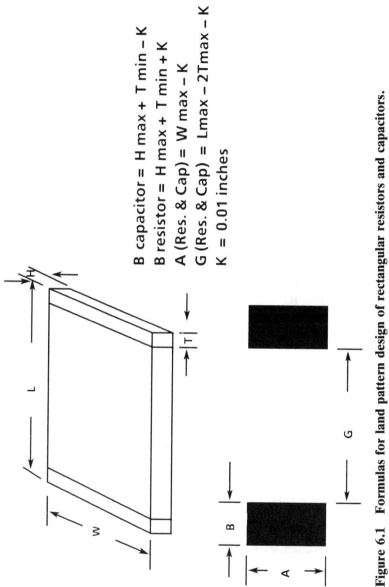

B capacitor = H max + T min – K

B resistor = H max + T min + K

A (Res. & Cap) = W max – K

G (Res. & Cap) = Lmax – 2Tmax – K

K = 0.01 inches

**Figure 6.1** Formulas for land pattern design of rectangular resistors and capacitors.

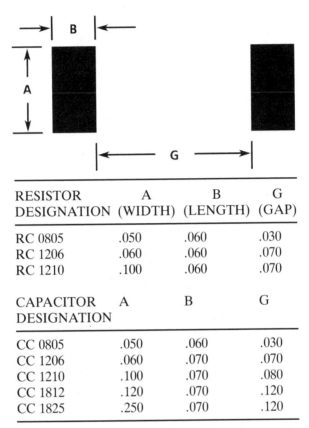

| RESISTOR DESIGNATION | A (WIDTH) | B (LENGTH) | G (GAP) |
|---|---|---|---|
| RC 0805 | .050 | .060 | .030 |
| RC 1206 | .060 | .060 | .070 |
| RC 1210 | .100 | .060 | .070 |

| CAPACITOR DESIGNATION | A | B | G |
|---|---|---|---|
| CC 0805 | .050 | .060 | .030 |
| CC 1206 | .060 | .070 | .070 |
| CC 1210 | .100 | .070 | .080 |
| CC 1812 | .120 | .070 | .120 |
| CC 1825 | .250 | .070 | .120 |

**Figure 6.2 Land pattern dimensions (inches) for rectangular ceramic resistors and capacitors. See Figures 3.2 and 3.5, respectively, for component dimensions.**

When using the formulas shown in Figure 6.1 for molded plastic tantalum capacitors, use $w$ instead of $W$ and $h$ instead of $H$ to indicate the land pattern dimensions. The applicable formulas and land pattern dimensions for tantalum capacitors are shown in Figures 6.3 and 6.4, respectively.

As discussed in Chapter 3, there are two types of tantalum capacitor: the molded plastic type and the kind with a welded stub on one side. (Refer to Chapter 3 for dimensions and differences.) As with the ceramic capacitors, in the welded stub type tantalum device the solderable termination width is equal to the component body width and the solderable termination height is equal to the component height. Thus the formulas for the land pattern for welded stub tantalum capacitors and ceramic capacitors are

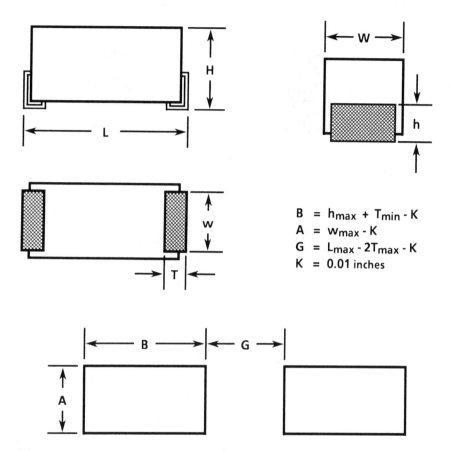

$$B = h_{max} + T_{min} - K$$
$$A = w_{max} - K$$
$$G = L_{max} - 2T_{max} - K$$
$$K = 0.01 \text{ inches}$$

**Figure 6.3    Formulas for land pattern design of tantalum capacitors.**

similar (see Figure 6.1). The land pattern dimensions for welded stub tantalum capacitors are shown in Figure 6.5.

## 6.3    LAND PATTERNS FOR CYLINDRICAL PASSIVE (MELF) DEVICES

Metal electrode leadless face (MELF) devices are used for capacitors, resistors, and diodes. See Chapter 3 for a description and component dimensions. MELF devices are soldered by both wave and reflow processes. This is one component that should have different land patterns for reflow and wave soldering. The land pattern for reflow soldering will also work for wave soldering, but the reverse is not true, because the MELF

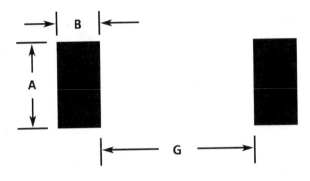

| STANDARD CAPACITANCE RANGE | | | |
|---|---|---|---|
| PACKAGE DESIGNATION | A (WIDTH) | B (LENGTH) | G (GAP) |
| Molded A (3216) | .050 | .060 | .040 |
| Molded B (3528) | .090 | .060 | .050 |
| Molded C (6032) | .090 | .090 | .120 |
| Molded D (7243) | .100 | .100 | .160 |

| EXTENDED RANGE | | | |
|---|---|---|---|
| | A (WIDTH) | B (LENGTH) | G (GAP) |
| 3518 | .070 | .080 | .040 |
| 3527 | .100 | .080 | .040 |
| 7227 | .100 | .120 | .120 |
| 7257 | .230 | .120 | .120 |

**Figure 6.4   Land pattern dimensions (inches) for molded tantalum capacitors. See Figure 3.7 for component dimensions.**

components, being cylindrical, have a tendency to roll either during handling or during reflow soldering.

Hence a notch is provided in the land pattern to keep the device in place during reflow soldering. The notch (dimensions $D$ and $E$ in Figure 6.6) can be omitted for wave soldering, since the adhesive will prevent the component from rolling sideways. The depth of the notch is determined by the following formula:

$$D = B - \left(\frac{2B + A - L_{max}}{2}\right)$$

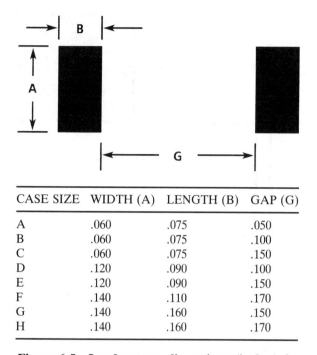

| CASE SIZE | WIDTH (A) | LENGTH (B) | GAP (G) |
|-----------|-----------|------------|---------|
| A | .060 | .075 | .050 |
| B | .060 | .075 | .100 |
| C | .060 | .075 | .150 |
| D | .120 | .090 | .100 |
| E | .120 | .090 | .150 |
| F | .140 | .110 | .170 |
| G | .140 | .160 | .150 |
| H | .140 | .160 | .170 |

**Figure 6.5   Land pattern dimensions (inches) for tantalum capacitors with welded stub. See Figure 3.6 for component dimensions.**

where $L_{max}$ is the maximum body length of the tubular device (not shown in Figure 6.6. See chapter 3 for dimension.

The formulas for $A$, $B$ and $C$ are similar to those discussed earlier for ceramic resistors and capacitors (Figure 6.1). The land pattern dimensions for MELFs are shown in Figure 6.6.

## 6.4    LAND PATTERNS FOR TRANSISTORS

As discussed in Chapter 3, the commonly used types of small outline transistor are SOT 23, SOT 89, and SOT 143. There is no formula given for the land pattern design of SOTs.

Land pattern design for SOTs is simple: maintain the distance between the center of lands equal to the center distance between the leads, and provide at least 15 mils for internal and external pad extension. This concept is very similar to the concepts for PLCC and SOIC land pattern designs, discussed next. Based on these two simple concepts, the land pattern dimensions for SOT 23, SOT 89, and SOT 143 are shown in Figures 6.7, 6.8, and 6.9, respectively.

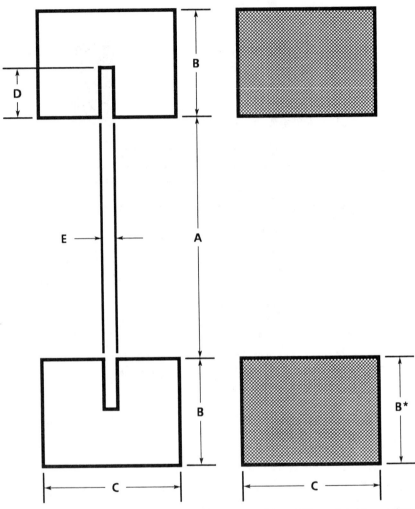

*Rectangular pads if components are to be wave soldered. The notch (dimensions D and E) are necessary for reflow soldering only to prevent cylindrical MELF devices from rolling sideways.

| PACKAGE | A | B | C | D | E |
|---|---|---|---|---|---|
| W Resistor | .170 | .075 | .100 | .035 | .010 |
| MLL 34 | .090 | .055 | .080 | .025 | .010 |
| SOD 80 | .090 | .060 | .080 | .030 | .010 |
| MLL 41 | .140 | .070 | .110 | .035 | .010 |

**Figure 6.6  Land pattern design for tubular MELF devices: resistors, capacitors, inductors, and diodes (dimensions in inches). See Figure 3.9 for component dimensions.**

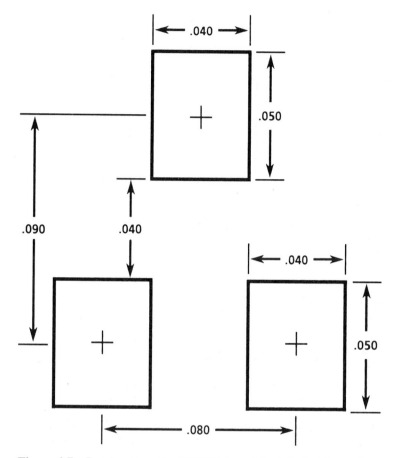

**Figure 6.7   Land pattern for SOT 23 transistor/diode (dimensions in inches). See Figure 3.16 for component dimensions.**

## 6.5    LAND PATTERNS FOR PLASTIC LEADED CHIP CARRIERS

Plastic leaded chip carriers are the commonly used packages for commercial applications for pin counts above 20. Various pad sizes for PLCCs have been found to provide reliable solder joints [3]. This is due to the compliancy (ductility) of J leads and the good CTE compatibility between the plastic package and the glass epoxy board.

For PLCCs, the total pad length is not sufficient. The land pattern design must also provide enough pad extension beyond the package lead to accommodate a good fillet. In PLCCs, the outside fillet is the major

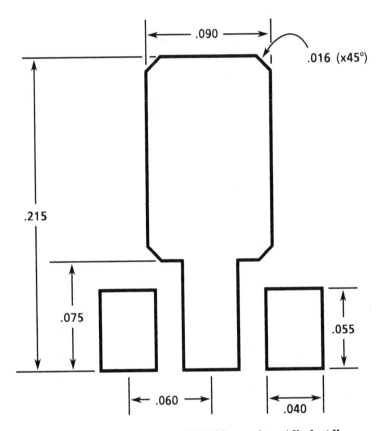

**Figure 6.8   Land pattern for SOT 89 transistor/diode (dimensions in inches). See Figure 3.17 for component dimensions.**

fillet and absorbs most of the stresses. Thus there must be enough outside pad extension for outside fillets to ensure solder joint reliability. A 0.015 inch pad extension is considered to be satisfactory for good outside fillets.

The land pattern design should also provide sufficient space for routing between the pads and should accommodate reasonable variation in component tolerances. As mentioned earlier, the JEDEC standards generally have loose component tolerances. PLCCs are no exception.

The formulas for the land pattern for PLCCs (Figure 6.10) take into account the considerations discussed above and result in the following dimensions:

$$\text{pad width} = 0.025 \pm 0.005 \text{ inch}$$
$$\text{pad length} = 0.075 \pm 0.005 \text{ inch}$$
$$A \text{ or } B = C + K$$

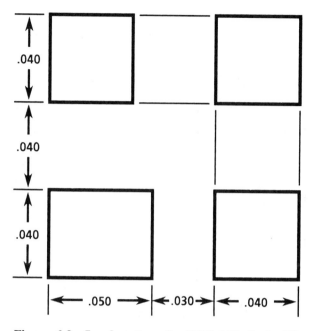

**Figure 6.9 Land pattern for SOT 143 diode (dimensions in inches). See Figure 3.18 for component dimensions.**

The 25 mil by 75 mil PLCC pad size has been selected to meet both reliability and manufacturability requirements. A wider pad tends to reduce the gap between the pads and to restrict the routing space and may cause solder bridging. For this reason, surface mount packages that have a 50 mil center lead pitch (such as PLCC, SOIC, and SOJ) use a 0.025 inch pad width.

The value of $K$ in Figure 6.10 is 0.030 inch for PLCC; this will allow a minimum pad extension of 0.015 inch, since dimensions $A$ and $B$ are based on the maximum value of $C$ and $K$. If the $C$ dimension is lower than maximum allowed in the JEDEC standard, the pad extension will be greater than 0.015 inch. This extension will be reduced on one side and increased on the opposite side by placement tolerance of the pick-and-place machine.

The land pattern dimensions for PLCCs shown in Figure 6.11 are based on the formulas in Figure 6.10. Figure 6.11 also shows the land pattern dimensions for leadless ceramic chip carriers. The reasons for adopting the same land pattern dimensions for PLCCs and LCCCs are discussed next.

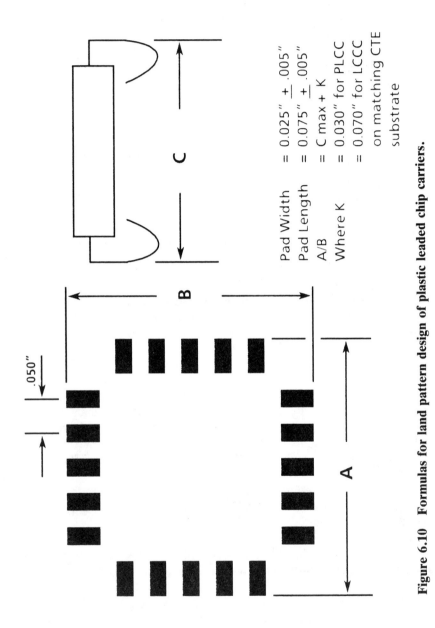

Pad Width = 0.025" + .005"
Pad Length = 0.075" + .005"
A/B = C max + K
Where K = 0.030" for PLCC
= 0.070" for LCCC
on matching CTE
substrate

**Figure 6.10 Formulas for land pattern design of plastic leaded chip carriers.**

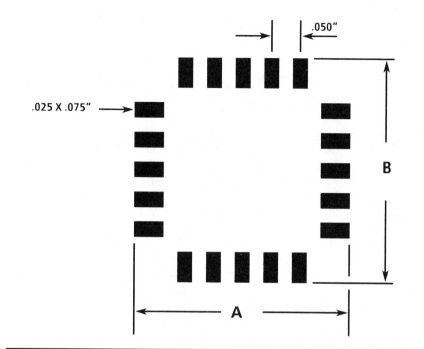

| PIN COUNT OF PLCC AND LCCC | LAND PATTERN DIMENSIONS | |
|---|---|---|
| | WIDTH (A) | LENGTH (B) |
| RECTANGULAR PACKAGES: | | |
| 18-short | .350 | .490 |
| 18-long | .350 | .550 |
| 22 | .350 | .550 |
| 28 | .420 | .620 |
| 32 | .520 | .620 |
| SQUARE PACKAGES: | | |
| 16 | .370 | .370 |
| 20 | .420 | .420 |
| 24 | .470 | .470 |
| 28 | .520 | .520 |
| 44 | .720 | .720 |
| 52 | .820 | .820 |
| 68 | 1.020 | 1.020 |
| 84 | 1.220 | 1.220 |
| 100 | 1.420 | 1.420 |
| 124 | 1.720 | 1.720 |
| 156 | 2.120 | 2.120 |

**Figure 6.11   Land pattern design for plastic leaded chip carriers and lead-less ceramic chip carriers (dimensions in inches). See Figure 3.11 for component dimensions.**

## 6.6 LAND PATTERNS FOR LEADLESS CERAMIC CHIP CARRIERS

The land pattern dimensions for LCCCs in Figure 6.11 are same as those for PLCCs. If we refer back to Chapter 3, we will find that the LCCC packages, for a given pin count, are about 0.040 inch smaller than the PLCC packages. To achieve the same land pattern for both LCCCs and PLCCs, the value of $K$ has to be different for LCCCs. Hence in Figure 6.10 we have $K = 0.070$ inch for LCCCs and $K = 0.030$ inch for PLCCs.

Why should we have the same land pattern for PLCCs and LCCCs? Duplication here allows a 0.015 inch outside pad extension for PLCCs but a 0.035 inch outside pad extension for LCCCs. The larger pad extension for LCCC is needed for better reliability because the solder fillet in these chip carriers is formed only on the outside of the package.

In PLCCs, solder fillets are formed both inside and outside the package, and also the PLCC leads take up some strain because they are compliant. Hence we can tolerate smaller outside pad extensions for PLCC. Having the same land pattern dimensions for both PLCCs and LCCCs simplifies the land pattern design and provides reliable solder joints when LCCCs are soldered to a substrate with a compatible coefficient of thermal expansion.

Larger pad extension for LCCCs may be needed for packages larger than 44 pins. Some companies use a larger pad size (25 mils × 100 mils) and a larger pad extension when the CTEs of package and substrates are not compatible, such as when using glass epoxy substrates for LCCCs. Use of glass epoxy boards for LCCCs is not recommended, for reasons detailed in Chapter 4.

## 6.7 LAND PATTERNS FOR SMALL OUTLINE INTEGRATED CIRCUITS AND R-PACKS

There are some differences between gull wing packages (SOICs or R-packs) and J-lead packages (PLCCs, SOJ). For example, the gull wing leads of SOICs, are more compliant than the J leads of PLCCs because the former are not as work hardened during lead forming. Also, the SOIC packages are relatively small in comparison to PLCC packages, hence induce less stress on solder joints. For these reasons, the reliability of solder joints for SOICs is not of serious concern. However, manufacturability is a concern for SOICs because the leads of SOICs and R-packs are prone to damage during handling.

There are other differences as well. For example, as discussed earlier,

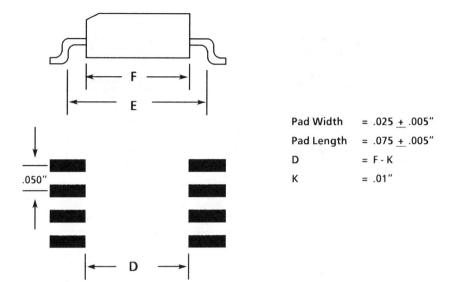

Pad Width    = .025 + .005"
Pad Length   = .075 + .005"
D            = F - K
K            = .01"

**Figure 6.12    Formulas for land pattern design of small outline integrated circuits and R-packs.**

in PLCCs the major fillet is on the outside of the package; but in the gull wing packages such as SOICs, the major fillet is on the inside.

The formulas for SOICs take into account the location of the major fillets to ensure proper fillet formation, as summarized below and shown in Figure 6.12:

$$\text{pad width} = 0.025 \pm 0.005 \text{ inch}$$
$$\text{pad length} = 0.075 \pm 0.005 \text{ inch}$$
$$D = F - K$$

where $K$ is a constant and is recommended to be 0.01 inch.

The $F$ dimension is used for the land pattern design to ensure formation of major fillets and solder joint reliability. Also, the $F$ dimension for SOICs (Figure 6.12) has as tight a tolerance as the dimension $C$ (Figure 6.10) in PLCCs.

SOICs and R-packs appear physically the same, although they may differ in body width. For example, as discussed in Chapter 3, the body widths of SOICs and of R-packs are about 150 and 300 mils, and 220 mils, respectively.

Currently, the EIA committee on R-packs is working to standardize these components in the same body width as the narrow body (150 mil) SOICs. Since SOICs and R-packs physically look alike and have the same lead configuration (gull wing), I have combined them for land pattern design.

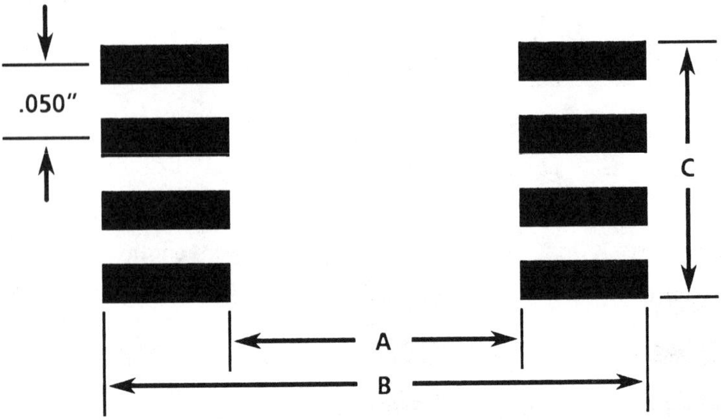

| PACKAGE— LEAD COUNT | A | B | C |
|---|---|---|---|
| SO-8 | .140 | .290 | .175 |
| SO-14 | .140 | .290 | .325 |
| SO-16 | .140 | .290 | .375 |
| SOL-14 | .300 | .450 | .325 |
| SOL-16 | .300 | .450 | .375 |
| SOL-20 | .300 | .450 | .475 |
| SOL-24 | .300 | .450 | .575 |
| SOL-28 | .300 | .450 | .675 |
| R-PACK-14 | .200 | .350 | .325 |
| R-PACK-16 | .200 | .350 | .375 |

**Figure 6.13  Land pattern dimensions (inches) for small outline integrated circuits and resistor networks. See Figure 3.19 for SOIC dimensions and Section 3.2.2 for resistor network dimensions.**

The land pattern dimensions for SOICs and R-packs shown in Figure 6.13 are based on the formulas discussed above and shown in Figure 6.12.

## 6.8    LAND PATTERNS FOR SOJ (MEMORY) PACKAGES

The small outline, J lead devices are similar in design to PLCCs except that the leads are on only two sides of the package. SOJ packages are primarily used for 1 and 4 MB memory devices.

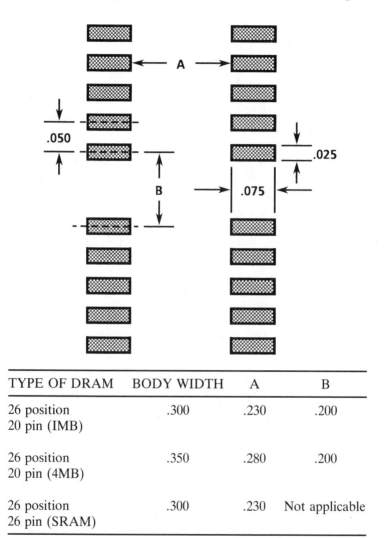

| TYPE OF DRAM | BODY WIDTH | A | B |
|---|---|---|---|
| 26 position 20 pin (IMB) | .300 | .230 | .200 |
| 26 position 20 pin (4MB) | .350 | .280 | .200 |
| 26 position 26 pin (SRAM) | .300 | .230 | Not applicable |

**Figure 6.14   Land pattern dimensions (inches) for SOJ memory (1 and 4 MB) and SRAM devices. See Figure 3.22 and Section 3.4.4 for package dimensions.**

Static RAMs, also available in SOJ package, have the same body width but they have all 26 pins (as opposed to 20 pins for DRAM packages). The land pattern dimensions for these devices are shown in Figure 6.14. The land pattern dimensions for 4 MB memory devices are tentative. It is possible, but less likely that the body width of 4 MB DRAM will shrink to 300 mils in order to be compatible with the land pattern design of 1 MB DRAMS.

## 6.9 LAND PATTERNS FOR DIP (BUTT MOUNT) PACKAGES

Since all packages are not available in surface mount, it is sometimes desirable to shear DIP leads for butt mounting. For example, in an assembly having active surface mount devices on both sides but only a few DIPs on the top side, using DIPs with butt leads may cut down on the number of soldering and cleaning steps. Use of DIPs with butt leads is not always desirable, however. Chapter 7 (Section 7.4) discusses cases in which butt-mounted leads may save on process steps to cut cost.

Even when it is cost effective to use butt-mounted DIP leads, the reliability of a butt solder joint may be questionable. Refer to pros and cons of different lead configurations in Chapter 3, Section 3.7.2. Also, care should be taken when shearing leads to prevent damage to components.

Figure 6.15 shows the land pattern dimensions for butt-mounted DIP leads. Some users prefer bending DIP leads into the gull wing form. The land pattern dimensions for gull wing DIP leads are not shown, but the concept discussed in connection with SOIC can be used to derive the land pattern dimensions.

## 6.10 LAND PATTERNS FOR FINE PITCH, GULL WING PACKAGES

Land patterns for gull wing 0.25 inch center, fine pitch packages are based on the same concepts discussed for SOICs. The land pattern dimensions for 0.25 inch center, fine pitch packages are shown in Figure 6.16. There are many other lead pitches such as 15 and 20 mils that are available in fine pitch. The same concept can be used for them as well, except that the distance between the pads should be equal to the lead pitch or lead center dimensions. However, 20 and 25 mil pitch devices may use the same pad width (12 mils) for ease of screen printing on 20 mil pitch land patterns.

## 6.11 LAND PATTERNS FOR SOLDER PASTE AND SOLDER MASK SCREENS

The relationship between the pieces of artwork used for solder paste film, solder mask film, and the top layers of PCB film should be closely controlled. Generally the supplier modifies the film to achieve the user's

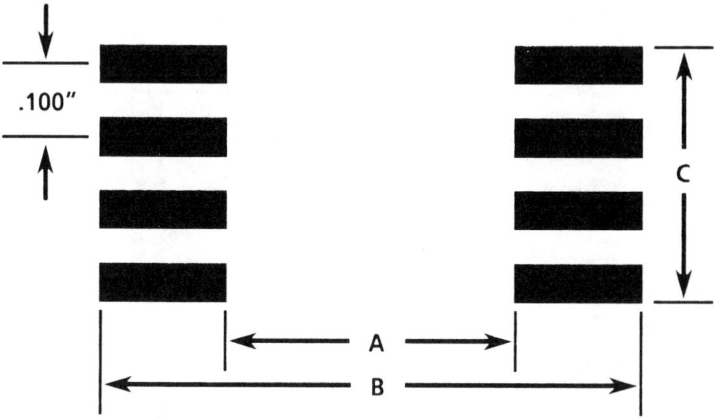

| DIP LEAD COUNT | A | B | C |
|---|---|---|---|
| 8 | .200 | .350 | .300 |
| 14 | .200 | .350 | .600 |
| 16 | .200 | .350 | .700 |
| 18 | .200 | .350 | .800 |
| 20 | .200 | .350 | .900 |
| 22 | .300 | .450 | 1.000 |
| 24 | .500 | .650 | 1.100 |
| 28 | .500 | .650 | 1.300 |

**Figure 6.15 Land pattern design for butt lead DIP packages (dimensions in inches). The body widths for DIPs are 300 mils (14–20 pins), 400 mils (22 pins) and 600 mils (24 and above) and the lead pitch is 100 mils for all pin counts.**

requirement. As a general guideline, the width and length of the film pad for the solder paste screen should be 0.002 inch *smaller* than the solder pads on the PCB. Similarly, the width and length of the film pad for the solder mask screen should be 0.002 inch *bigger* than the solder pads on the PCB. The 0.002 inch dimenlsion is valid for solder masks such as dry film or photoimageable solder masks that provide good resolution. For wet film solder masks, however, 0.005 inch should be used. The latter guideline is adequate to prevent the solder mask from getting on the pad and to prevent bridging of paste before reflow.

The area around the fine pitch land pattern (B − A in Figure 6.16) should be free of solder mask, if a solder mask is not available to achieve the required resolution between the fine pitch pads. However, this does increase the potential for solder bridging.

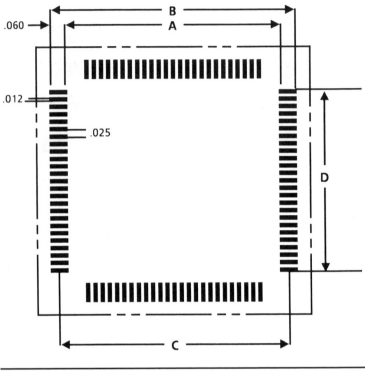

| TERMINAL COUNT | A | B | C | D |
|---|---|---|---|---|
| 84 | .690 | .810 | .750 | .500 |
| 100 | .785 | .905 | .845 | .600 |
| 132 | .985 | 1.105 | 1.045 | .800 |
| 144* | 1.15 | 1.27 | 1.21 | .875 |
| 164 | 1.190 | 1.310 | 1.250 | 1.000 |
| 196 | 1.385 | 1.505 | 1.445 | 1.200 |
| 244 | 1.555 | 1.675 | 1.615 | 1.500 |

**Figure 6.16 Land patterns for U.S. version of fine pitch, 25 mil center gull wing miniature packages (dimensions in inches). See Figure 3.23 for component dimensions.**
***Japanese (Toshiba Package)**

## 6.12 SUMMARY

Surface mount land patterns define the sites at which components are soldered for needed electrical connection and mechanical strength. They play a critical role in the reliability of surface mount solder joints. Certain basic land pattern design parameters are applicable to all components for achieving this goal; these are solderable lead/termination width and height, lead contact points or areas on the printed circuit board, and extension of pads beyond contact points or areas (both inside and outside the package).

The foregoing parameters determine the size and shape of the solder fillets, which provide the needed strength for solder joint reliability. Keeping these parameters and component tolerances in mind, formulas for land pattern design should be used. For every formula, a constant $K$ is defined to accommodate variations in component tolerances. The basic variables in the land pattern design guidelines can be summarized as follows:

1.  The center-to-center distance between adjacent pads should be equal to the center-to-center distance between adjacent component leads.
2.  The pad width should equal the component lead or termination width plus or minus a constant.
3.  The pad length is determined by the height and width of solderable terminations or leads and the contact points or areas of components. Pad length plays a more critical role than pad width in solder joint reliability.
4.  The gap between opposite pads of the land pattern depends on the component body width and dimensional tolerances and is most critical in properly locating the components on the pads. This gap has an impact on both reliability and solder defects.

# REFERENCES

1.  ANSI/IPC-SM-782. Surface Mount Land Patterns: Configurations and Design Rules). March 1987 IPC, Lincolnwood, IL.

2.  News, IPC and EIA cooperate on outlines and Land Patterns, IPC Technical Review, July 1988, Page 9.

3.  Prasad, R.P. "Surface mount land patterns and design rules. Surface mount land patterns play a key role in the reliability of SMT." *Printed Circuit Design,* December 1986, pp. 13-20.

4.  IPC-SM-782. Surface Mount Land Pattern Round Robin Test Plan. IPC, Lincolnwood, IL, August 1988.

5.  Giordano, Jerry and Khoe, David, Bourns Inc., Chip Resistor Design helps prevent tombstoning, Surface Mount Technology, August, 1988, pp 23-25.

# Chapter 7
# Design for Manufacturability, Testing, and Repair

## 7.0 INTRODUCTION

Design for manufacturability is the practice of designing board products that can be produced in a cost-effective manner using existing manufacturing processes and equipment. The benefits of a manufacturable design are better quality, quicker time to market, lower labor and material costs, shorter throughput time, and fewer design iterations. The quickest way to waste thousands of dollars is to produce an SMT design that cannot be assembled, repaired, or tested using the existing equipment. Design for manufacturability is essentially a yield issue, hence a cost issue. It plays a critical role in reducing defects in printed circuit assemblies.

Design for manufacturability (DFM) is gaining more recognition as it becomes clear that cost reduction of printed circuit assemblies cannot be controlled by manufacturing engineers alone. The printed circuit board designer also plays a critical role in cost reduction. The days of throwing the design over the fence to the manufacturing engineer are gone—if indeed they ever really existed.

Companies planning surface mount products generally work out manufacturable designs by trial and error and at considerable expense. Getting decent yield can be a source of constant irritation and often total frustration.

How should one go about establishing design guidelines for manufacturability? In this chapter we aim to minimize the frustration of producing a manufacturable SMT board by presenting some major DFM guidelines, suitable for use in developing in-house design rules.

These guidelines have been validated on thousands of products built in production environments. Certain guidelines are equipment and process dependent—they may not apply equally to all manufacturing facilities. Other guidelines are generic, hence will apply to every company regardless of the equipment or processes used.

Any DFM guidelines, including the ones presented in this chapter, must be verified on prototype assemblies before an item is released for high volume production. Validation of DFM should not pose a major problem because most designs go through a few revisions, at least in the beginning stages, to fine-tune the electrical performance. During those revisions, one should also be on the lookout for manufacturing defects.

We must keep in mind that design for manufacturability alone cannot eliminate all defects in surface mount assemblies. Defects fall into three major categories: design-related problems, vendor-related problems, and problems related to manufacturing processes and equipment. Each defect should be analyzed for its source, to permit appropriate corrective action to be taken.

Simply put, all major issues in design, materials, and process must be fine-tuned to achieve zero defects. Materials and process issues are covered in Chapters 8 to 14 of this book. In this chapter we discuss DFM guidelines to prevent design-related problems.

There are two major design-related issues: land pattern design and design for manufacturability. The first is very critical because it determines the strength of the solder joint, hence its reliability; it also has an impact on solder defects. Land pattern design, which was covered in Chapter 6, should be considered to be an integral part of design for manufacturability.

The generic issues in design for manufacturability, which were covered in Chapter 5, are important in determining whether SMT or another packaging option should be used. They should also be considered an integral part of DFM. In this chapter we concentrate on DFM issues that affect manufacturability from the manufacturing yield viewpoint, in order to minimize board assembly, testing, and rework costs.

## 7.1 DESIGN GUIDELINES, RULES, THE ROLE OF THE DESIGNER

For the designer to be able to design a manufacturable product, it is important to establish guidelines and rules. The distinction between guidelines and rules is very important. Rules are necessary for compatibility with the planned manufacturing equipment and processes. If the design rules are not followed, the product either cannot be manufactured or must be built manually. In the case of SMT, having to build manually is tantamount to not being able to build at all. Guidelines, on the other hand, are nice to have and may make life easier for the manufacturing engineer, but they are not critical in getting the product out the door. This means that rules cannot be violated, but the guidelines may be overruled if there are good and sufficient reasons.

Depending on the product and on marketing constraints, even the rules can be violated as long as the members of the product team (manufacturing engineer, test engineer, CAD designer, electrical designer, component engineer, reliablity engineer, and marketing representative responsible for the product) understand and agree on the consequences of each violation.

It is not enough, however, to establish design rules and guidelines for manufacturability. A corporate culture must exist to support adherence to guidelines and rules. In this book we have called everything "guidelines" because a book cannot be a substitute for an in-house specification, whereas it can be used as a basis for developing in-house guidelines and rules.

It is not an exaggeration to say that the designer plays a critical role in the success and financial viability of any product using SMT. This has been true in the through-hole mount design as well, but not to the same degree. There is only one way to assemble a through-hole mount board: stuff and wave solder it. In SMT, however, the designer has many options, depending on the type of assembly. For example, in SMT, the same component can be placed on either the top side or the bottom side, with different manufacturing consequences in each case. These issues are discussed later on. Clearly the designer needs to be thoroughly familiar with the surface mount manufacturing processes.

This is not to belittle the importance of DFM in through-hole assemblies, because DFM for autoinsertion is important. But unlike through-hole mount, SMT does not have the option of manual placement, and the rules for testability are very different because in SMT we no longer have the ends of the devices that also served as the test nodes. In a nutshell, DFM and the role of the designer in the success of SMT cannot be overemphasized.

We discuss various issues the PCB designer must confront in designing a manufacturable SMT assembly.

## 7.2 STANDARD FORM FACTOR CONSIDERATIONS

To reduce manufacturing costs, products should fit into a standard form factor, that is, a standard board shape and size and a standard tooling hole location and size. (See Chapter 11, Figure 11.4.) This is the first item of importance in design for manufacturability, and it means that even with the variety of new products, it is possible to have similarities that allow the use of standard manufacturing processes.

If the products do not fit into a standard form factor, manual setups are required, which leads to longer throughput time and higher tooling and manufacturing costs. Even when cost is not a major factor, especially for prototype volumes, the schedule may be important. The lead time required for fixtures and tooling for the screen printer, placement, and soldering machines may be unusually long if a nonstandard form factor is used. This issue must be taken into account before committing to a nonstandard form factor.

Board size should not be confused with standard form factor. Even if it is not possible to have a standard board size (which is generally the case in industry because of multiple product designs for different applications), standard form factor strategy can be implemented in manufacturing. If it is not possible to have one standard form factor, it is desirable to have a limited number rather than many form factors in a manufacturing line. Limitation can be achieved by selecting only a few standard panel sizes that accommodate various small boards in a multipack design. The individual boards within the multipack panel must be on a standard grid with respect to each other and with the tooling holes on the panel. The grid chosen must be the same as the one used to create the board. Any multipack design must allow for easy depaneling in individual boards.

The maximum size of the standard panel or board should be selected with the capabilities of the machines in mind, as well as potential warp and twist problems in the board. Limiting the size of the panel to a maximum of 12 inches by 12 inches is a good idea, even though panels of 12 inches by 18 inches are not uncommon. Any board or panel size larger than the 12 inch square may require either permanent or temporary stiffeners to prevent warpage when it goes through the soldering processes.

## 7.3   COMPONENT SELECTION CONSIDERATIONS FOR MANUFACTURABILITY

The objective of this section is not to discuss surface mount component issues in general but only to point out some dos and don'ts of component selection for easy manufacturing. Surface mount components were covered in detail in Chapter 3. For easier manufacturing, the following guidelines should be used.

1.    Use only surface mount components that solve the real estate constraints or reduce the number of manufacturing steps (from Type II SMT to full surface mount Type I SMT) and bear no cost premium. Unless surface mount components meet at least one of these requirements, their use will not be cost effective.

2.    All surface mount components should be autoplaceable. On the surface it seems that all surface mount components have to be autoplaceable. This is not true. For example, tantalum capacitors with welded stubs cause problems during placement. Instead use encapsulated tantalum

capacitors, which are more reliable and easy to place, as well. Connectors and sockets also fall into a difficult to autoplace category heat sinks.

3.     Component terminations with solder plating are better than solder-dipped terminations because of the "dog bone" effect that occurs during dipping. The uneven surfaces can prevent the placement jaws from centering components.

4.     Select only components that can withstand the soldering temperature anticipated. Many through-hole components have been converted to surface mount by cutting or bending the leads, but not all of them can withstand the reflow temperatures. For example, delay lines that have internal connections soldered with eutectic solder (63% Sn and 37% Pb, mp = 183°C) will melt during reflow soldering at 215 to 220°C.

5.     Use 1206 size passive components (resistors and capacitors) whenever possible. Smaller sizes such as 0805 should be used only when the required value or dielectric is unavailable in another size. The 0805 sizes are harder to place than the larger sizes.

6.     Use marked components if there is no cost premium. Parts without marking do create problems if they get mixed up or have to be repaired in the field, since the documentation (schematic or parts list) must be referred to when making replacements. Use of components on tape and reel eliminates part mixup.

7.     For secondary side attachment in Type II or Type III SMT, low profile SOT devices are preferred over the high profile SOTs because the former are easier to bond with adhesive. For reflow soldering, there is no difference. However, the higher profile SOTs are better for cleaning, for both wave and reflow soldering.

8.     Avoid using metal electrode leadless face passive components unless the product is targeted for a very cost-sensitive market. MELFs are cheaper than their rectangular counterparts. However, they tend to be less reliable and harder to place. Also they tend to roll off during reflow soldering. Rolling off, which can be prevented by using the land pattern featuring a notch in the pad (see Chapter 6), is not an issue in wave soldering, since the components are glued. For diodes, MELFs can also be replaced with SOTs.

9.     To prevent leaching (dissolution of the solderable termination), use nickel barriers on resistors and capacitors. Leaching exposes nonsolderable ceramic dielectric, resulting in poor solder joints. This requirement for a nickel barrier underplating is more critical for wave soldering than for reflow soldering, where the limited amount of solder paste does not cause leaching. Now it is almost an industry standard to have nickel barriers in chip components. See Chapter 10 for a detailed discussion of this subject.

10.     Avoid using surface mount connectors and sockets unless there

is good reason. Both tend to take up more real estate than their conventional counterparts. This situation is changing with the offering of connectors on 50 mil centers. All connectors should have tooling holes for mechanical fastening to the board to prevent solder joint cracking.

    11.    Use only qualified components. Some components appear to be surface mountable but are not, as discussed in item 4 above. Qualification requirements are defined by their application and the manufacturing processes (placement, soldering, cleaning, and repair, etc.) that the components must withstand. By defining the qualification requirements and using only qualified components, many problems in manufacturing and in field service can be avoided. For example, during the qualification stage it can be determined whether the component requires baking before reflow to prevent package cracking. See Chapter 5 (Section 5.7) for details on package cracking.

    12.    The last but most important guideline is to minimize the different types of component and to prevent component proliferation. This is especially critical now, while the industry has yet to settle on component standards. The benefits of component commonality or standardization are many. It is easier, for example, to establish standard land patterns and to limit dealings to a few vendors.

Dealing with the smallest possible number of vendors provides improved purchasing leverage and reduced material overhead. Also, depending on the input slot capability of the pick-and-place machine, most of the commonly used component types may be line loaded and available right on the placement machine. This certainly provides quick response or throughput time in manufacturing by eliminating setup time. It is one of the key elements of the current industry trend toward a just-in-time manufacturing philosophy. This all adds up to improved quality and reliability and reduced cost.

Using application-specific integrated circuit devices whenever possible is the ultimate in reducing the usage of different part types. Substituting one ASIC for various programmable logic arrays (PLAs) is an example. One will not only save considerably in placement time, but also in programming, inventory, and feeder requirements. An added benefit is that thermal problems will be alleviated, since PLAs are the biggest culprit in compounding the thermal problems on a board. Because of nonrecurring development costs, however, ASICs are feasible only in high volume applications. Also one must allow 4 to 6 months of development time for the ASIC in the product schedule. Nevertheless, when the ASIC route is feasible, the savings in both component and manufacturing costs can be substantial.

## 7.4 SOLDERING CONSIDERATIONS

When converting to SMT, one of the first questions generally asked is what type of SMT should be used. Should it be Type III, Type II, or Type I? Sometimes the answer is very clear, a function of component availability and real estate constraints. At other times one must weigh various factors before making a decision. The initial tradeoff must include options other than SMT. Refer to Chapter 5 for the considerations to be balanced in making the final decision for or against SMT. Here we assume that the board must be SMT. Thus the first order of business is to complete a preliminary part placement to determine the real estate constraint, hence the type of SMT assembly. Refer to Section 5.3 in Chapter 5 for a rough estimate of real estate requirements.

For Type III and Type I SMT, the soldering issues are clear-cut. But for Type II, the same components, depending on whether they are placed on the top side or the bottom side, have different soldering implications. For example, in Type II assemblies, the discrete devices such as resistors, capacitors, transistors, and diodes can be mounted either on the secondary (wave soldering), or the primary (reflow soldering) side.

Other surface-mounted devices such as R-packs and SOICs should be mounted on the primary side for reflow soldering. Many companies place them on either side, but this is not recommended because of the potential problems in solder defects (shadowing, bridging, etc.) and adverse impact on package reliability. The risks of wave soldering any leaded surface mount devices are not worth taking. Since surface mount packages are susceptible to package cracking due to moisture absorption, wave soldering of plastic surface mount devices should be avoided unless the reliability data indicate otherwise.

Many companies wave solder SOICs and butt mount and reflow solder DIPs. Now we examine the pros and cons of this approach to determine when and if it is wise.

Let us discuss the soldering options for an assembly like that of Figure 7.1, which has only a few through-hole devices on top and active surface mount devices on both sides, due to real estate constraints. (We assume that the board cannot grow any larger and that we need all the functions.) This puts the designer in a quandary. If both sides of the boards are reflow soldered, the through-hole components must be hand soldered; this will increase cost. If the leaded surface mount devices on the bottom side (along with the through-hole a devices on top side) are wave soldered, package reliability may be degraded. However, if we butt mount the through-hole devices and reflow both sides, the butt solder joints may not be reliable.

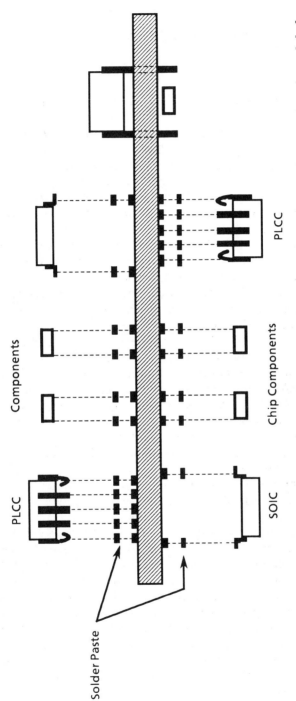

**Figure 7.1a  A Type II SMT packaging case in which it is definitely desirable to convert the last few through-hole components to surface mount to achieve Type I SMT assembly.**

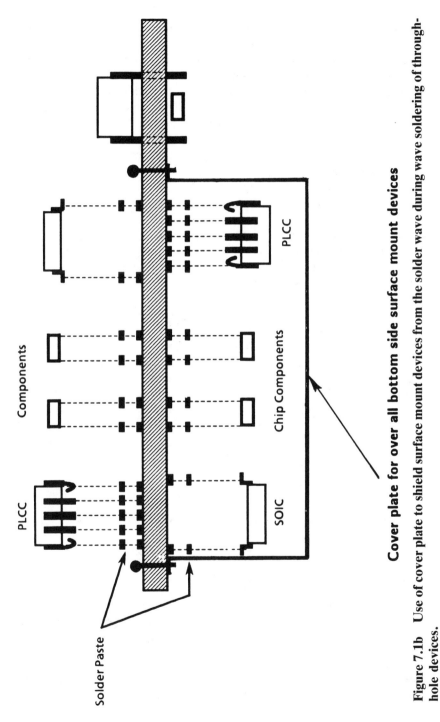

**Figure 7.1b**   Use of cover plate to shield surface mount devices from the solder wave during wave soldering of through-hole devices.

Some studies have indicated that butt joints may be satisfactory but the results require further evaluation.

If the DIP leads are bent to form either a J or a gull wing, care must be taken to prevent package cracking. Nevertheless, if one must choose among these difficult options for designing an assembly, such as the one shown in Figure 7.1a, a butt mount or a gull wing mount, achieved by cutting or bending the leads, may be desirable. Such a design will eliminate the reliability concerns associated with wave soldering of leaded surface mount devices and will reduce soldering steps, hence cost. Appropriate tooling is available.

Another option for soldering an assembly, as in Figure 7.1a, is to use a cover plate over the surface mount devices on the bottom side. (See Figure 7.1b.) If such an option is used, the process sequence is as follows: components are reflow soldered on one side and then cleaned. A cover plate is put on over these devices. Then the second side of the assembly is reflow soldered. During this operation the bottom side solder joint does not reflow. After second cleaning, the through hole devices are inserted on the side opposite of the cover plate and the assembly is wave soldered. Since the bottom side surface mount devices are covered with the cover plate, no flux or solder touches these devices. This prevents any reliability problem associated with wave soldering of active surface mount devices.

The cover plate material can be made of either stainless steel or titanium. Titanium is an expensive and difficult material to work with, but will hold its shape better after repeated use. If the cover plate is made of some other metals, such as copper or aluminum, the solder pot will get contaminated. High temperature plastic may also be an option. In addition to material selection, one must also be aware that the cover plate must be custom made for each product, and the through-hole components must be

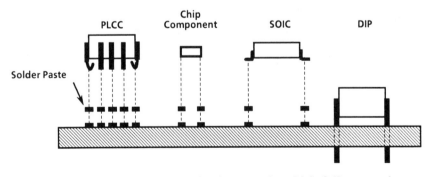

**Figure 7.2 A Type II SMT packaging case in which full conversion to SMT will save soldering and cleaning process steps.**

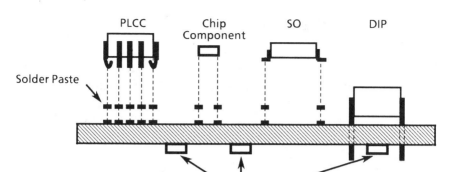

**Figure 7.3 A Type II SMT packaging case in which full conversion to SMT will not save soldering and cleaning process steps.**

segregated in one corner or end of the board. Hence, this may not be a feasible option if certain through-hole components must be located next to certain surface mount components. Also, the wave solder pot may have to be lowered to allow sufficient clearance for the cover plate. Despite these technical challenges, many well known companies are using cover plates for boards that must have surface mount active devices on both sides and many through-hole devices on one side.

The case illustrated in Figure 7.2 may also justify modification of DIP leads. Since this assembly does not have any components on the secondary side, by converting the few through-hole devices to surface mount, the wave soldering step can be eliminated, reducing cost. However, the through-hole devices must withstand reflow soldering temperatures.

If passive devices are added to the secondary side (Figure 7.3), the soldering cost advantage disappears even if the through-hole devices are converted, because the number of soldering steps is not reduced. At least two soldering operations are necessary no matter what happens to the through-hole devices. In such a case it is better to begin by reflow soldering the top side surface mount devices and then wave soldering the through hole on top and surface mount passive on the bottom, in a single step.

The foregoing discussion, as it relates to Figures 7.1, 7.2, and 7.3, suggests that butt mounting of through-hole devices may not be desirable because the reliability of the solder joints is compromised. It is also not desirable to wave solder surface mount devices such as SOICs and PLCCs because the reliability of the package is compromised. But if one must choose, butt mounting may be preferable to wave soldering of active surface mount devices.

Let us take another case to illustrate soldering considerations in design: a full surface mount, double-sided board (Type I SMT assembly). In a double-sided Type I assembly, it is important to decide which side should bear the larger components (PLCCs above 84 pins or smaller PLCCs attached to heat sink), which may fall off if the assembly is inverted. Obviously larger or heavier components (those attached to heat sinks) should be placed on one side only, and the side containing them should be reflow soldered last. There is an added benefit to this option. The smaller components mounted on the opposite side, which are reflowed twice, tend to self-align more during the second reflow cycle. Self-alignment is generally not seen in larger devices.

Finally let us discuss a third design involving soldering considerations. Some double-sided Type I assemblies containing only passive devices on the secondary side have both wave and reflow soldering options for the secondary side. Wave soldering the secondary side instead of reflow soldering is not any cheaper, but the former may be desirable for many reasons. For example, via holes can be fully filled only during wave soldering. Plugging of vias is necessary to achieve the vacuum required for ATE work. If wave soldering is not used, one can fill the vias during reflow by dispensing solder paste over them. The screen or stencils can be ordered with this requirement in mind. Either option is acceptable, but a choice must be made at the design stage to permit time for the artwork for screens or stencils to be generated if the reflow option is taken. Also, if wave soldering option for the secondary side is chosen, only wet film solder mask should be used as discussed later on in Section 7.8.

## 7.5 COMPONENT ORIENTATION CONSIDERATION

One of the important considerations for manufacturing is proper alignment of components on the PCB. The guidelines for component orientation vary depending on type; that is, components of similar types should be aligned in the same orientation for ease of component placement, inspection, and soldering.

Orienting similar components in the same direction is very desirable for pick-and-place equipment because not all machines have head rotational capability. Uniform orientation also facilitates the inspection of components for misplacement and is very helpful for troubleshooting during testing. Hence the pin 1 ends of polarized tantalum capacitors and active devices should be aligned as shown in Figure 7.4.

For wave soldering, proper alignment is necessary to prevent uneven solder fillets or solder skips (see Figure 7.5). The recommended component orientation for wave soldering to prevent these defects is shown in Figure 7.6.

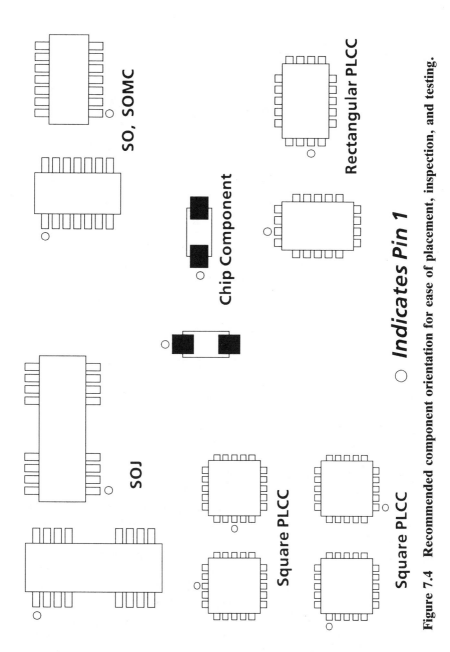

**Figure 7.4** Recommended component orientation for ease of placement, inspection, and testing.

## • Uneven fillets

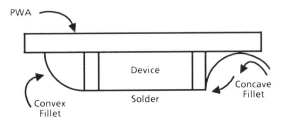

## • Solder skips can occur on trailing termination

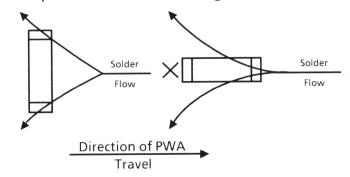

**Figure 7.5** **Uneven fillets or solder skips will result if both component terminations are not soldered simultaneously.**

It is incorrectly thought that any orientation for passive components is acceptable if dual-wave or one of the various other "chip wave" geometries is used. A dual-wave solder pot is certainly very effective for minimizing solder skips, but it does nothing to prevent uneven fillets. The orientation shown in Figure 7.6 is the only one that will yield uniform solder fillets on both terminations. Otherwise, trailing terminations generally have excess solder fillets (Figure 7.5, top), hence unnecessary excess stress on the solder joint, which may crack ceramic chip capacitors; cracking of chip components also may be caused by thermal shock, as discussed in Chapter 3 (Section 3.2.3) and Chapter 14 (Section 14.3.2).

It should be also noted from Figure 7.6 that when small and large components are adjacent to each other *and the spacing between them* is less than 0.100 inch, the smaller component must be on the leading edge during wave soldering. Otherwise the larger component may shadow the trailing smaller component. This effect is similar to the case of staggered

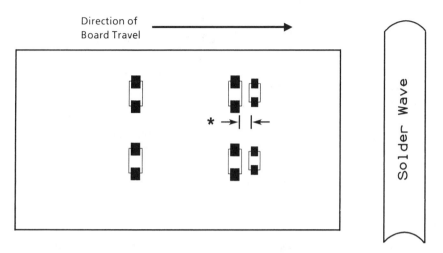

\* When pad-to-pad dimension is less than .100, smaller component must enter
wave first, because of shadowing

**Figure 7.6 Recommended component orientation for wave soldering.**

components, shown in Figure 7.7 where the leading component may
shadow the trailing component. In other words, both unequal and staggered
components may shadow the trailing component during wave soldering if
the spacing between them is less than 0.100 inch. Other cases of interpackage spacing are discussed in the next section.

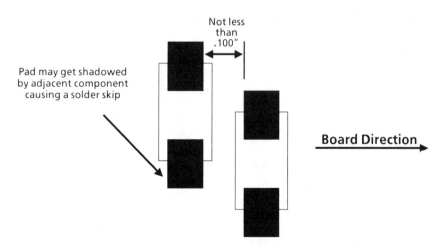

**Figure 7.7 Minimum distance between staggered components to prevent
shadowing of the trailing component in wave soldering.**

## 7.6 INTERPACKAGE SPACING CONSIDERATIONS

Interpackage spacing is the most important issue in design for manufacturability; it controls the cost effectiveness of placement, soldering, testing, inspection, and repair. A minimum interpackage spacing is required to satisfy the various manufacturing requirements, but there is no maximum limit—the more the better. We all know that designers like to pack surface mount components in as tightly as possible. After all, isn't that one of the reasons for going to SMT? Therefore, to prevent excessively tight packing with a view to keeping the assembly manufacturable in a cost-effective manner, we must develop parameters for interpackage spacing. This is not the place to make compromises, and every company should establish rules (not guidelines) for this purpose.

There are three types of spacing: pad to pad, lead to lead, and body to body. For manufacturing purposes, the lead-to-lead or termination-to-termination spacing is what matters. But for the CAD designer, the pad-to-pad spacing is the important rule for placement. Thus interpackage spacings should be specified between adjacent land patterns of different packages for ease of CAD layout. It might be more appropriate to use the term "interpad spacings", but this could cause confusion with the interpad spacings of the same package; hence we will continue to refer to interpackage spacings.

### 7.6.1 Assumptions in Interpackaging Spacing Requirements

The interpackage spacing guidelines presented in this chapter make the following assumptions:

1. The land patterns shown in Chapter 6 are used.
2. All the spacings are shown between the land patterns of adjacent packages.
3. For mixed assemblies, autoinsertion equipment requires about 0.100 inch clearance for through-hole devices.
4. About 0.075 to 0.085 inch of clearance is needed for automated and hand probe testing.
5. The interpackage spacing must allow for test probe access at ATE. Hand probe access at ATE and test clip access must also be available for troubleshooting.
6. Automated cleaning with adequate pressure nozzles (>30 psi) is used.

7.    Component shift during reflow soldering and placement may be up to 25% of pad width (IPC 815 requirement for Class 3 military applications). For commercial applications, a shift of 50% is allowed by IPC, but this would be too liberal for the assumptions made in this section. In addition, the spacing between two isolated conductors should not be less than 0.005 inch. Since there is no shift or float during wave soldering, the component shift requirement is not considered for this process. For this and other reasons, interpackage spacing is process dependent, being different for reflow and for wave soldering. The requirements specified later are identified as such.

8.    Solder joint melting of adjacent components is allowed; hence the number of times a part can be replaced and repaired is minimized (two times maximum). The assumption is that solder joints can be reflowed four times (twice in reflow for a double-sided assembly) and twice in rework. This is a very important point. Much larger (>0.5 inch) interpackage spacing is required with hot air reflow if melting of the solder joints of the adjacent components is to be completely prevented.

9.    Inspection is visual, and the assemblies can be tilted at various angles for inspection of solder joints.

## 7.6.2    Interpackage Spacing Requirements

Before mentioning specific numbers for interpackage spacing, we need to compare the requirement for through-hole assemblies. It should be pointed out that the interpackage spacing specified in the literature [1] can be as high as 0.150 to 0.200 inch. There is no problem with these numbers, because as discussed earlier, the more the better.

The interpackage spacing used in conventional through-hole assemblies is 0.100 inch. The spacing for surface mount must be less than 0.100 inch. Otherwise the advantages of SMT are lost to a great degree. Therefore all the spacings specified in this Section are less than 0.100 inch between the leads (and lands) of adjacent packages. Thousands of actual boards, meeting all the manufacturability requirements in a production environment, have been built using the interpackage spacing specified in this section. Some of those boards are shown in Chapter 1. One of them has been built by three different manufacturers in production quantities in this country and overseas. However, this figure (<0.100 inch for SMT) is not the sole basis for the guidelines presented here. The important thing to keep in mind is that the actual interpackage spacing (lead to lead or termination to termination) will be 25 to 30 mils greater than what is shown in the figures in this section. For actual distance, refer to the component package dimensions shown in Chapter 3 and the land pattern dimensions shown in Chapter 6.

As a general rule of thumb, the pad-to-pad spacing of adjacent components should be 0.050 to 0.070 inch minimum, depending on component type and orientation [2]. For PLCCs, this means 0.085 inch between leads of adjacent components. For components such as SOJ and passives, where leads or terminations are on only two sides of the packages, the interpackage spacing should be 0.040 to 0.050 inch, as indicated in Figure 7.8.

For mix-and-match boards, the spacing between the pads of through-hole mount devices and the pads of surface mount components should be at least 60 mils (see Figure 7.9). The idea is to ensure 100 mils clearance for through-holes components for the autoinsertion equipment.

The interpackage dimensions shown in Figures 7.8 and 7.9 apply only to reflow soldering for surface mount components and wave soldering of

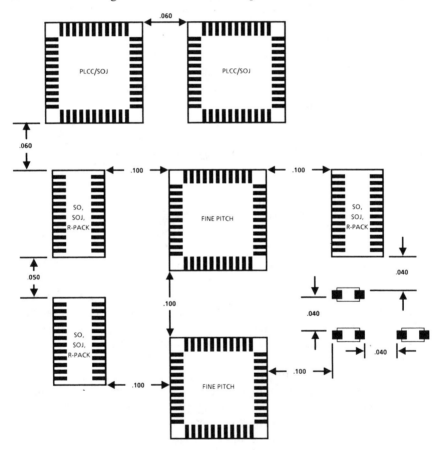

**Figure 7.8 Minimum pad-to-pad clearances between surface mount pads of active devices for reflow soldering. At least 100 mils of clearance must be provided all around the fine pitch land patterns to allow multilevel paste printing (see Figure 9.15 in Chapter 9).**

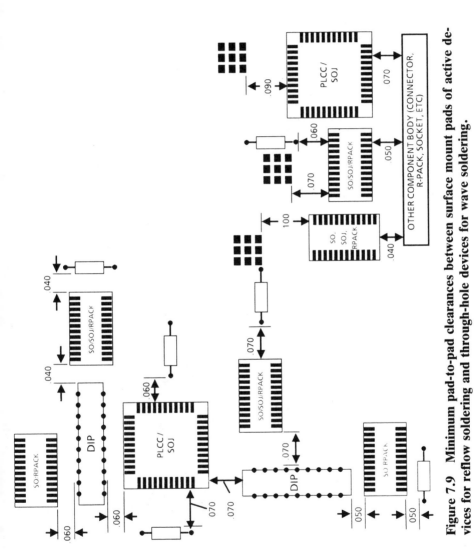

**Figure 7.9** Minimum pad-to-pad clearances between surface mount pads of active devices for reflow soldering and through-hole devices for wave soldering.

through-hole components. These dimensions can be about 10 mils smaller for memory devices because room for test accessories such as test clips and hand probing may not be needed for memory arrays.

If the secondary side passive components are wave soldered, the requirements are different. Those for staggered and unequal adjacent components were discussed earlier (Figures 7.6 and 7.7). When the adjacent chip components for wave soldering are of the same size (height and length), however, the possibility of shadowing does not arise and the interpackage distance can be reduced to 0.040 inch (Figure 7.10).

If the heights of adjacent components as shown in Figure 7.6 and 7.7 are different but their lengths are equal (e.g., the 1206 chip resistor and capacitor), the interpackage spacing needs to be only 0.040 inch. Either component can be on the trailing side.

Note also from Figure 7.10 that the external distance between chip component and DIP pad is 0.060 inch but the internal distance is only 0.040, because it is assumed that the DIPs are clinched outward. The internal distance between a DIP and a passive device is only 0.040 inch but the internal distance between a pin grid array (PGA) and a passive device is 0.060 (Figure 7.11). Greater distance for PGAs is required because the numerous pins make these devices susceptible to bridging. Figure 7.11

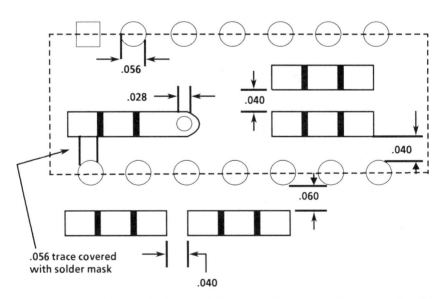

**Figure 7.10 Minimum pad-to-pad clearances between surface mount pads of passive devices and DIPs for wave soldering.**

## INSIDE A PGA PATTERN:

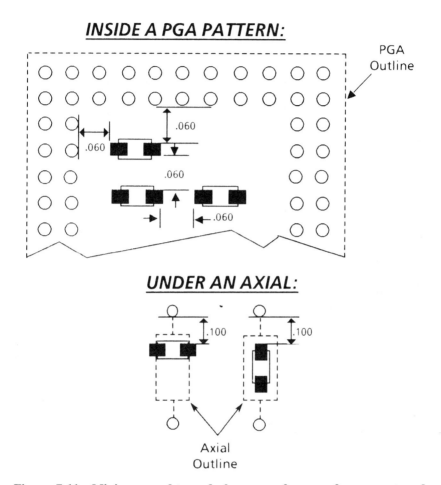

## UNDER AN AXIAL:

**Figure 7.11** **Minimum pad-to-pad clearances for a surface mount and through-hole PGA and axial component for wave soldering.**

also shows a 0.100 inch internal distance to pads of axial devices, and it is assumed that these leads are clinched inward.

Table 7.1 summarizes the *minimum* interpackaging dimensions among various types of surface mount and through-hole components. The 70 mil dimensions in Table 7.1 are intended to allow sufficient space for autoinsertion and for PLCC/DIP clips for testing. There should also be a minimum of 0.050 inch of clear space all around the edges of the PCB if boards are to be tested from the connector. This requirement is increased to 0.100 inch minimum if a vacuum-actuated, bed-of-nails fixture is used.

It is worth repeating that the requirements discussed above are *absolute* minima.

**Table 7.1 A summary of MINIMUM interpackage spacings among various types of surface mount and through-hole components**

| | MINIMUM SPACING (MILS) | | | | |
|---|---|---|---|---|---|
| | PLCC/SOJ PAD, SIDE | SO/R-PACK PAD, SIDE | SO/SOJ/ R-PACK BODY, END | CHIP COMPO-NENT PAD, SIDE | CHIP COMPONENT PAD, END |
| PLCC/SOJ side, pad to | 60[a] | — | — | — | — |
| SO/R-Pack side, pad to | 60[a] | 50 | — | — | — |
| SO/SOJ/R-pack end, body to | 70[a] | 50 | 50 | — | — |
| Chip component side, pad to | 70[a] | 50[a] | 40 | 40 | — |
| Chip component end, pad to | 40 | 40 | 40 | 40 | 40 |
| DIP side, pad to | 60 | 60 | 70 | 60 | 40 |
| DIP end, body to | 70[a] | 50 | 40 | 40 | 40 |
| Axial side, body to | 70[a] | 50 | 40 | 40 | 40 |
| Axial end, pad to | 60 | 60 | 70 | 70 | 50 |
| Stake pin, pad to | 90 | 70 | 100 | 70 | 40 |
| Other component, body to | 70[a] | 50 | 40 | 40 | 40 |

[a] A 10 mils smaller gap may be used between components in a memory array.
NOTE: Allow at least 100 mils of clearance all around the land pattern of a fine pitch (under 25 mil pitch) device.

## 7.7  VIA HOLE CONSIDERATIONS

It is mistakenly thought that holes simply disappear when using SMT. This is obviously untrue, however, since via holes are needed for interconnection. The vias are relatively quite narrow in diameter (14 to 18 mils), versus about 38 mils for the plated through holes used for component insertion.

The total number of via holes can be considerable. Table 7.2 compares plated through holes and via holes in a board before and after its conversion to Type II SMT (80% surface mount components). See Figure 1.12 (SBC 286/100) in Chapter 1 for a photograph of this board.

As shown in Table 7.2, even after conversion to SMT, a large number of via holes is required. The vias also serve as test points and must be filled to achieve adequate vacuum for the bed-of-nails (ATE) test. The solder mask requirements for via holes are discussed in Section 7.8

Via holes within the surface mount pads should be avoided because if solder flows inside such vias during reflow soldering, insufficient solder fillets may result. Figure 7.12 indicates that the recommended minimum distance between via holes and solder pads for reflow soldered assemblies is 0.025 inch. This distance can be reduced to 0.015 inch if solder mask is used between the via and the pad. Any smaller gap, however, will increase the risk of the solder mask getting on the lands, causing insufficient solder fillet. Insufficient solder fillets also may result when flush vias are connected to surface mount lands. Figure 7.13 illustrates bad design practices for reflow soldering, including flush vias and wider conductors. Designing wide traces connected to solder pads has the same effect as flush vias on solder pads (i.e., insufficient solder fillets).

Wide traces should either be covered with solder mask (preferably solder mask over bare copper to prevent cracking of solder mask) or necked down to 0.015 inch maximum.

**Table 7.2    Via hole and plated through-hole comparisons in an Intel single-sided computer board (SBC 286/100, Figure 1.12 in Chapter 1) board before and after its conversion to SMT**

|  | THROUGH-HOLE MOUNT VERSION | SURFACE MOUNT VERSION |
|---|---|---|
| Component holes | 3150 | 1300 |
| Via holes | 800 | 2200 |
| Total holes | 3950 | 3500 |

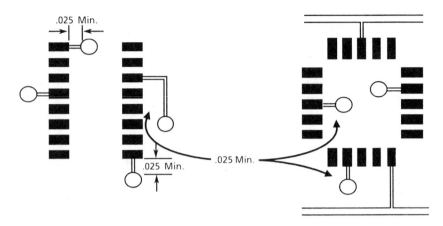

**Figure 7.12  Good design practice for via hole location in surface mount boards.**

As a general rule, vias within a pad should be avoided unless they can be filled by the board fabricator. That not only takes care of any potential flux entrapment problem but is highly desirable for attaining a good vacuum seal during in-circuit, bed-of-nails testing.

Note that via hole location is a concern only for reflow soldering (Type I SMT). In wave soldering (Type II and III SMT), the unlimited supply of solder ensures that solder fillets are not robbed of solder by the vias. In Type II and III assemblies all the vias are filled and there is no problem in ATE testing.

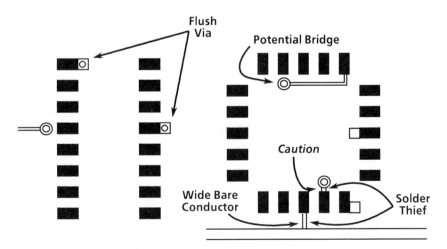

**Figure 7.13  Poor design practice for reflow soldering.**

Vias within pads of passive components glued to the underside (Type III SMT) are acceptable or even preferred because they may provide escape holes for outgassing. This feature may prove to be especially helpful if the flux used in wave soldering fails to dry sufficiently during preheat.

Whenever possible, via holes should be located underneath surface mount packages in Type I SMT assemblies. However, in Type II and Type III SMT, via holes underneath surface mount packages should be minimized or avoided because during wave soldering of these assemblies, flux may be trapped underneath the packages. As a practical matter, vias are placed both inside and outside components because of real estate constraint. This is not a real problem if the cleaning process is properly controlled.

## 7.8    SOLDER MASK CONSIDERATIONS

Solder mask is used for tenting (covering) vias, as shown in Figure 7.14. Tenting is good for Type I SMT boards because it can cover the via holes and prevent vacuum leakage during testing. Of course the test pads should not be tented, since solder mask, being nonconductive, will prevent electrical contact between the test probe and the test nodes.

Generally dry film solder mask is used for fine line boards, but it should not be used in the gap between the pads of a chip component. The thick dry film solder mask acts as a pivot between the pads and compounds the tombstoning problem.

Wet film solder mask is not a problem because its coating is relatively thin. In wave soldering, moreover, a wet film mask allows routing of traces between pads of chip components, increases adhesive board strength, and minimizes formation of voids during adhesive cure. Since it cannot tent vias, its use is limited to Type II and III assemblies, where tenting is not an issue.

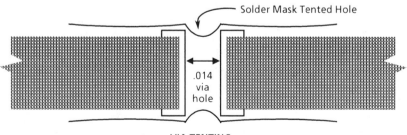

Solder Mask Tented Hole

.014
via
hole

VIA TENTING

**Figure 7.14    Tented via hole with solder mask.**

Because of the very high resolution requirements, very few vendors have the capability to mask the small spaces (10-12 mils) between adjacent pads when fine pitch components are used. Therefore, it may be necessary to require no solder mask around the land patterns of fine pitch packages. Unfortunately this is where one needs solder mask, especially in surface mount components that are to be reflow soldered. During solder paste application and reflow soldering, the fine pitch leads (<25 mils pitch) have a tendency to bridge. As solder mask technology improves, the vendors will be able to provide solder mask where it is needed most in surface mounting.

If solder mask is used over tin-lead, the solder melts and crinkles the mask, which may crack and cause entrapment of flux and moisture. This damage may eventually lead to circuit failure due to corrosion.

To prevent crinkling of solder masks, all masks should be applied over bare copper.

## 7.9 REPAIRABILITY CONSIDERATIONS

The interpackage spacing guidelines presented earlier may or may not cause problems depending on the type of equipment needed for repairs. For example, if the repair tool does not use a vacuum tip for component removal and replacement, there may be access problems. If overly thick conductive tips (Figure 7.15 left) or mechanical tips in conjunction with hot air (Figure 7.15 right) are used for repair, they may not fit into the tight interpackage spacing discussed earlier. This problem may disappear if the repair/rework tool manufacturers succeed in making hot air tools with vacuum pickup (removal and replacement) capability available for passive components.

Access is not a problem for active components, because the hot air tools used for their removal and replacement blow hot air from the top (see Chapter 14). However, the nozzles should have tight tolerances to minimize the reflow of solder joints of adjacent components.

With the best of tools available at this time, it is very difficult to prevent totally the reflow of adjacent solder joints unless the interpackage spacing is more than 0.5 inch. This dimension, which is not feasible due to real estate constraints, is not absolutely necessary either, from a metallurgical standpoint.

As long as the number of times a component can be removed and replaced is controlled, the intermetallic thickness due to repair will not grow to the point of causing failure. For this reason, a maximum of two repairs should be allowed, assuming that the solder joint has already seen

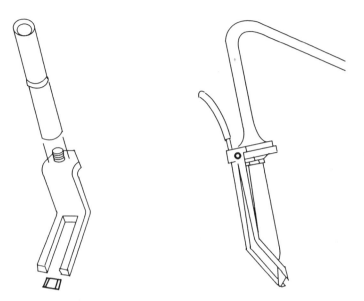

**Figure 7.15    Equipment for the repair of passive devices:
(left) conductive and (right) hot air with mechanical tweezers.**

two reflows if it is on a double-sided board. As we said earlier, the manufacturability requirements are process and equipment dependent, and one should be mindful of the manufacturing situation before setting requirements.

# 7.10    CLEANLINESS CONSIDERATIONS

Given the smaller standoff heights between surface mount components and the board, the cleaning of surface mount assemblies is of some concern. When a relatively active flux is used, which may be the case with wave soldering, cleaning problems are compounded. The elements in designing for cleaning are component orientation and solder mask use.

The guidelines recommended for component orientation and inter-package spacing in Sections 7.5 and 7.6 apply for cleaning as well. Use of solder mask is recommended whenever possible. It promotes cleanliness and helps in other areas, as discussed earlier. The most important element of designing for cleaning is the use of surface insulation resistance (SIR) patterns to monitor the process. Refer to Chapter 13 (Section 13.11: Design for Cleaning) for details.

## 7.11 TESTABILITY CONSIDERATIONS

In this section we discuss only the design issues in testing that must be taken into account when the board is being designed. The guidelines for procurement of test fixtures, test probes, and accessories are discussed in Chapter 14. When boards are designed for through-hole technology, board testability comes "free," at least insofar as testability can be defined as the ability to probe for in-circuit testing. This is because in through-hole boards, the lead ends of the devices also serve as test nodes and can be accessed by test probes from the bottom side. This is not always the case in Type I and Type II SMT boards. Type III SMT boards are essentially the same as through-hole boards, since all the active devices are through hole.

Design for testability of SMT boards boils down to whether to provide a test pad for each test node. In addition to taking up real estate, furnishing test pads raises cost. Not all test pads, however demand their own real estate on the board. Most of the vias required for interconnection also serve as the test pads.

The number of test nodes needed depends on the level of defects in manufacturing. If the defect rate is really low, we can tolerate a lower number of test nodes and accept a poor diagnostic capability. However, if the defect rate is high, we need adequate diagnostic capability, which means that we must provide access to all test nodes.

The alternative to ATE testing is functional testing, which can be much more expensive [3]. Also the solder joints in SMT boards are hard to inspect (inside fillet of SOICs, for example). The higher density of the SMT boards puts an additional constraint on visual inspection.

In addition, many of the components do not have part markings, which makes visual checking of the values of the devices impossible. Improperly trained operators can easily make mistakes in identifying polarity and pin 1.

This is not to suggest that ATE is the only right answer for SMT assemblies. There is no right or wrong approach in designing for testability. Rather, one must determine the appropriate approach. With process under control, one can tolerate access to less than all components or even depend on some form of self-test or functional test.

Companies just starting out in SMT, however, still need to get process under control. Until they do, access to all test nodes must be provided, to fully utilize the troubleshooting capability of ATE.

Thus in the short term, we need to depend on the full diagnostic capability of ATE and should provide full nodal access. In the long term, we need to get the processes under control and not waste the space and money on conventional (ATE) testing. Instead, we should strive for a combination of appropriate process control and test approaches for a cost-effective solution.

## 7.11.1   Guidelines for ATE Testing

Designing for testability revolves around whether all the test nodes must be accessible by automated test equipment (ATE). In any event, the test pads must be accessible from both sides: the bottom bottom access is needed for ATE test probes and the top access is needed for hand probing while the board is under test in ATE. In such a situation the bottom test pads cannot be accessed manually. Providing the access on both sides requires additional real estate because the design must include adequate interpackage spacing.

Interpackage spacing also has an impact on the selection of test probes and the size of the test pads (outer diameter of via pads). The two types of test probe generally used for in-circuit or bed-of-nails testing in ATE are 50 and 100 mil probes. The smaller probes allow the use of smaller (25-40 mils) test pads.

The smaller the test pad, the higher the need for fixture accuracy to guarantee good probe contact. Use of 25 mil test pads is rare, and a 32 mil test pad size may be a good compromise between the 25 and 40 mil sizes to meet the conflicting requirements of real estate constraints and fixture accuracy for good probe contact. The 100 mil probes require larger test pads, but this is seldom troublesome because the lead ends of conventional devices serve the purpose.

Figure 7.16, which should be used as a guideline for establishing the spacing between test pads and the adjacent package, is based on the diameter of the test probes. Since most boards contain both surface mount and through-hole components, both types of test probe are used. One should keep the probe density per unit area to a minimum to make sure that the fixture can be pulled down with the available vacuum.

Since the test pads are very small, there is no room for overlap of solder mask. Additional space for probe contact may be provided by making the test pads square instead of round. In addition to providing additional space for probe contact on the pad corners, a square test pad is readily distinguished from other (nontest) via pads.

As many via holes as possible should be filled or sealed with solder mask to prevent air leakage when the fixture is being pulled down. As mentioned earlier, this is not a problem in Type II and III SMT assemblies, since all the via holes are filled during wave soldering. The problem is encountered in Type I assemblies, which do not use wave soldering.

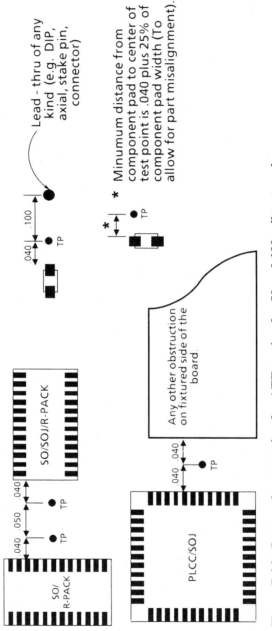

**Figure 7.16**  Interpackage spacing for ATE testing for 50 and 100 mil test probes.

The via holes that do not serve as test pads in Type I assemblies should be tented with solder mask. If there is still a problelm with air leakage, the via holes for test pads can be filled with solder paste. This is not a separate step but is accomplished when the paste is screened onto other solder pads. If possible, screening of paste into vias under devices should be avoided, to minimize entrapment of flux and solder balls under the components.

## 7.12   SUMMARY

Designing for manufacturability, testing, and repair is very important for yield improvement, hence cost reduction. Design for manufacturability, especially repairability and testability, has an essential impact on the real estate savings that SMT provides. If we get our processes under control to achieve a zero or very low defect rate, DFM will become much critical with respect to the needs to design for repair and testing. In other words we can pack components tightly and have not to worry about finding or repairing defects since they will be essentially non-existent.

However, the industry has not been able to accomplish zero defects even in conventional assemblies, which have been around for more than 30 years. Thus because SMT is in its infancy and will take time to achieve a very low defect rate, we need to give up some of the real estate benefits that SMT provides and design surface mount boards that can be manufactured, inspected, tested, and repaired in a cost-effective manner.

The consequences of designing boards that do not meet these requirements can be serious to the extent of exerting adverse impacts on revenues or schedules or both. This means that a product may not succeed in the marketplace, and that could be a serious problem indeed.

## REFERENCES

1.   Solberg, Vern. "SMT Design guidelines for automation—the robotic assembly process." Proceedings of the Technical Program, August 1988, pp. 271–280, available from surface mount technology association, Edina, Mn.

2.   Prasad, R. P. "Designing surface mount for manufacturability." *Printed Circuit Design,* May 1987, ppo. 8–11.

3.   Kunin, D., and Desai, N. "Design put to test." *Circuits Manufacturing,* March 1987, pp. 52–58.

# Part Three
# Manufacturing with Surface Mounting

# Chapter 8
# Adhesive and Its Application

## 8.0 INTRODUCTION

An adhesive in surface mounting is used to hold passive components on the bottom side of the board during wave soldering. This is necessary to avoid the displacement of these components under the action of the wave. When soldering is complete, the adhesive no longer has a useful function.

The types of components most commonly glued to the bottom side of Type III and Type II SMT boards are rectangular chip capacitors and resistors, the cylindrical transistors known as metal electrode leadless face (MELFs), and small outline transistors (SOTs). These components are wave soldered together with the through-hole mount (THM) devices.

Adhesive is also used to hold multileaded active devices such as small outline integrated circuits (SOICs) and plastic leaded chip carriers (PLCCs) on the bottom side for wave soldering. Wave soldering of active devices is generally not recommended for reasons of reliability. In another uncommon application, an adhesive holds both active and passive components placed on solder paste on the bottom or secondary side of the board during reflow soldering, to allow the simultaneous reflow soldering of surface mount components on both sides.

This chapter focuses on adhesive types, and on dispensing methods and curing mechanisms for adhesives. Some of the important technical considerations in the selection and qualification of such materials for surface mounting.

## 8.1 IDEAL ADHESIVE FOR SURFACE MOUNTING

Many factors must be considered in the selection of an adhesive for surface mounting. It is important, in particular, to keep in mind three main areas: precure properties, cure properties, and postcure properties.

## 8.1.1    Precure Properties

One-part adhesives are preferred over two-part adhesives for surface mounting because it is a nuisance to have to mix two-part adhesives in the right proportions for the right amount of time. One-part adhesives, eliminating one process variable in manufacturing, are easier to apply, and one does not have to worry about the short working life (pot life) of the mixture. The single-part adhesives have a shorter "shelf life", however. The terms "shelf life" and "pot life" can be confusing. "Shelf life" refers to the usable life of the adhesives as it sits in the container, whereas "pot life," as indicated above refers to the usable life of the adhesive after the two main components (catalyst and resin) have been mixed and catalysis has begun.

Two-part adhesives start to harden almost as soon as the two components are mixed, hence have a short pot life even though each component has a long shelf life. Elaborate metering and dispensing systems are generally required for automated metering, mixing in the right proportion, and then dispensing two-part adhesives.

Colored adhesives are very desirable because they are easy to spot if applied in excess amount, such that they contact the pads. Adhesive on pads prevents the soldering of terminations, hence is not allowed. For most adhesives it is a simple matter to generate color by the addition of pigments. In certain formulations, however, pigments are not allowed because they would act as catalysts for side reactions with the polymers, perhaps drastically altering the cure properties. Typical colors for surface mount adhesives are red or yellow, but any color that allows easy detections can be used.

The uncured adhesive must have sufficient green strength to hold components in place during handling and placement before curing. This property is similar to the tackiness requirement of solder paste, which must secure components in their places before reflow. The adhesive should have sufficient volume to fill the gap without spreading onto the solderable pads. It must be nontoxic, odorless, environmentally safe, and nonflammable.

Some consideration must also be given to storage conditions and shelf life. Most adhesives will have longer shelf life if refrigerated. Finally the adhesive must be compatible with the dispensing method to be used in manufacturing. This means that it must have the proper viscosity. Adhesives that require refrigeration must be allowed to equilibrate to ambient temperature before use to assure accurate dispensing. The issue of changes in viscosity with temperature is discussed in Section 8.5.3.

## 8.1.2    Cure Properties

The cure properties relate to the time and temperature of cure needed to accomplish the desired bond strength. The shorter the time and the lower the temperature to achieve the desired result, the better the adhesive. The specific times and temperatures for some adhesives are discussed later (Section 8.6).

The surface mount adhesive must have a short cure time at low temperature, and it must provide adequate bond strength after curing to hold the part in the wave. If there is too much bond strength, reworking may be difficult; too little bond strength may cause loss of components in the wave. However, high bond strength at room temperature does not mean poor reworkability, as discussed in Section 8.1.3.

The adhesive should cure at a temperature low enough to prevent warpage in the substrate and damage to components. In other words, it is preferable that the adhesive be curable below the glass transition temperature of the substrate (120°C for FR-4). However, a very short cure time above glass transition temperature is generally acceptable. The cured adhesive should neither increase strength too much nor degrade strength during wave soldering.

A short cure time is desired to ensure sufficient throughput. Low shrinkage during cure, to minimize the stress on attached components, is another cure property. Finally, there should be no outgassing in adhesives, because this phenomenon will entrap flux and cause serious cleaning problems. Voids also may result from rapid curing of adhesive. This issue is addressed in Section 8.6.1.2.

## 8.1.3    Postcure Properties

Although the adhesive loses its function, after wave soldering, it still must not degrade the reliabilty of the assembly during subsequent manufacturing processes such as cleaning and repair/rework. Among the important postcure properties for adhesives is reworkability. To ensure reworkability, the adhesive should have a relatively low glass transition temperature. The cured adhesives soften (i.e., reach their $T_g$) as they are heated during rework. For fully cured adhesives $T_g$ range of 75 to 95°C is considered to accommodate reworkability.

Temperatures under the components often exceed 100°C during rework because the terminations must reach much higher temperatures (>183°C) for eutectic tin-lead solder to melt. As long as the $T_g$ of the cured

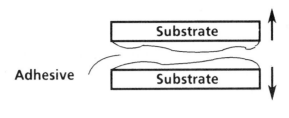

## Cohesive Failure

**Figure 8.1 Schematic representation of a "cohesive" failure in adhesive.**

adhesive stays below 100°C, and the amount of adhesive is not excessive, reworkability should not be a problem. As discussed later in Section 8.7, differential scanning calorimetry (DSC) can be used to determine the $T_g$ of a cured adhesive.

Another useful indicator of reworkability is the location of the shear line after rework. If the shear line exists in the adhesive bulk, as shown in Figure 8.1, it means that the weakest link is in the adhesive at the reworking temperature. The failure would occur as shown in Figure 8.1 because one would not want to lift the solder mask or pad during rework. In the failure mechanism shown in Figure 8.2, on the other hand, there is hardly any bond between the substrate and adhesive. This can happen due to contamination or undercuring of the adhesive.

Other important postcure properties for adhesives include nonconductivity, moisture resistance, and noncorrosivity. The adhesive should also have adequate insulation resistance and should remain inert to cleaning solvents. Insulation resistance is generally not a problem because the building blocks of most adhesives are insulative in nature, but insulation re-

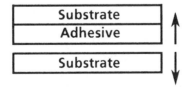

## Adhesive Failure

**Figure 8.2 Schematic representation of an "adhesive" failure at the adhesive-substrate bond line.**

sistance under humidity should be checked before final selection of an adhesive is made.

The test conditions for checking surface insulation resistance (SIR) are discussed in Chapter 13 (Cleaning). It should be noted that SIR test results can flag not only poor insulative characteristics but also an adhesive's voiding characteristics. See Section 8.6.1.2.

## 8.2    GENERAL CLASSIFICATION OF ADHESIVES

Adhesives can be based on electrical properties (insulative or conductive), chemical properties (acrylic or epoxy), curing properties (thermal or UV/thermal cure) or physical characteristics after cure (thermoplastic or thermosetting). Of course these electrical, chemical, curing, and physical properties are interrelated.

Based on conductivity, adhesives can be classified as insulative or conductive. They must have high insulation resistance, since electrical interconnection is provided by the solder. However use of conductive adhesives as a replacement for solder for interconnection purposes is also being suggested. Silver fillers are usually added to adhesives to impart electrical conduction. We discuss conductive adhesives briefly in Section 8.4. This chapter focuses on nonconductive (insulative) adhesives because they are most widely used in the wave soldering of surface mount components.

Surface mount adhesives can be classified as elastomeric, thermoplastic, or thermosetting. Elastomeric adhesives, as the name implies, are materials having great elasticity. These adhesives may be formulated in solvents from synthetic or naturally occurring polymers. They are noted for high peel strength and flexibility, but they are not used generally in surface mounting.

Thermoplastic adhesives do not harden by cross-linking of polymers. Instead, they harden by evaporation of solvents or by cooling from high temperature to room temperature. They can soften and harden any number of times as the temperature is raised or lowered. If the softening takes place to such an extent that they cannot withstand the rough action of the wave and may be displaced by the wave, however, thermoplastic adhesives cannot be used for surface mounting.

Thermosetting adhesives cure by cross-linking polymer molecules, a process that strengthens the bulk adhesive and transforms it from a rubbery state (Figure 8.3, left) to a rigid state (Figure 8.3, right). Once such a material has cured, subsequent heating does not add new bonds. Thermosetting adhesives are available as either one- or two-part systems.

# Before Curing

# After Curing

Polymer chains with
no cross-linking

Cross-linked polymer chains
(simplified view)

**Figure 8.3 Schematic representation of the curing mechanism in a thermosetting adhesive.**

## 8.3 ADHESIVES FOR SURFACE MOUNTING

Both conductive and nonconductive adhesives are used in surface mounting. Conductive adhesives are discussed in Section 8.4. The most commonly used nonconductive adhesives are epoxies and acrylics. Sometimes urethanes and cyanoacrylates are chosen.

### 8.3.1 Epoxy Adhesives

Epoxies are available in one- and two-part systems. Two-part adhesives cure at room temperature but require careful mixing in the proper proportions. This makes the two-part systems less desirable for production situations. Single-component adhesives cure at elevated temperatures, with the time for cure depending on the temperature. It is difficult to formulate a one-part epoxy adhesive having a long shelf life that does not need high curing temperature and long cure time. Epoxies in general are cured thermally and are suitable for all different methods of application. The catalysts for heat curing adhesives are epoxides. An exposide ring contains an oxygen atom bonded to two carbon atoms which are bonded to each other. The thermal energy breaks this bond to start the curing process.

The shelf life of adhesives in opened packages is short, but this may not be an important issue for most syringing application since small (5 gram) packages of adhesives are available. Any unused adhesive can be thrown away without much cost impact if its shelf life has expired. For most one-part epoxies, the shelf life at 25°C is about 2 months. The shelf life can be.prolonged, generally up to 3 months, by storage in a refrigerator at low temperature (0°C). Epoxy adhesives, like almost all adhesives used in surface mounting, should be handled with care because they may cause skin irritation. Good ventilation is also essential.

## 8.3.2 Acrylic Adhesives

Like epoxies, acrylics are thermosetting adhesives that come as either single- or two-part systems, but with a unique chemistry for quick curing. Acrylic adhesives harden by a polymerization reaction like that of the epoxies, but the mechanism of cure is different. The curing of adhesive is accomplished by using long wavelength ultraviolet light or heat. The UV light causes the decomposition of peroxides in the adhesive and generates a radical or odd-electron species. These radicals cause a chain reaction to cure the adhesive by forming a high molecular weight polymer (cured adhesive).

The acrylic adhesive must extend past the components to allow initiation of polymerization by the UV light. Since all the adhesive cannot be exposed to UV light, there may be uncured adhesive under the component. Not surprisingly, the presence of such uncured adhesive will pose reliability problems during subsequent processing or in the field in the event of any chemical activity in the uncured portion. In addition, uncured adhesives cause outgassing during solder and will form voids. Such voids may entrap flux.

Total cure of acrylic adhesives is generally accomplished by both UV light and heat to ensure cure and to also reduce the cure time. These adhesives have been widely used for surface mounting because of faster throughput and because they allow in-line curing between placement and wave soldering of components. Like the epoxies, the acrylics are amenable to dispensing by various methods.

The acrylic adhesives differ from the epoxy types in one major way. Most but not all acrylic adhesives are anaerobic (i.e., can cure in the absence of air). To prevent natural curing, therefore, they should not be wrapped in airtight containers. To avoid cure in the bag, the adhesive must be able to breathe. The acrylic adhesives are nontoxic but are irritating to the skin. Venting and skin protection are required, although automated dispensing often can eliminate skin contact problems. Acrylic adhesives can be stored for up to 6 months when refrigerated at 5°C and for up to 2 months at 30°C.

## 8.3.3 Other Adhesives for Surface Mounting

These are Two other kinds of nonconductive adhesives are available: urethanes and cyanoacrylates. The urethane adhesives are typically moisture sensitive and require dry storage to prevent premature polymerization. Nor are they resistant to water and solvents after polymerization. These materials seldom are used in surface mounting.

The cyanoacrylates are fast bonding, single-component adhesives generally known by their commercial names (Instant Glue, Super Glue, and Crazy Glue, etc.). They cure by moisture absorption without application of heat. The cyanoacrylates are considered too fast bonding for SMT, and they require good surface fit. An adhesive that cures too quickly is not suitable for surface mounting because some time lapse between adhesive placement and component placement is necessary. Also, cyanoacrylates are thermoplastic and may not withstand the heat of wave soldering.

## 8.4 CONDUCTIVE ADHESIVES FOR SURFACE MOUNTING

Conductive adhesives have been proposed for surface mounting as a replacement for solder to correct the problem of solder joint cracking [1]. It is generally accepted that leadless ceramic chip carriers (LCCCs) soldered to a glass epoxy substrate are prone to solder joint cracking problems due to mismatch in the coefficients of thermal expansion (CTE) between the carriers and the substrate. This problem exists in military applications, which require hermetically sealed ceramic packages.

Most electrically conductive adhesives are epoxy-based thermosetting resins that are hardened by applying heat. They cannot attain a flowable state again, but they will soften at their glass transition temperature. Nonconductive epoxy resin serves as a matrix, and conductivity is provided by filler metals. The metal particles must be present in large percentage so that they are touching each other, or they must be in close proximity, to allow electron tunneling to the next conductive particle through the nonconductive epoxy matrix. Typically it takes 60 to 80% filler metals, generally precious metals such as gold or silver, to make an adhesive electrically conductive. This is why these adhesives are very expensive. To reduce cost, nickel-filled adhesives are used. Copper is also used as filler metal, but oxidation causes this metal to lose its conductivity.

It is suggested that in surface mounting, it is cheaper to use conductive epoxy instead of solder [1]. The cost saving is derived form the reduction in process steps from nine typical in reflow soldering (Figure 8.4) to five (Figure 8.5) if conductive adhesive is used. The process for applying the conductive adhesive is very simple. The adhesive is screen printed onto circuit boards to a thickness of 2 mils, using 200 mesh stainless steel screen with a 1.8 mil emulsion thickness [1]. Nonconductive adhesive is also used

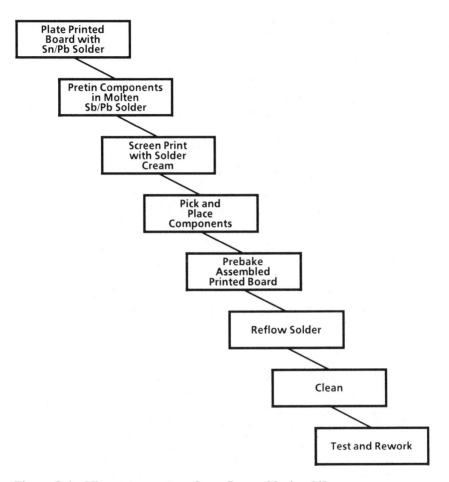

**Figure 8.4   Nine process steps for reflow soldering [1].**

to provide the mechanical strength for larger components. For smaller chip components, nonconductive adhesive is not needed for mechanical strength.

After the adhesive has been applied and the components placed, both conductive and nonconductive adhesives are cured. Depending on the adhesive, the heat for curing is provided by heating in a convection oven, by exposing to infrared or ultraviolet radiation, or by vapor phase condensation. Curing times can vary from a few minutes to an hour, depending on the adhesive and the curing equipment. High strength materials cure at 150 to 180°C., lower strength, highly elastomeric materials cure at 80 to 160°C [2].

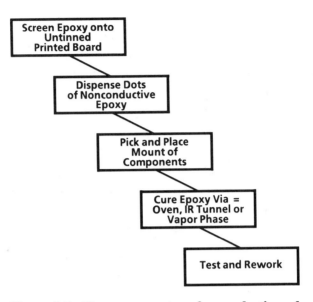

**Figure 8.5    Five process steps for conductive adhesive [1].**

Conductive adhesives have been ignored by the surface mounting industry for many reasons. The military market is still struggling with the solder joint cracking problem, and the exotic substrates selected thus far instead of adhesives have not been completely successful. However, there is not much reliability data for adhesives, either, and since plastic packages with leads are used in commercial applications, the cracking problem does not exist, in that market sector. The higher cost of conductive adhesives is also a factor.

In one study [3], neither eutectic solder nor conductive epoxy could prevent joint cracking, so indium solder was used. Thus the claim of preventing solder joint cracking with adhesives needs more supporting data. Moreover, the process step savings provided by conductive adhesive may be exaggerated. The only meaningful process step saving between Figure 8.4 (solder reflow) and Figure 8.5 (conductive epoxy) is elimination of cleaning when adhesive is used. Solder plating cannot be eliminated because it is required for interconnection, and tinning of component leads and adhesive application during reflow are not common practices.

There are other reasons for not using conductive adhesives for surface mounting. As indicated throughout this book, very few assemblies are entirely surface mount (Type I); most are a mixture of surface mount and through-hole mount (Type II). Conductive adhesives, however, do not work for interconnection of through holes and therefore cannot be used for mixed assemblies. In any event, since the electrical conductivity of these adhesives is lower than that of solder, they cannot replace solder if high electrical conductivity is critical. When conductive adhesives can be used, component placement must be precise, for twisted components cannot be corrected without the risk that they will smudge and cause shorts.

Repairs may also be difficult with conductive adhesives. A conductive adhesive is hard to remove from a conductive pad, yet complete removal is critical because the new component must sit flush. Also, it is not as convenient to just touch up leads with conductive adhesive, as can be done with solder. Probably the biggest reason for the very slight usage of conductive adhesives, however, is unfamiliarity for mass application. The conductive adhesives do have their place in applications such as hybrids and semiconductors. It is unlikely, however, that they will ever replace solder, a familiar and relatively inexpensive material.

## 8.5 Adhesive Application Methods

The commonly used methods for applying adhesives in surface mounting are pin transfer, screening, and syringing or pressure transfer. Proper selection depends of a method on a great number of considerations such as type of adhesive, volume or dot size, and speed of application. No matter which method is used, the following guidelines should be followed when dispensing adhesives.

1. Adhesives that are kept refrigerated should be removed form the refrigerator and allowed to come to room temperature before their containers are opened.

2. The adhesive should not extend onto the circuit pads. Adhesive that is placed on a part should not extend onto the component termination.

3. Sufficient adhesive should be applied to ensure that when the component is placed, most of the space between the substrate and the component is filled with adhesive. For large components, more than one dot may be required.

4. It is very important that the proper amount of adhesive be placed. As mentioned earlier, too little will cause loss of components in the solder wave and too much will either cause repair problems or flow onto the pad under component pressure, preventing proper soldering.

5. Figure 8.6 can be used as a general guideline for dot size requirements.

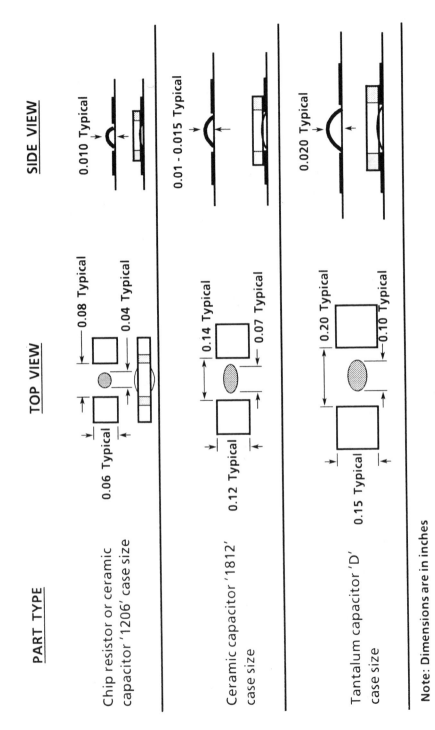

**Figure 8.6   Guidelines for adhesive dot sizes for surface mount components.**

6. Unused adhesives should be discarded.
7. If two-part adhesives are used, it will be necessary to properly proportion the "A" (resin) and "B" (catalyst) materials and mix them thoroughly before dispensing. This can be done either manually or automatically. Two-part adhesives are not commonly used for surface mounting because they introduce additional process and equipment variables.

## 8.5.1    Screening

Like solder paste application, screen printing uses a squeegee to push adhesive through the holes in the screen onto the substrate where adhesive is required. The screens are made using an artwork film of the outer layer showing the locations at which adhesive needs to be deposited. Chapter 9 covers the screen printing process and lists equipment for paste application; this process applies to the screening of adhesives as well. Screen printing is a very fast process. It allows the deposition of adhesives on all locations in one stroke. Thickness and size of adhesive dots are determined by the thickness of the wire mesh and the emulsion on the screen.

Screening of adhesive is cumbersome, hence is not a very common production process. Aligning the screens before printing and cleaning them after printing are difficult tasks. Also, care is necessary to prevent smudging of adhesives onto adjacent pads, to preserve solderability. Screening can be a problem if deposits of varying thicknesses are required for components with different standoff heights since screens can deposit the same thickness of adhesive for all component sizes.

## 8.5.2    Pin Transfer

Pin transfer, like screening, is a very fast dispensing method because it applies adhesive en masse. Viscosity control is very critical in pin transfer to prevent tailing, as it is for good printing in screening. The pin transfer system can be controlled by hardware or by software.

In hardware-controlled systems, a grid of pins, which is installed on a plate on locations corresponding to adhesive locations on the substrate, is lowered into a shallow adhesive tray to pick up adhesive. Then the grid is lowered onto the substrate. When the grid is raised again, a fixed amount of adhesive sticks to the substrate because the adhesive has greater affinity for the nonmetallic substrate surface than for the metallic pins. Gravity ensures that an almost uniform amount of adhesive is carried by the pins each time. Hardware-controlled systems are much faster than their software-controlled counterparts, but not as flexible. Some control in the size of dots can be exercised by changing the pin sizes, but this is very difficult.

Software-controlled systems offer greater flexibility at a slower speed, but there are some variations. For example, in some Japanese equipment, such as TDK's, a jaw picks up the part, the adhesives is applied to the part (not the substrate), with a knife rather than with a pin, and then the part is placed on the substrate. This software-controlled system applies adhesive on one part at a time in fast succession. Thus when there is a great variety of boards to be assembled, software-controlled systems are preferable.

For prototyping, pin transfer can be achieved manually, using a stylus, shown in Figure 8.7. Manual pin transfer provides great flexibility. For example, as shown in Figure 8.8, a dot of adhesive may be placed on the body of a large component, if this is necessary to ensure an adequate adhesive bond.

### 8.5.3   Syringing

Syringing, the most commonly used method for dispensing adhesive, is not as fast as other methods, but it allows the best flexibility. Adhesive placed inside a syringe and is dispensed through a hollow needle by means of pressure applied pneumatically, hydraulically, or by an electric drive. In all cases, the system needs to control the flow rate of the adhesive to ensure uniformity.

As discussed in Chapter 11, adhesive dispensing systems that can place

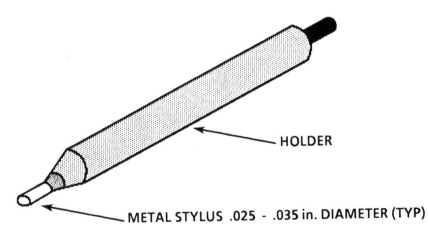

**Figure 8.7   A stylus for manual pin transfer of adhesive.**

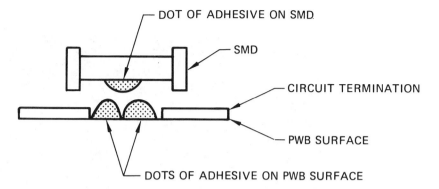

**Figure 8.8 Guideline for adhesive dots on large components.**

two to three dots per second are generally an integral part of the pick-and-place equipment. Either dedicated adhesive dispensing heads with their own X-Y table can be used, or the component placement X-Y table can be shared for adhesive placement. The latter option is cheaper but slower because the placement head can be either dispensing adhesive or placing components, not both.

The adhesive dispensing head, as an integral part of the pick-and-place system is shown in Figure 8.9 which shows the syringe out of its housing. This figure shows a typical size of syringe with 5 to 10 grams of adhesive capacity. Refer to Chapter 11, Figure 11.5, which shows the syringe inside its housing, ready for dispensing adhesive onto the substrate. The dispensing head can be programmed to dispense adhesive on desired locations. The coordinates of adhesive dots are generally downloaded from the CAD systems not only to save time in programming the placement equipment but also to provide better accuracy. This is a very effective method for varying dot size or even for placing two dots, as is sometimes required.

If the adhesive requires ultraviolet cure, the dot of adhesive should be placed or formed so that a small amount extends out from under the edges of the component, but away from the terminations and the leads, as shown in Figure 8.10. The exposed adhesive is necessary to initiate the ultraviolet cure.

The important dispensing parameters, pressure and timing, control the dot size and to some extent tailing, which is a function of adhesive viscosity. By varying the pressure, the dot size can be changed. Stringing or tailing, the dragging of adhesive's "tail" to the next location over components and substrate surface, can cause serious problems of solder skips on the pads. Stringing can be reduced by making some adjustments to the dispensing system. For example, smaller distance between the board and the nozzle, larger diameter nozzle tips and lower air pressure help reduce

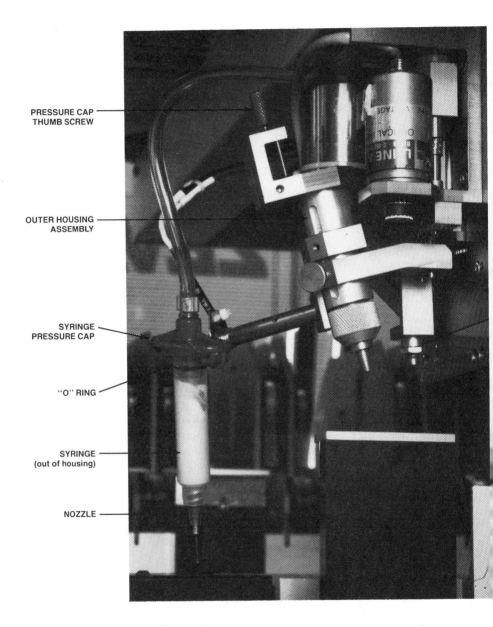

PRESSURE CAP
THUMB SCREW

OUTER HOUSING
ASSEMBLY

SYRINGE
PRESSURE CAP

"O" RING

SYRINGE
(out of housing)

NOZZLE

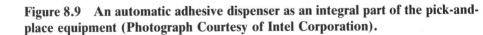

**Figure 8.9   An automatic adhesive dispenser as an integral part of the pick-and-place equipment (Photograph Courtesy of Intel Corporation).**

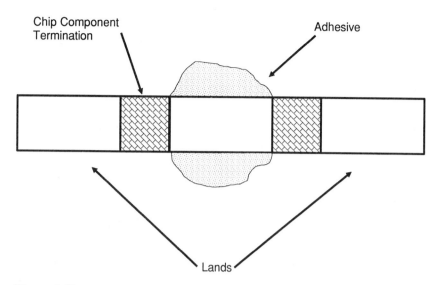

**Figure 8.10**  **The adhesive extension required on the sides of components using UV curing adhesive.**

the incidence of stringing. If pressure is used for dispensing, which is commonly the case, any change in viscosity and restriction to flow rate will cause the pressure to drop off, resulting in a decrease in flow rate and a change in dot size.

The viscosity of adhesive also plays a role in stringing. For example higher viscosity adhesives are more prone to stringing than lower viscosity adhesives. However, a very low viscosity may cause dispensing of excessive amount of adhesive. Since viscosity changes with temperature, a change in ambient temperature can have a significant impact on the amount of adhesive that is dispensed. From a study conducted at IBM by Meeks [4], and as shown in Figure 8.11, only a 5°C change in ambient temperature can influence the amount of adhesive dispensed almost by 50% (from 0.13 gram to 0.190 gram). All other dispensing variables, such as nozzle size, pressure, and time, are the same. Temperature controlled housing should be used to prevent variation in dot size due to change in ambient temperature.

Skipping of adhesive is another common problem in adhesive dispensing. The likely causes of skipping are clogged nozzles, worn dispenser tips, and circuit boards that are not flat [5]. The nozzles generally clog if adhesive is left unused for a long time (from a few hours to a few days, depending on the adhesive). To avoid clogging of nozzles, either the syringe should be discarded after every use or a wire can be put inside the nozzle

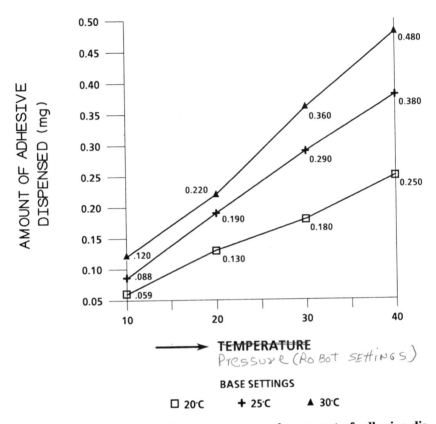

**Figure 8.11 The impact of temperature on the amount of adhesive dispensed [4].**

tip. A very high viscosity can also cause skipping. When using automatic dispensers in pick-and-place systems, care should be exercised to keep gripper tips clean of the adhesive. If contact with adhesive occurs during component placement, the gripper tips should be cleaned with isopropyl alcohol or naphtha. Also, when using the pick-and place machine for dispensing adhesive, a minimum amount of pressure should be used to bring the component leads or terminations down onto the pads. Once the components have been placed on the adhesive, lateral movement should be avoided. These precautions are necessary to ensure that no adhesive gets onto the pads. Dispensing can also be accomplished manually for prototyping applications using a semiautomated dispenser (Figure 8.12). Controlling dot size uniformity is difficult with semi-automated dispensers.

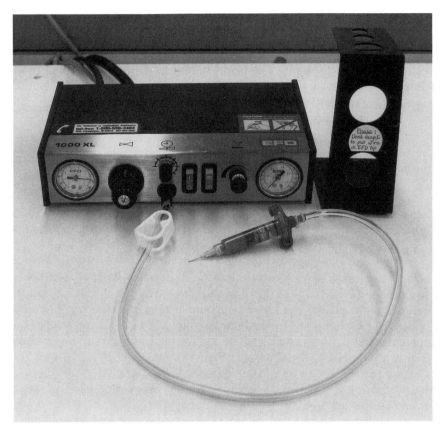

**Figure 8.12   A semiautomatic dispenser for adhesive application.**

## 8.6   CURING OF ADHESIVES

Once adhesive has been applied, components are placed. Now the adhesive must be cured to hold the part through the soldering process. There are two commonly used methods of cure: thermal cure and a combination of UV light and thermal cure. We discuss these curing processes in turn.

### 8.6.1   Thermal Cure

Most of the epoxy adhesives are designed for thermal cure, which is the most prevalent method of cure. Thermal cure can be accomplished simply in a convection oven or an infrared oven, without added investment in a UV system. The infrared oven can also be used for reflow soldering. This is one of the reasons for the popularity of thermal cure, especially in IR ovens. (See Chapter 12 for details on the heat transfer mechanism in IR ovens.) The single-part epoxy adhesives require a relatively longer cure time and higher temperatures. When using higher temperature, care should be taken that boards do not warp and are properly held.

### 8.6.1.1 Thermal Cure Profile and Bond Strength

As shown in Figure 8.13, the adhesive cure profile depends on the equipment. Convection ovens require longer, but the temperature is lower; infrared ovens provide the same result in shorter time, since the curing is done at higher temperatures.

Different adhesives give different cure strengths for the same cure profile (time and temperature of cure), as shown in Figure 8.14, where strength is depicted as the force required to shear off a chip capacitor, cured in a convection oven, from the board at room temperature. For each adhesive, the cure time is fixed at 15 minutes. The cure temperature in a convection oven was varied, and strength was measured with a Chatillon pull test gauge. Graphs similar to Figure 8.14 should be developed for evaluating the cure profile and corresponding bond strength of an adhesive for production applications. The curing can be done either in a convection oven or in an infrared oven.

In adhesive cure, temperature is more important than time. This is shown in Figure 8.15. At any given cure temperature, the shear strength shows a minor increase as the time of cure is increased. However, when the cure temperature is increased, the shear strength increases significantly at the same cure time. For all practical purposes, the recommended min-

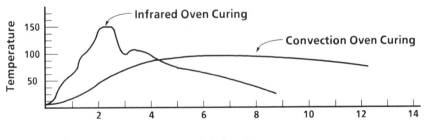

Time in minutes

**Figure 8.13  Adhesive cure profiles in IR and convection ovens.**

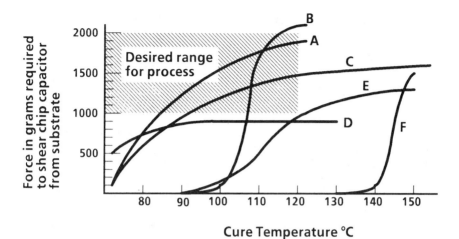

**Figure 8.14   Cure strength for adhesives A to F at different temperatures for 15 minutes in a convection oven.**

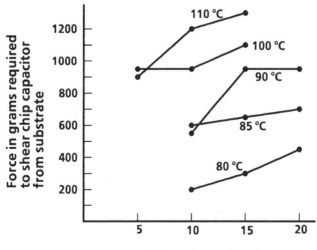

**Figure 8.15   Cure strength of adhesive C, an epoxy adhesive, at different times and temperatures.**

imum and maximum shear strengths for cured adhesive are 1000 and 2000 grams, respectively. However, higher bond strengths (up to 4000 grams) have been found not to cause rework problems, since adhesive softens at rework temperatures.

This point about cure temperature being more important than time is also true for infrared curing, as shown in Table 8.1. As the peak temperature in the IR oven is raised by raising panel temperatures, the average shear strength increases drastically. Table 8.1 also shows that additional curing takes place during soldering. Thus an adhesive that is partially cured during its cure cycle will be fully cured during wave soldering.

Most of the adhesive gets its final cure during the preheat phase of wave soldering. Hence it is not absolutely essential to accomplish the full cure during the curing cycle. Adequate cure is necessary to hold the component during wave soldering, however and if one waits for the needed cure until the actual soldering, it may be too late. Chips could fall off in the solder wave.

It should be kept in mind that the surface of the substrate plays an important role in the determining the bond strength of a cured adhesive. This is to be expected, because bonding is a surface phenomenon. For example, a glass epoxy substrate surface has more bonding sites in the polymer structure than a surface covered by a solder mask.

As illustrated in Figure 8.16, different solder masks give different bond strengths with the same adhesive using identical curing profiles. Hence an

**Table 8.1  Impact of curing temperature and after cure in the wave on the bond strength of epoxy adhesive. The table shows that additional curing takes place during soldering. (Intel Corporation.)**

| BELT SPEED (FT/MIN)[a] | PEAK TEMPERATURE (°C) | MEANSHEAR STRENGTH (GRAMS) | | APPROXIMATE REWORK (REMOVAL) TIME (SECONDS) |
|---|---|---|---|---|
| | | AFTER IR CURE | AFTER WAVE SOLDERING | |
| 4.0 | 150 | 3000 | 3900 | 4–6 |
| 5.0 | 137 | 2000 | 3900 | 4–5 |
| 6.0 | 127 | 1000 | 3700 | 3–5 |

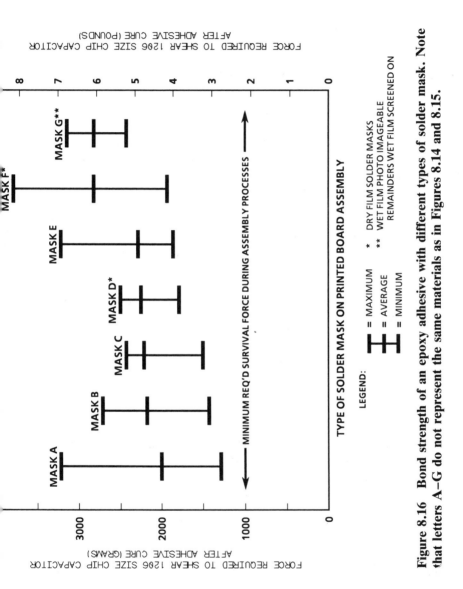

**Figure 8.16** Bond strength of an epoxy adhesive with different types of solder mask. Note that letters A–G do not represent the same materials as in Figures **8.14** and **8.15**.

adhesive should be evaluated on the substrate surface that is to be used. If there is a new kind of solder mask or if the solder mask is changed, an adhesive that was acceptable before may have only marginal shear strength under the new conditions and may cause loss of many chip components in the wave.

### 8.6.1.2 Adhesive Cure Profile and Flux Entrapment

One additional requirement for adhesive thermal cure is also very important. The cure profile should be such that voids are not formed in the adhesive. If belt speed is increased to meet produciton throughput requirements, the rapid ramp rate during cure may cause voiding in the adhesive.

Voiding may not be caused by rapid ramp rate alone, however. Some adhesives are more susceptible to voiding characteristics than others. For example, entrapped air in the adhesive may cause voiding during cure. Voids in adhesive also may be caused by moisture absorption in the bare circuit boards during storage. Similarly, susceptibility of moisture absorption increases if a board is not covered with solder mask. During adhesive cure, as well, the evolution of water vapor may cause voids. Whatever the cause, if voids in adhesive are formed during the cure cycle, they will entrap flux, which is almost impossible to remove during cleaning.

Baking boards before adhesive application, using an adhesive that has been centrifuged to remove any air after packing of the adhesive in the syringe, certainly helps to prevent the formation of voids, as does the use of solder mask. Nevertheless, the most important way to prevent flux entrapment due to formation of voids in the adhesive during cure is to characterize the adhesive and the cure profile. This must be done before an adhesive with a given cure profile is used on products. How should an adhesive be characterized? We discussed earlier the precure, cure, and postcure properties of an ideal adhesive. Characterization of the adhesive cure profile should be added to the list.

There are two important elements in the cure profile for an adhesive, namely initial ramp rate (the rate at which temperature is raised) and peak temperature. The ramp rate, or the rate at which the adhesive is cured, determines its susceptibility to voiding, whereas the peak temperature determines the percentage cure and the bond strength after cure. Both are important, but controlling the ramp rate during cure is the more critical.

Figure 8.17 shows the recommended cure profile for an epoxy adhesive in an oven. Naturally this profile will vary from oven to oven, but it can be used as a general guideline. A safe ramp rate for an in-line infrared adhesive cure oven for an epoxy adhesive is 0.5°C/second. For comparison

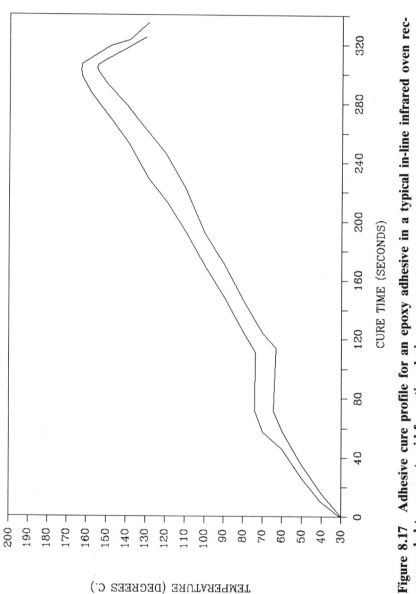

**Figure 8.17** **Adhesive cure profile for an epoxy adhesive in a typical in-line infrared oven recommended to prevent void formation during cure.**

purposes, the ramp rate in a typical batch convection oven is 0.1°C/second for the peak cure temperature of 100°C for 15 minutes. Such a low ramp rate in a batch oven may be ideal for preventing void formation, but it is not acceptable for production.

A 0.5°C/second ramp rate may conflict with the required throughput (belt speed) in some manufacturing environments. For example, for a typical oven, depending on the heating zones in the oven and its overall length, ramp rate of 0.5°C/second translates into a belt speed of about 30 inches/minute. This may or may not be acceptable to meet the throughput requirement. Increasing the belt speed, however, will increase the ramp rate beyond the acceptable limit. For example, a belt speed of 42 inches/minute will push the ramp rate to 0.8°C/second for most ovens, an unacceptable limit.

When the belt speed is raised above 30 inches/minutes for typical infrared ovens currently on the market, the risk for void formation in the adhesive increases significantly. Since the ramp rate may exceed 0.5°C/second at belt speeds above 30 inches/minute, does this mean that higher belt speeds cannot be used? Higher belt speeds are certainly possible, but the adhesive must be fully characterized.

Cleanliness tests, including the surface insulation resistance test (see Chapter 13), should be an integral part of the adhesive characterization process. Cleanliness tests other than SIR tests are necessary because SIR tests, which are generally valid only if water-soluble fluxes are used, may not flag rosin flux entrapment problems. Another way to determine the voiding characteristics of an adhesive for a given cure profile is to look for voids visually during the initial cure profile development and adhesive qualification phases.

Figure 8.18 offers accept/reject criteria for voids after cure; again, these are only guidelines. The SIR or other applicable cleanliness tests should be the determining factors for the accept/reject criteria applied to voids, ramp rate, and belt speed. Additional data that must be collected on the final profile are adhesive to be used and acceptable bond strength to prevent loss of chips in the wave without compromising reworkability. The bond strength requirement was discussed in the preceding section. It should be pointed out that the cure profile and the voiding characteristics of thermal and UV adhesive will differ, since the UV adhesive is intended for faster cure.

## 8.6.2   UV/Thermal Cure

The UV/thermal cure system uses, as the name implies, both UV light and heat. The very fast cure that is provided may be ideal for the high

Gross Voiding:        (40% to 70%)

*(unacceptable)*

Moderate Voiding:  (25% to 40%)

*(unacceptable)*

Gross Porosity:       (5% to 20%)

*(unacceptable)*

Moderate Porosity:  (2% to 5%)

*(minimum)*

*(acceptable)*

Minor Porosity:       (0% to 2%)

*(acceptable)*

**Figure 8.18   Accept/reject criteria for voids in adhesive (to prevent flux entrapment during wave soldering).**

throughput required in an in-line manufacturing situation. The adhesives used for this system (i.e., acrylics) require both UV light and heat for full cure and have two cure "peaks," as discussed later in connection with differential scanning calorimetry (DSC).

The UV/thermal adhesive must extend past the components to allow initiation of polymerization by the UV light, which is essentially used to "tack" components in place and to partially cure the adhesive. Final cure is accomplished by heat energy from IR or convection or a combination of both.

It is not absolutely necessary to cure the UV/thermal adhesives by both UV light and heat. If a higher temperature is used, the UV cure step can be skipped. A higher cure temperature also may be necessary when the adhesive cannot extend past the component body (as required for UV cure: see Figure 8.10, discussed earlier) because of lead hindrance of components such as SOTs or SOICs.

For UV/thermal systems, it is important to have the right wattage, intensity, and ventilation. The lamp power of 200 watts/inch at a 4 inch distance using 2 kW lamp generally requires about 15 seconds of UV cure. Depending on the maximum temperature in an IR or convection oven, a full cure can be accomplished in less than 2 minutes.

## 8.7 EVALUATION OF ADHESIVES WITH DIFFERENTIAL SCANNING CALORIMETRY

Adhesives are received in batches from the manufacturer, and batch-to-batch variations in composition are to be expected even though the chemical ingredient are not changed. Some of these differences, however minute they may be, may affect the cure properties of an adhesive. For example, one particular batch may not cure to a strength that is adequate to withstand forces during wave soldering after it has been exposed to the standard cure profile. This can have a damaging impact on product yield.

Adhesives can be characterized by the supplier or the by user, as mutually agreed, to monitor the quality of adhesive from batch to batch. This is done by determining whether the adhesive can be fully cured when subjected to the curing profile. The equipment can be programmed to simulate any curing profile. It also can measure the glass transition temperature of the adhesive after cure, to determine whether the reworkability properties have changed.

In sections that follow, we discuss the results of adhesive characterization of epoxy and acrylic adhesives based on evaluations done at Intel [6]. The thermal events of interest for surface mount adhesives are the glass transition temperature and the curing peak.

## 8.7.1 Basic Principles of DSC Analysis

Differential scanning calorimetry is a thermal analysis technique used for material characterization, such as ascertaining the curing properties of adhesives. Typical DSC equipment is shown in Figure 8.19. The output from a DSC analysis is either an isothermal DSC curve (heat flow versus time at a fixed temperature) or an isochronal DSC curve (heat flow versus temperature at a fixed heating rate).

Basically, in the DSC method of analysis, the heat generated by the sample material (e.g., an adhesive) is compared to the heat generated in a reference material as the temperature of both materials is raised at a predetermined rate.

The DSC controller aims to maintain the same temperature for both the sample and the reference material. If heat is generated by the sample material at a particular temperature, the DSC controller will reduce the heat input to the sample in comparison to that to the reference material, and vice versa. This difference of heat input is plotted as the heat flow to or from the sample (Y axis) as a function of temperature (X axis).

The reference material should undergo no change in either its physical or chemical properties in the temperature range of study. If the sample material does not evolve heat to or absorb heat from the ambient, the plot will be a straight line. If there is some heat-evolving (exothermic) or heat-absorbing (endothermic) event, the DSC plot of heat flow versus temperature will exhibit a discontinuuity.

As seen in Figure 8.20, the glass transition temperature is represented

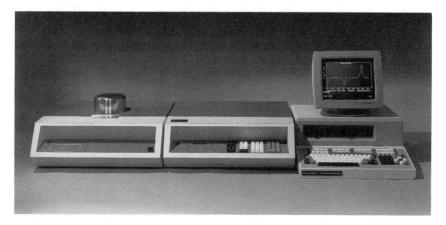

**Figure 8.19 A differential scanning calorimeter. (Photograph courtesy Perkin-Elmer.)**

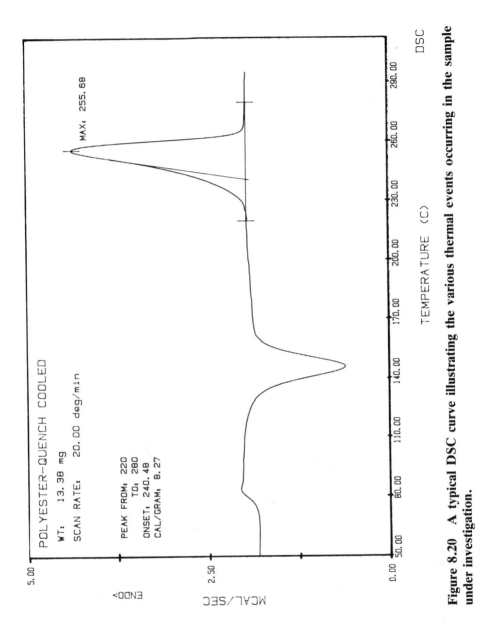

**Figure 8.20 A typical DSC curve illustrating the various thermal events occurring in the sample under investigation.**

as a change in the value of the baseline heat flow. This occurs because the heat capacity (the quantity of heat needed to raise the temperature of the adhesive by 1°C) of the adhesive below the glass transition temperature is different from the heat capacity above $T_g$.

As also seen in Figure 8.20, curing of the adhesive is represented by an exothermic peak. That is, the adhesive gives off heat when it undergoes curing. This cure peak can be analyzed to give the starting temperature of the cure and the extent of cure for a particular temperature profile. A fusion peak is also shown in Figure 8.20. However, adhesives do not undergo fusion or melting at the temperatures of use because these temperatures are too low.

The DSC curve shown in Figure 8.20 is an isochronal curve; that is, the heat flow is measured as the temperatures of the sample and the reference material are increased at a constant rate. Isothermal DSC curves can also be generated at any particular temperature. Isothermal curves depict heat flow versus time at a constant predetermined temperature.

Both isochronal and isothermal DSC curves are very useful in characterizing surface mount adhesive curing rates and glass transition temperatures. Results on the characterization of an epoxy adhesive (thermal cure) and an acrylic adhesive (UV/thermal cure) are presented next.

## 8.7.2. DSC Characterization of an Epoxy Adhesive

Figure 8.21 shows an isochronal curve for an uncured epoxy adhesive (adhesive A, Figure 8.14) subjected to a heating profile in the DSC furnace from 25°C to 270°C at a heating rate of 75°C/minute. This heating rate is very similar to that which the adhesive sees during the IR oven cure. The results from Figure 8.21 indicate the following:

1.  There is an exothermic peak corresponding to the curing of the adhesive. The onset temperature of this peak is 122°C, and it reaches a minimum at 154°C.
2.  The heat liberated during the curing of the adhesive is 96.26 calories per gram of adhesive.
3.  The adhesive starts curing before it reaches the maximum temperature of 155°C (the peak temperature for most epoxy adhesives in an infrared oven).

When the same adhesive in the uncured state is cured in the IR oven and analyzed with DSC, the results are different, as shown in Figure 8.22, which compares the isochronal DSC curves of the adhesive in the uncured and cured states. Before cure, the adhesive is in a fluid state. Its $T_g$ is

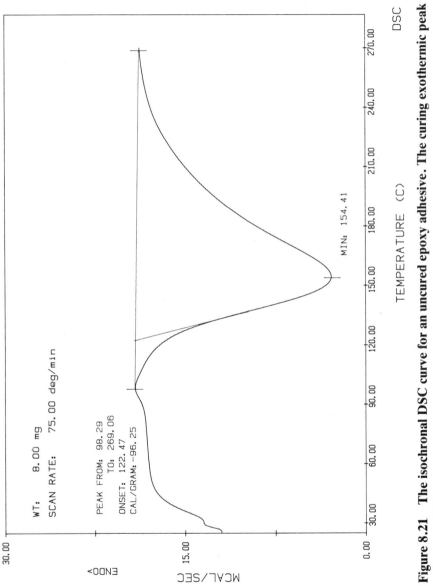

**Figure 8.21** The isochronal DSC curve for an uncured epoxy adhesive. The curing exothermic peak is clearly visible.

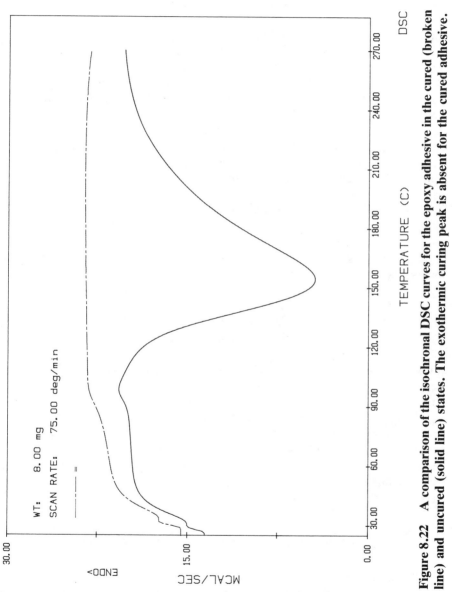

**Figure 8.22   A comparison of the isochronal DSC curves for the epoxy adhesive in the cured (broken line) and uncured (solid line) states. The exothermic curing peak is absent for the cured adhesive.**

below room temperature because by definition this is the point at which the adhesive transforms from a rigid state to a glassy or fluid state. After cure, the $T_g$ of the adhesive will increase because of cross-linking in the carbon chains as discussed earlier (see Figure 8.3).

An increase in heat flow occurs during the glass transition of an adhesive. Since the magnitude of this change is quite small, the $Y$ axis in the DSC curves shown in Figure 8.22 must be expanded, as shown in Figure 8.23, to reveal the $T_g$ effect. Figure 8.23 is a portion of the DSC curve from Figure 8.22 (broken curve) replotted with the $Y$ axis scaled up. From Figure 8.23, it is apparent that the onset of the $T_g$ occurs at 73°C and the midpoint of the $T_g$ occurs at 80°C. It is appropriate to characterize the $T_g$ from DSC curves as the temperature value at the midpoint of the transition range. (The onset value depends on the baseline slope construction, whereas the midpoint value is based on the inflection point on the DSC curve and is therefore independent of any geometric construction.)

Since the $T_g$ in the curved state is quite low, it will be very easy to rework the adhesive after IR cure. After IR cure, however, the adhesive and the SMT assembly must be subjected to another heat treatment, namely, wave soldering. This further heating will increase the $T_g$ of the adhesive, is evident from Figure 8.24, which shows a higher $T_g$ (85°C).

The DSC curve in Figure 8.24 for the same epoxy adhesive characterized in Figure 8.23 but after wave soldering. The 5°C increases in the $T_g$ after wave soldering implies that additional physical curing of the adhesive occurs during the wave soldering step. A small increase in the $T_g$ after wave soldering is not a real problem because the adhesive still can be reworked. The requirement for SMT adhesives is that after all soldering steps the $T_g$ be below the melting point of the solder (i.e., 183°C).

Figure 8.24 also shows that the adhesive starts decomposing at about 260°C, as evidenced by the small wiggles at that temperature. The DSC curve shown in Figure 8.24 was run with the encapsulated sample placed in a nitrogen atmosphere. In other atmospheres, the decomposition temperature might be different.

## 8.7.3  DSC Characterization of an Acrylic Adhesive

As mentioned earlier, acrylic adhesives require UV light and heat, but they can be cured by heat alone if the temperature is high enough. However, as mentioned earlier in Section 8.3.2, the catalysts for UV curing are the peroxides (photoinitiators). Figure 8.25 shows the isochronal DSC curve of an acrylic adhesive (adhesive G) that had been cured in DSC equipment with a simulated IR oven curing profile. Also shown in Figure

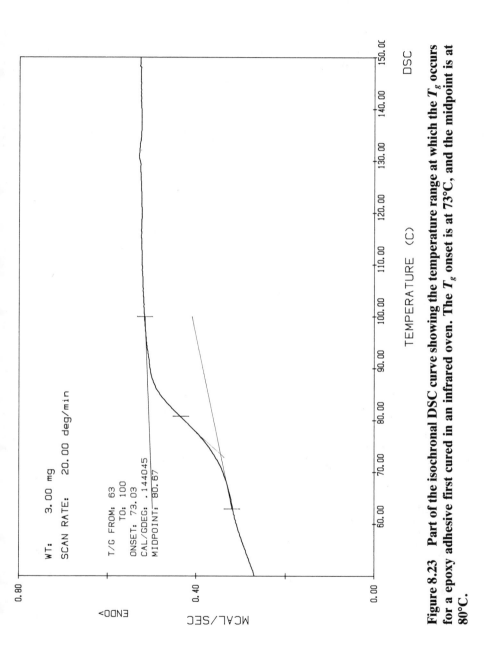

**Figure 8.23** Part of the isochronal DSC curve showing the temperature range at which the $T_g$ occurs for a epoxy adhesive first cured in an infrared oven. The $T_g$ onset is at 73°C, and the midpoint is at 80°C.

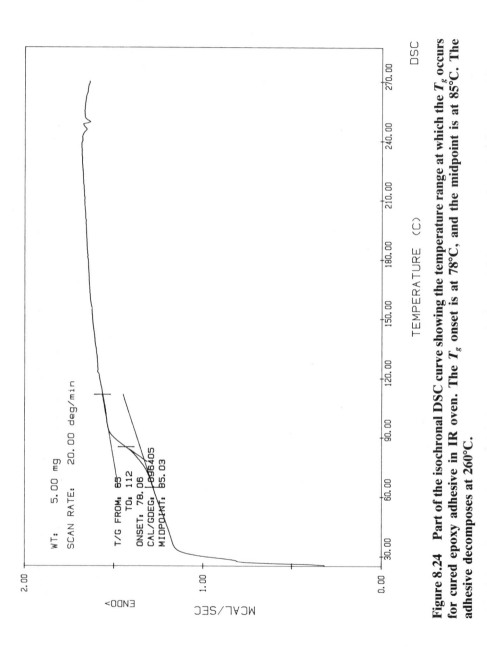

**Figure 8.24** Part of the isochronal DSC curve showing the temperature range at which the $T_g$ occurs for cured epoxy adhesive in IR oven. The $T_g$ onset is at 78°C, and the midpoint is at 85°C. The adhesive decomposes at 260°C.

8.25 for comparison is the isochronal DSC curve for the same adhesive in an uncured state. From Figure 8.25 we can draw two conclusions.

1. The uncured adhesive shows two cure peaks, one at 150°C and another at 190°C (solid line in Figure 8.25). The lower temperature peak is caused by the curing reaction induced by the photoinitiator catalyst in the UV adhesive. In other words, the photoinitiators added to the UV adhesive start the curing reaction at ambient temperature. They reduce the curing cycle time in the UV adhesives by "tacking" components in place. For complete cure, thermal energy is required.

2. DSC analysis offers two ways to distinguish a UV adhesive from a thermal adhesive. An uncured UV adhesive will show two curing peaks (solid line in Figure 8.25). If however, it is first thermally cured and then analyzed by DSC, it will show only one peak (broken line in Figure 8.25). This characteristic is similar to the curing characteristics of an uncured thermal adhesive as discussed earlier. In other words, a low temperature thermally cured UV adhesive will look like an uncured thermal adhesive.

Can an acrylic UV adhesive be fully cured without UV? Yes, but as shown in Figure 8.26, the curing temperature must be significantly higher than that used for other adhesive cures. Such a high temperature may damage the temperature-sensitive through-hole components used in mixed assemblies.

Figure 8.26 compares the isochronal DSC curves for an adhesive in the uncured state and in the fully cured state. The adhesive was fully cured by heating it to 240°C, which is much higher than the 155°C maximum curing temperature in the IR oven. Figure 8.26 shows that the adhesive sample cured at 240°C does not exhibit any peak (broken line). This indicates that to achieve full cure of this adhesive, the material must be heated above the peak temperature of about 190°C. Again, as in Figure 8.25, the solid line in Figure 8.26 is for the same adhesive in an uncured state.

When UV acrylic adhesives are subjected to heat only, they do not fully cure. The partial cure can meet the bond strength requirement for wave soldering, however. Does this mean that the UV adhesives can be thermally cured and one does not need to worry about the undercure as long as the bond strength requirements are met? Not necessarily. A partially cured adhesive is more susceptible to absorption of deleterious chemicals during soldering and cleaning than a fully cured adhesive, and unless the partially cured adhesive meets all the other requirements, such as insulation resistance, it should not be used.

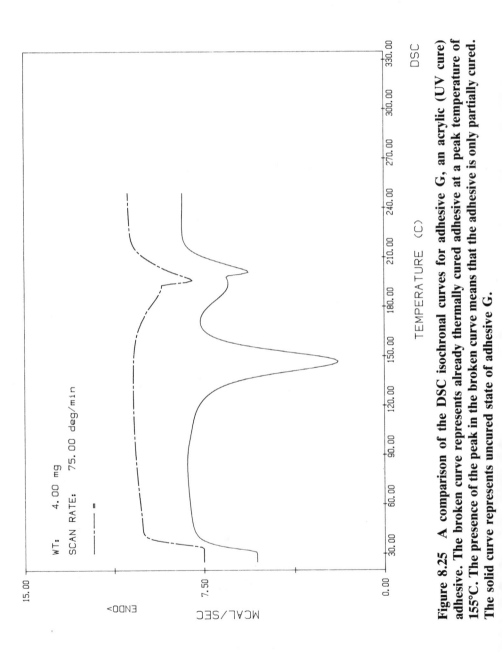

**Figure 8.25** A comparison of the DSC isochronal curves for adhesive G, an acrylic (UV cure) adhesive. The broken curve represents already thermally cured adhesive at a peak temperature of 155°C. The presence of the peak in the broken curve means that the adhesive is only partially cured. The solid curve represents uncured state of adhesive G.

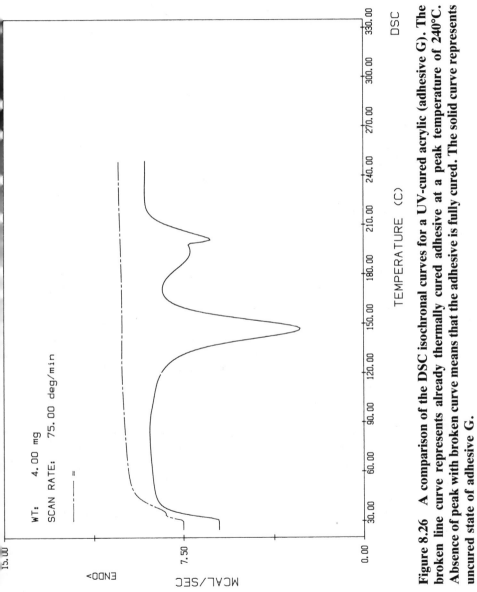

**Figure 8.26 A comparison of the DSC isochronal curves for a UV-cured acrylic (adhesive G). The broken line curve represents already thermally cured adhesive at a peak temperature of 240°C. Absence of peak with broken curve means that the adhesive is fully cured. The solid curve represents uncured state of adhesive G.**

## 8.8 SUMMARY

Adhesive plays a critical role in the soldering of mixed surface mount assemblies. Among the considerations that should be taken into account in the selection of an adhesive are desired precure, cure, and postcure properties. Dispensing the right amount of adhesive is important, as well. Too little may cause loss of devices in the wave, and too much may be too hard to rework or may spread on the pad, resulting in solder defects. Adhesive may be dispensed in various ways, but syringing is most widely used method.

Epoxies and acrylics are most widely used types of adhesive. Adhesives for both thermal land UV/thermal cures are available, but the former are more prevalent. The cure profile selected for an adhesive should be at the lowest temperature and shortest time that will produce the required bond strength. Bond strength depends on the cure profile and the substrate surface, but the impact of temperature on bond strength is predominant.

Bond strength should not be the sole selection criterion, however. Consideration of voiding characteristics may be even more significant to ensure that flux entrapment and cleaning problems after wave soldering are not encountered. One way to prevent voiding is to control the ramp rate during the cure cycle of the adhesive and to fully characterize the adhesive cure profile before the adhesive is used on products. An integral part of that characterization process should be to visually look for gross voids and to confirm their impact by conducting cleanliness tests such as the test for surface insulation resistance. In addition, the cure profile should not affect the temperature-sensitive through-hole components used in mixed assemblies.

Minor variations in an adhesive can change its curing characteristics. Thus it is important that a consistent quality be maintained from lot to lot. One of the simplest ways to monitor the curing characteristics of an adhesive is to analyze samples using differential scanning calorimetry (DSC). This can be performed either by the user or the supplier, as mutually agreed. DSC also serves other purposes. For example, it can be used to distinguish a thermal cure adhesive from an UV/thermal cure adhesive and fully cured adhesive from a partially cured or uncured adhesive. DSC is used for characterizing other materials needed in the electronics industry, as well.

## REFERENCES

1.    Estes, R.H., and Kulesza, F.W., "Surface mount technology—The epoxy alternative." *Proceedings, NEPCON West,* February 1984, pp. 219–234.

2.    Pound, Ronald., "Conductive epoxy is tested for SMT solder replacement." *Electronic Packaging and Production,* February 1985, pp. 86–90.

3.    Holt, Kenneth D., et al. "Use of indium alloy solder for microwave component attachment." IPC Technical IPC-Paper TP-361, April 1982.

4.    Meeks, S. "Application of surface mount adhesive to hot air leveled solder (HASL) circuit board evaluation of the bottom side adhesive dispense." Paper IPC-TR-664, presented at the IPC fall meeting, Chicago, 1987.

5.    Kropp, Philip and Eales, S. Kyle, Trouble shooting guide for surface mount adhesive, surface mount technology, August 1988, pp. 50–51.

6.    Aspandiar, R. Intel Corporation, internal report, February, 1987.

# Chapter 9
# Solder Paste and Its Application

## 9.0 INTRODUCTION

In the reflow soldering of surface mount assemblies, solder paste is used for the connection between the leads or terminations of surface mount components and the lands. Solder paste is applied to the surface mount lands by screening, stenciling, or dispensing. Each process has its pros and cons.

Although the imperfections found after soldering are generally termed "soldering defects," this can be misleading, since the soldering process alone does not control product quality. Rather, the quality of a finished surface mount assembly is determined by many process and design factors. Screen printing, for example, is one of the processes that plays a very important role in the quality of the final assembly.

In this chapter we cover the three major elements of the screening process—namely, solder paste quality, equipment variables, and printing process variables and discuss their impact on print quality.

We begin with the basics of solder paste and the important considerations in its selection and proceed to the selection of screen printing equipment, concluding the chapter with the pros and cons of various solder paste application methods.

## 9.1 SOLDER PASTE PROPERTIES

Solder paste is essentially comprised of metal powder particles in a thickened flux vehicle, as shown in Figure 9.1. The actual process steps in making of metal powder and the paste formulations are mostly proprietary, but in general, molten solder of desired composition is quickly chilled on a rotating wheel to form fine solder particles. The process takes place in an inert environment to minimize the oxidation of the particles.

Solder paste serves multiple critical purposes in soldering of Type I

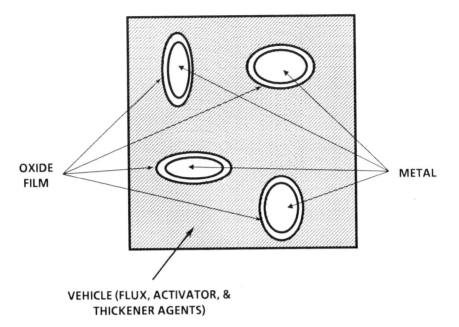

**Figure 9.1    Different constituents of solder paste.**

(full surface mount) and Type II (mixed) surface mount assemblies. Since it also contains flux necessary for effective soldering, there is no need to add flux separately and to worry about control of flux activity and density as in through-hole mount assemblies. The flux also acts as a temporary adhesive for holding surface mount components in place during placement and handling before reflow soldering. Clearly proper selection of solder paste is very important in producing defect-free, reliable surface mount assemblies. The properties of the solder paste that should be carefully controlled and evaluated before selecting a paste for surface mount assembly are discussed in the sections that follow.

There are various ways to evaluate solder paste. Vendor data and some industry specifications, such as IPC-SP-819 [1], along with the recommendations discussed below, can be used to establish an in-house solder paste specification for vendor qualification.

Certain simple tests should be conducted by the user regularly to monitor the quality of solder paste. These tests, such as the solder ball test and viscosity measurement, can prevent building of lot of assemblies that may have to be reworked, recleaned, or even scrapped.

## 9.1.1    Metal Composition

Table 9.1 [2] shows some of the commonly used solder compositions that are available for various applications in the electronics industry. The critical factors that determine the composition of the solder in the solder paste are as follows.

1.    Materials of the substrate and the components of the surface mount assemblies: high temperature solders (>225°C) can be used to attach ceramic packages to ceramic substrates. However, for the attachment of plastic surface mount packages to FR-4 (epoxy glass) substrates, lower melting solders are used to avoid degradation of the substrate and package materials during soldering.

2.    Compatibility of the solder with the metallizations on the substrate and the component leads: indium-lead alloy solders are recommended for soldering to gold terminations, since the popular tin-lead alloy solders form deleterious intermetallic compounds with gold.

3.    Strength of solder: if high tensile and shear strength at elevated temperatures is a requirement, tin-antimony alloy solders are recommended.

4.    Cost: silver is usually added to tin-lead solders to diminish the dissolution of the silver from component leads. A tin-lead solder containing 2% silver is recommended over a tin-solder containing 4% silver because the former is cheaper and has an equivalent effect in diminishing silver dissolution from the leads.

For soldering plastic surface mount components to FR-4 substrates, eutectic tin-lead (63% Sn-37% Pb, mp = 183°C) or tin-lead-silver (62% Sn-36% Pb-2% Ag, mp = 179°C) solders are widely used. However, other solder alloys are available in paste form [3].

## 9.1.2    Metal Content

The metal content in solder paste determines the solder fillet size. Fillet size increases with an increase in the percentage of metal, but the tendency for solder bridging also increases with increase in metal content at a given viscosity. A higher metal content will result in higher thickness of the reflowed solder, as shown in Table 9.2 [4].

As Table 9.2 indicates, the final thickness of reflowed solder can vary from 50% of the paste thickness for 90% metal content to as low as 22%

**Table 9.1  Melting point of different compositions of solder for the electronics industry. [2]**

| COMPOSITION | RANGE[a] | DIFFERENCE | PROPERTIES AND USES |
|---|---|---|---|
| 75 Pb/25 In | 250S-264L | 14 | Less gold leaching, more ductile than Sn/ Pb alloys; die attachment, closures, and general circuit assembly |
| 50 Pb/50 In | 180S-209L | 29 | |
| 25 Pb/75 In | 156S-165L | 9 | |
| 37.5 Sn/37.5 Pb/ 25 In | 134S-181L | 47 | Good wettability; not recommended for gold |
| 80 Au/20 Sn | 280E | 0 | Highest quality for gold surfaces; die attachment and closures |
| 63 Sn/37 Pb[b] | 183E | 0 | Widely used tin-lead solders for surface mounting and general circuit assembly; low cost and good bonding properties; not recommended for silver and gold soldering because of high leaching rate |
| 60 Sn/40 Pb[b] | 183S-188L | 5 | |
| 50 Sn/50 Pb | 183S-216L | 33 | |
| 10 Sn/90 Pb | 268S-302L | 34 | |
| 5 Sn/95 Pb | 308S-312L | 4 | |

**Table 9.1** (*Continued*)

| COMPOSITION | RANGE<sup>a</sup> | DIFFERENCE | PROPERTIES AND USES |
|---|---|---|---|
| 62 Sn/36 Pb/2 Ag<sup>b</sup> | 179E | 0 | Tin-lead solders |
| 10 Sn/88 Pb/2 Ag | 268S-290L | 22 | containing small |
| 1 Sn/97.5 Pb/1.5 Ag | 309E | 0 | amounts of silver to minimize leaching of silver conductors and leads; not recommended for gold; 3701 (62/36/2) is strongest tin lead solder |
| 96.5 Sn/3.5 Ag | 221E | 0 | Widely used tin- |
| 95 Sn/5 Ag | 221S-240L | 19 | silver solders providing very strong, lead-free joints; minimizes silver leaching; not recommended for gold |
| 42 Sn/58 Bi | 138E | 0 | Low temperature eutectic with high strength |

<sup>a</sup>S = solidus, L = liquidus, E = eutectic.
<sup>b</sup>Most commonly used soldering alloys.

of the paste thickness for 75% metal content. Thus only a minor variation in the metal content of the paste from lot to lot can have a significant impact on the quality of solder joints. For example a 10% variation in metal content can change an excess solder joint to an insufficient solder joint for the same paste thickness.

Typically, solder pastes for surface mount assembly contain 88 to 90% metal. A method for verifying the metal content of a solder paste is given in IPC-SM-819 [1].

**Table 9.2    For a given solder paste thickness, impact of metal content on reflowed solder thickness [4]**

|  | THICKNESS (INCH) | |
| --- | --- | --- |
| METAL CONTENT (%) | WET SOLDER PASTE | REFLOWED SOLDER |
| 90 | 0.009 | 0.0045 |
| 85 | 0.009 | 0.0035 |
| 80 | 0.009 | 0.0025 |
| 75 | 0.009 | 0.0020 |

### 9.1.3    Particle Size and Shape

Powder particle shape determines the oxide content of the powder and as well as the paste's printability. Spherical powders are preferred over elliptical powders. The lower the surface area, the lower the oxidation. Hence the finer and irregularly shaped powders, with their larger surface area, have a higher metal oxide content than spherical powders.

Solder pastes containing powders of irregular shape are prone to clog screens and stencils. Figure 9.2 shows acceptable and unacceptable solder particles for use in solder paste. It should be noted that the pastes do not have to be as perfect, as shown in Figure 9.2a. Some variation in particle sizes and shapes is only to be expected.

Printing problems are encountered wth solder pastes containing powder particles that are large in diameter, which readily clog screen and stencils. On the other hand, a solder paste with powder particles that are too small is prone to form solder balls during reflow.

The commonly used powder size is $-200/+325$ mesh; that is, at least 99% by weight of the powder particles will pass through a 200 (holes/square inch) mesh and less than 20% of the powder particles by weight will pass through a 325 mesh.

### 9.1.4    Flux Activators and Wetting Action

The flux is one of the main constituents of the solder paste vehicle. Solder pastes are available with three different flux types: R (rosin flux), RMA (mildly activated rosin), and RA (fully activated rosin). The activators in the RMA and RA fluxes promote wetting of the molten solder to the surface mount lands and component terminations or leads by removing oxides and other surface contaminants.

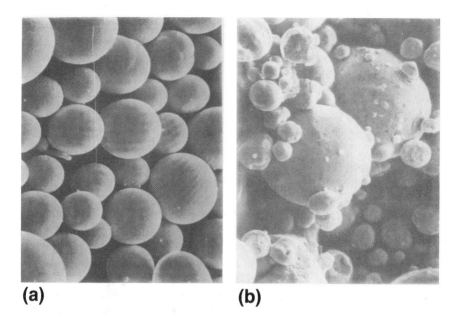

**(a)**          **(b)**

**Figure 9.2** **(a) Perfectly spherical and (b) unacceptable solder particles for solder paste.**

Activators used in solder paste range from rosin to halide-containing compounds and organic acids (OA). The R fluxes, which do not contain any activators, may not give acceptable solderability during reflow soldering. However the OA or RA fluxes may be too aggressive, calling for extra precautions during cleaning of the assemblies to avoid flux residue contamination.

The type of flux has a direct impact on the cleanliness of the assembly. In surface mounting, because of the lower standoff between component and substrate, the flux may be trapped, leading to reliability problems. See Chapter 13 on fluxes and cleaning. For this reason, RMA fluxes are generally used in solder pastes for surface mount applications. Specifically, an RMA flux has sufficient activator to clean the solder-coated or plated lands and component terminations or leads, thereby enabling the molten solder to wet these areas. Rosin fluxes generally require solvents containing chlorinated fluorocarbons (CFCs), however, and these compounds have been determined to cause depletion of the ozone layer. The 1987 Montreal Protocol signed by various nations including the United States restricts CFC use (see Chapter 13), and this environmental concern has given rise to the development of solder pastes that contain water-soluble fluxes.

Among the main problems associated with water-soluble pastes are poor tackiness and cleanability after reflow. If these technical problems are solved, water-soluble solder pastes may be widely used in the future. They certainly have good wetting action (ability to remove oxides and prepare a clean surface for soldering). The wetting action of the paste is determined by the activity of its flux. Refer to IPC-SP-819 for the wetting action test procedure [1].

## 9.1.5 Solvent and Void Formation

The solvent dissolves the flux and imparts the pasty characteristics to the metal powder in the solder paste. It controls the tackiness of the paste by its evaporation rate under ambient conditions. The solvent should not be hygroscopic. It should have a high flash point and should be compatible with the activator and the rheological modifiers.

The two most important factors that control the formation of voids in the solder fillets are the solvent in the solder paste and the reflow profile. Not only do voids formed during reflow lower the strength of the fillet, but because they represent an obstacle to the efficient transfer of heat from the device to the substrate, they could result in the overheating of the product. Voids in the solder fillets of reflowed assemblies can be determined by x-ray analysis of the assemblies. The total area of voids calculated as a percentage of the total solder fillet area should not exceed 10%.

## 9.1.6 Rheological Properties

The rheological properties of solder paste—viscosity, slump, tackiness, and working life—are controlled by the addition of rheological modifiers, also called thickening agents or secondary solvents. Rheological modifiers are generally high boiling solvents, since they have to function at temperatures up to the melting point. However, some amount of the high boiling solvents is trapped in the solder joint after solidification, since there is insufficient time for these modifiers to boil away completely.

### 9.1.6.1 Viscosity

Solder pastes are thixotropic fluids. When deformed at a constant rate of shear stress or strain rate, the solder paste viscosity will decrease over time, thus implying that its structure breaks down progressively [5]. Furthermore, the viscosity of the solder paste decreases as the shear stress on the solder paste increases. Explained simply, the paste is thin when a shear stress is applied (as with a squeegee), but thick when no stress is applied. This property is highly desirable for printing, since as suggested by Figure

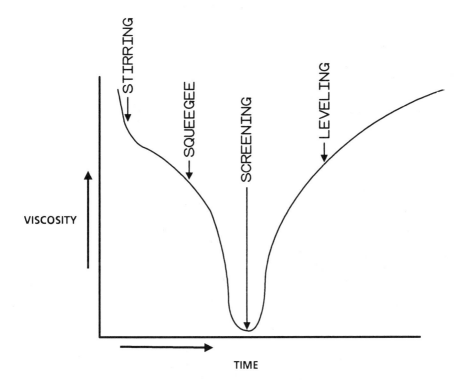

**Figure 9.3 Typical solder paste viscosity change during screen printing operation.**

9.3, the solder paste will stay on top of the open areas of the screen or stencil, flowing only when the squeegee induces a stress on it [6].

Once the paste has been deposited on the lands, the squeegee-induced shear stress is removed and the paste returns to its highly viscous form, thus staying on the lands and not flowing onto the nonmetallized surface of the board. In addition to shear force, particle size and ambient temperature have an effect on solder paste viscosity [6]. Figure 9.4 represents the decrease in viscosity that occurs as shear force or temperature is *increased* or particle size is *decreased* (finer particles).

It should be noted that Figure 9.4 is intended to show the effect on viscosity (on the Y axis) of only one variable (on the X axis) at a time. The reason for combining them is to simply illustrate a similar effect of these three variables on viscosity.

Just prior to application, the solder paste viscosity should be measured at 25°C using a Brookfield viscometer, which has almost become a de facto standard. Figure 9.5 shows the model RTV. There are various sizes of spindles (TA, TB, TC, TD, TE, TF, etc.); the TF spindle is the industry standard for solder paste viscosity measurement at 5 RPM.

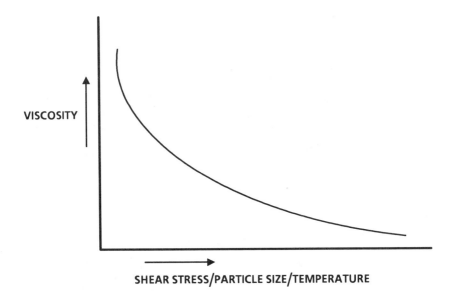

VISCOSITY

SHEAR STRESS/PARTICLE SIZE/TEMPERATURE

**Figure 9.4   Impact of increasing shear force and temperature and decreasing particle size on solder paste viscosity with identical metal content and flux vehicle.**

The solder paste should be inspected visually for appearance: a homogeneous, light to medium gray paste is desirable. A thin layer of flux that makes the contour of solder particles still perceptible is acceptable. It should not exhibit a dry crusty appearance. To inspect for foreign substances, lumpiness, or a crusted surface, the solder paste may be stirred with a clean spatula. See Section 9.5.1 for desired viscosity ranges and the impact on print quality of this variable.

### 9.1.6.2   Slump

Slump is the ability of the paste to spread out after being deposited on the lands. For good soldering results, solder paste slump must be minimized. Slump also depends on the percentage of metal in the solder paste. The most reliable way to control slump is to establish bounds within which slumping is not a problem and ensure that the paste remains within those bounds. Excessive slump of a paste can cause bridging. A general procedure to test slump using dummy board print patterns is given in IPC-SP-819 [1].

### 9.1.6.3   Working Life and Tackiness

There is some confusion about the definition of "working life." The working life of a solder paste is generally defined as *the length of time* the

**Figure 9.5    The de facto industry standard, a Brookfield viscometer. (Photograph courtesy of Intel Corporation.)**

paste can be left without degradation of its rheological properties on the screen/stencil before printing or on the substrate after printing. Both parts of the definition may be correct, but they are not equally useful. For example, the length of time the paste can be left on the screen before printing does not help us much.

However, the length of time the paste can be left on the substrate after printing is useful information. It defines the maximum time the placement machine has to place the parts. Therefore a more useful definition of "working life" is: the maximum time that can elapse between opening of the solder paste jar to paste reflow without degradation of the paste's rheological properties. This includes the total time needed for printing, placement, baking, and handling between operations.

Tackiness is the ability of the solder paste to hold the surface mount components in place after placement but before reflow soldering. The tackiness of a paste is an indicator of whether the working life of a paste has elapsed. If a tackiness check reveals that a paste has been pressed into service past its working life, the paste is no longer useful to hold components in place during placement and in handling before reflow.

Refer to IPC-SP-819 for a standard test for tackiness [1]. Morris [7] also discusses solder paste tackiness measurement. Tackiness of a paste is important for determining whether the paste on the screen or board will have to be changed in the event of a time delay for any reason (e.g., change of shifts) during a production run. Devising a controlled production planning and scheduling may be preferable to having to conduct the tackiness test. Different manufacturers, depending on their operation, will have different requirements for tackiness.

## 9.1.7   Solder Balls

Solder balls are small spherical particles of solder, usually 2 to 5 mils in diameter, which reside on the nonmetallic surfaces of the board. Solder balls, especially the mobile ones, are a reliability hazard because potentially they can short metallic conductors at any time during the life of the substrate. There are two mechanisms by which solder balls are formed.

1.   Solder balls are caused by very fine powder particles in the solder paste. They are carried away from the main solder deposit as the flux melts and flows before the solder itself melts. This happens especially when the paste is deposited outside the land area either by design (screen opening is larger than the land area) or by misregistration. These small powder particles then lose contact with the larger solder paste deposit and when the solder melts, each particle becomes a small solder ball at the periphery of the original paste deposit. A collection of small solder balls around the main solder deposit is called a "halo."

2.   Solder balls are also formed when the oxide layer on the surface of the solder powder particles is so thick that the rosin flux and any activator in the paste is not sufficient to remove it. Since oxides cannot melt at soldering temperatures, it is pushed aside as solder ball by surrounding oxide free molten solder. Solder balls formed in this manner are usually

larger than those formed by the first mechanism because of presence of surface oxide which is less dense than the metal.

Oxidation of the solder powder particles is accelerated by improper storage, or by baking the paste at an excessively high temperature before reflow. Solder particle oxidation is also promoted by fretting, or the production of oxides by the mutual abrasion of powder particle surfaces during sieving to separate the particles into various size fractions [8].

The tendency of a solder paste to form solder balls can be determined with a simple test. A few 6 to 10 mil thick solder paste patterns are deposited by stencil or screen printing on a nonmetallic substrate, such as frosted glass, ceramic, or even FR-4, and then reflowed, preferably using a reflow profile close to the one used in production.

The resulting solder deposit should be examined visually. If the deposit is one large shiny ball with a smooth surface, the solder paste is acceptable. One or two discrete solder spheres are acceptable, provided they are no larger than 10% of the main ball or more than 5 mils in diameter. A halo of solder balls is unacceptable. Figure 9.6 shows examples of acceptable and unacceptable solder balls [1]. Naturally absence of solder balls is preferred, but a minor occurrence of very fine solder balls is also acceptable. Clustered solder balls are unacceptable, as is a halo ring consisting of numerous solder balls.

Oxidation of the solder paste metal alloy leads to the formation of solder balls and the loss of metal as the oxide. Even though the oxidation of the metal takes place on the surface of the metal particles, a good gauge of the tendency of a paste to form solder balls is the bulk oxide content.

In a manufacturing environment, it is not feasible to conduct the bulk oxide content test. It is more common and useful to test for solder balls, as described above. As a matter of fact, the solder ball test should be conducted regularly to monitor the quality of the solder paste and its tendency to ball.

## 9.1.8   Printability

The printability of a solder paste is reflected in the accuracy and reproducibility of the screened or stenciled solder paste pattern onto the land patterns. The printability of the solder paste, which is one of the most critical tests, should be determined according to the following experiments. Dummy boards can be used for test substrates.

1.   Weigh five cleaned dummy board substrates before ($W_1$) and after ($W_2$) the paste is screened or stenciled on. Use five substrates for each method of application, to get average data.

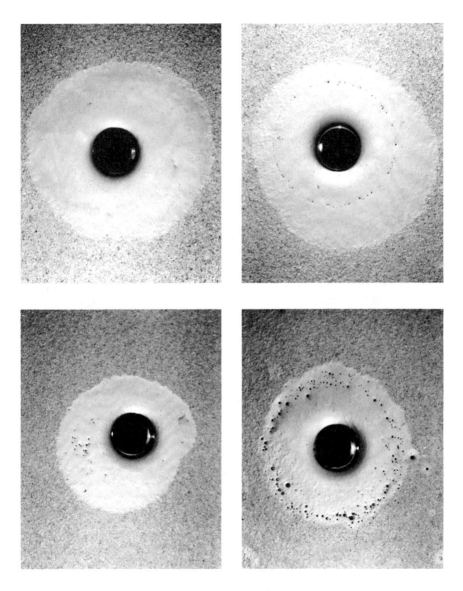

**Figure 9.6** Solder ball formation during solder ball tests for determining the suitability of solder paste. No solder ball (top left) is preferred, but minor occurrence of very fine solder balls (top right) is also acceptable. Clustered solder balls (bottom left) and a solder ball halo ring with numerous solder balls (bottom right) are unacceptable [1].

2.  Determine the weight of the solder paste applied to each board by the formula $W_2 - W_1$.
3.  Measure and record the paste height at four predetermined points on each substrate, using a depth scope.
4.  Perform steps 1 to 3 for freshly removed solder paste and for solder paste exposed to the atmosphere for 4 hours.

The acceptance criteria for use of that particular solder paste should be as follows.

- The solder paste weight should not vary more than 10% among the average measurements taken on one substrate.
- The paste height should not vary more than ±1 mil among the average measurements taken on one substrate. The acceptable paste height range is 8-10 mils. Commonly an 8 or 10 mil paste thickness is used with stencils (4-6 mils for fine pitch packages). In addition to the stencil thickness, the condition of the printing equipment determines print height.
- The solder paste pattern should have uniform coverage, without stringing and without separation of flux and solder, and it should print without forming a peak.

## 9.2  SOLDER PASTE PRINTING EQUIPMENT

The screen printing process involves a series of interrelated variables, but the printer is crucial for the achievement of the desired quality of print. The solder paste screen printers available on the market today fall into two main categories: laboratory or prototype and production. Each category has further subdivisions, since companies expect different levels of performance from laboratory and production printers. For example, a laboratory application that is R&D for one company would be prototype for another. Moreover, the production requirement itself can vary widely depending on volume. Since a clear-cut classification of equipment is not possible, the best thing to do is to select a screen printer to match the desired application.

If the goal is to use the equipment entirely for R&D, the equipment controls do not have to match what will be used in manufacturing. However, if the goal is to use the equipment for process development and improvement, the basic features of the machine should closely match to the ones to be used in manufacturing. The only difference may be in the automation features such as automatic board handling. For laboratory use or in the hybrid industry, smaller benchtop models generally are used, like the screen printer shown in Figure 9.7. Production application usually call for larger equipment with semiautomatic or automatic board loading and unloading capability.

No matter what the application is, the main features of screen printing equipment are mechanical rigidity of structure, registration method, the squeegee velocity controller, and the squeegee pressure controller [9]. Mechanical rigidity is important in maintaining parallelism between the squeegee travel path and the substrate, to achieve good print uniformity. Squeegee pressure and speed control must be repeatable, for the sake of uniformity of print. A good registration control with repeatable X, Y, and θ adjustments expedites setup work.

Having considered these generic variables, how should one select a screen printer? The first item to consider is the maximum size of the substrate that will be used. The screen or stencil frame that will handle the substrate must be larger than the substrate, since in addition to the print area for the substrate, the frame must accommodate the area needed for squeegee travel. The equipment also must accommodate the outer

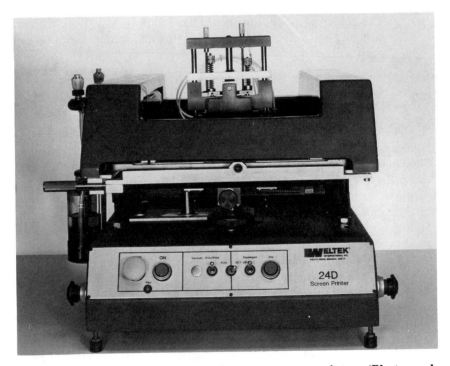

**Figure 9.7   Laboratory model solder paste screen printer. (Photograph courtesy of Weltek International.)**

frame of the screen/stencil. As a rough general rule, about 3 to 4 inches should be added on each side of the maximum print area to determine the maximum size of the machine. For example, to accommodate the printing on 7 inch × 12 inch and 9 inch × 9 inch boards in different orientations, a machine would have to be able to handle an outer frame size of 15 square inches.

The maximum number of boards that can be processed per hour is another consideration. It should be kept in mind that once the machine is set up, the screen printer will almost never be a bottleneck in an SMT manufacturing line. Throughput for a screen printer is a function of squeegee travel speed and board size; an average screen printer can print about 300 to 400 boards per hour regardless of the squeegee speed and method of operation [4].

Since the squeegee travels at about 1 to 6 inches per second (compared to 0.5 to 4 seconds per component during placement), hardly any pick-and-place equipment available can keep up with a screen printer unless there are very few components on the board. Unless the manufacturing line is totally automated, therefore, there is no need to go overboard on functions that increase the throughput of the screen printer.

The print mode of the equipment is another consideration. Printers are designed with a squeegee bar and a flood bar. In flood mode the flood bar spreads the paste, and in print mode the squeegee applies pressure, such that the actual printing takes place. When the flood bar is spreading the paste, the squeegee is up (riding), and when the squeegee is printing the flood bar is up (riding). Since the two bars are identical, if the design allows, their functions can alternate. Some equipment that can perform only in the flood/print mode, and some others offer both print/print and flood/print modes. Some of the older machines are not suitable for stencil application because they are not flexible and cannot provide the print/ print mode.

Only the print/print mode can be used with stencils, because the holes are completely etched and it is necessary to prevent the paste from oozing out of the holes and dripping on the substrate surface. For the screens, the flood/print mode is used. Naturally equipment is more desirable if it can function in various modes and can be used for both stencils and screens.

As mentioned earlier, print quality depends on the machine variables. Table 9.3 summarizes the equipment variables that determine print quality and repeatability [10]. These are generic variables and will apply to every screen printer. It is difficult to specify a specific control variable that will be applicable to different types of equipment. However, Table 9.4 suggests a range for printer settings that can be used as a general guideline for fine-tuning the settings of different machines [4].

**Table 9.3  Screen printing equipment variables [10]**

Structural
    Stiffness
    Parallelism
    Precision of mechanical parts (fit and movement)
Squeegee
    Velocity
    Acceleration
    Deceleration
    Pressure (down force)
    Stroke parallelism
    Parallelism in substrate
    Down stop
Modes of operation
    Contact/off-contact
    Bidirectional printing
    Flood Bar
    Multiple wet pass
Screen holder
    X axis
    Y axis
    Z axis
    Rotation (Theta)
    Peeloff
    Snapoff

## 9.3  SOLDER PASTE PRINTING PROCESSES

In printing solder paste, the substrate is placed on the work holder, held firmly by vacuum or mechanically, and aligned with the aid of tooling pins. Either a screen or stencil is used to apply solder paste. Figures 9.8 and 9.9 present cross-sectional and top views, respectively, that show the major differences between screens and stencils. The frame of the screens and stencils are similar; the differences lie in the construction of individual openings used for depositing the solder paste. Figure 9.9 shows schematically a screen and three different types of stencil. Figure 9.10 gives close-up photographic views corresponding to the diagrams of Figure 9.9. Figure 9.9 shows only one PLCC on a screen/stencil and Figure 9.10 shows openings for only a few components, in reality there are many components on the screens or stencils. The number of components on the screen/stencil

**Table 9.4   Recommended settings for screen printing equipment [4]**

| MACHINE VARIABLE | RECOMMENDED ADJUSTMENT |
|---|---|
| Squeegee pressure | |
|   Screens | 7–8 lb |
|   Stencils | 3–4 lb |
| Squeegee speed | |
|   Range | 0–20 inches/second |
|   Effective speed for SMT | 3–6 inches/second |
| Snap off distance | 0.020 to 0.030 inch |
| Angle of attack | 45° or 60° |
| Squeegee material | Polyurethane plastic with smooth edge, run flat and parallel to screen or metal mask |
| Leveling of screen/stencil (front/back and side to side) | Adjust so that it is parallel |
| Alignment | Repeated passes will bring the solder pad pattern of the screen in line with the solder pads on the substrate; align with micrometers or screws or visually |

matches the number of components on the board. Figure 9.11 is a photograph of a stencil that shows the openings for all the components on the board. Since each board is unique, stencils and screens are unique to particular boards and cannot be used for others. However, the frames can be used over and over for different board designs to save cost.

Typically, a screen will contain open wire mesh around which solder paste must flow to reach the substrate surface. A stencil opening is fully etched and does not obstruct paste flow. The screens or stencils have etched openings to match the land patterns on the substrate, where solder paste must be deposited for the electrical interconnections.

The screen/stencil is stretched in a metallic frame and is aligned above the substrate. The distance between the top of the board and the bottom of the stencil/screen is called the snapoff (see Figure 9.12). This gap or snapoff distance is a function of the equipment design and is about 0.020–0.030 inch.

Initially, as shown in Figure 9.13 the solder paste is manually placed on the stencil/screen with the print squeegee at one end of the stencil. During the printing process, the print squeegee presses down on the stencil

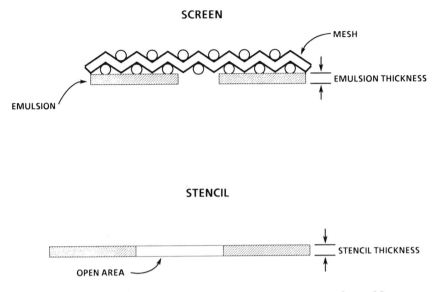

**Figure 9.8**  **Cross-sectional views of stencils and screens for solder paste printing.**

to the extent that the bottom of the stencil touches the top surface of the board. The solder paste is printed on the lands through the openings in the stencil/screen when the squeegee traverses the entire length of the image area etched in the metal mask. When screens are used, there is also the flood squeegee close to the print squeegee, whose function is to flood the screen with solder paste before the printing operation begins. The flood and print squeegees were discussed previously.

After the paste has been deposited, the screen peels away or snaps off immediately behind the squeegee and returns to its original position. (See Figure 9.12.) Thus snapoff distance and squeegee pressure are two important equipment-dependent variables for good quality printing. If there is no snapoff, the operation is called on-contact printing; it is used when an all-metal stencil is used. If there is a snapoff, the process is called off-contact printing. Off-contact printing is used with flexible metal masks and screens. As the squeegee traverses the board, this process continues, as illustrated in Figure 9.12, which shows the location of paste, screen, squeegee, substrate, and snapoff distance. The process sequence for applying solder paste is as follows:

- prepare solder paste.
- Set up and register screen/stencil.
- Print solder paste.
- Clean screen/stencil after printing.

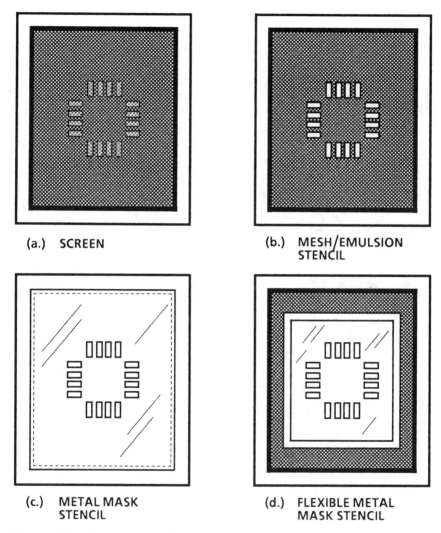

(a.)   SCREEN

(b.)   MESH/EMULSION
STENCIL

(c.)   METAL MASK
STENCIL

(d.)   FLEXIBLE METAL
MASK STENCIL

**Figure 9.9   Construction of screen (a), mesh/emulsion stencil (b), all-metal mask stencil (c), and flexible metal mask (d).**

When setting up the screen printer, the following guidelines should be observed for good registration of solder paste on the land patterns.

1.     No matter which method of application is used, be sure that the solder paste has been stored properly. A tightly sealed, unopened container of solder paste generally can be stored for 6 months at 4 to 29°C (40-84°F). If opened, store in a refrigerated environment.

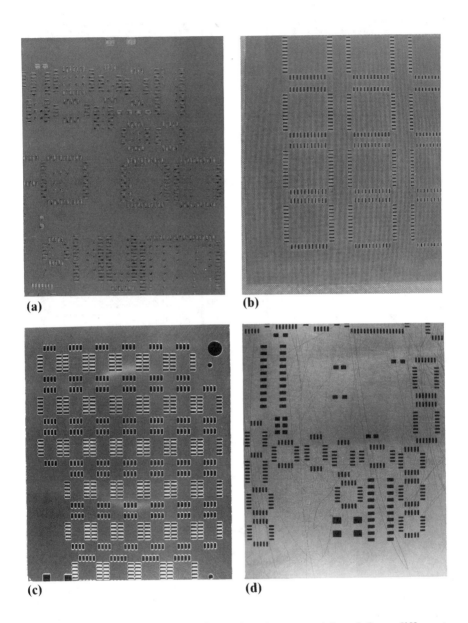

(a)　　　　　　　　　　(b)

(c)　　　　　　　　　　(d)

**Figure 9.10** **Close-up views of openings in screen (a) and three different stencil types: mesh/emulsion stencil (b), all-metal mask stencil (c), and flexible metal mask (d), which correspond to the diagrams of Figure 9.9. (Photographs courtesy of Intel Corporation.)**

**Figure 9.11** **A flexible metal mask stencil. (Photograph courtesy of Intel Corporation).**

2.  Allow a refrigerated container to reach room temperature before use.

3.  Check the solder paste for solder ball characteristics and viscosity.

4.  Place a clean sheet of Mylar over the board.

5.  Print solder paste onto the Mylar and inspect printed pattern for registration with the board, and for uniform coverage. If registration is poor, realign, using X-Y-Θ adjustments on the printer, and print onto a clean Mylar sheet again. If coverage is not uniform, examine squeegee, solder paste, and screen for evidence of areas with insufficient solder paste, and readjust the equipment controls to correct the deficiency.

6.  If the print is acceptable, remove the Mylar and proceed with

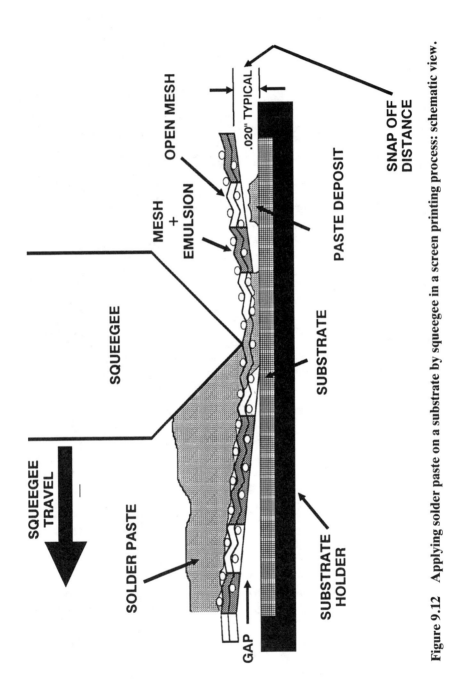

**Figure 9.12** Applying solder paste on a substrate by squeegee in a screen printing process: schematic view.

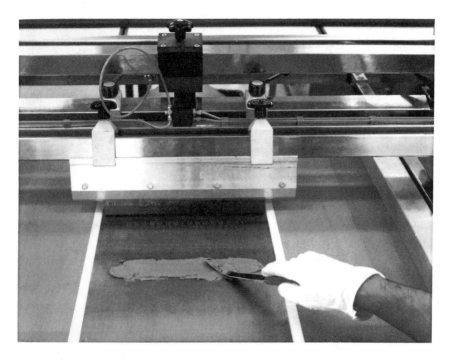

**Figure 9.13    Solder paste application before printing. (Photograph courtesy of Intel Corporation.)**

solder paste printing onto the board itself. Printing should provide adequate solder paste quantities for subsequent reflow processing. There should be no bridging between solder pads, and the solder paste must cover at least 75% of each solder pad, as shown in Figure 9.14. This item is discussed in more detail later in connection with print defects (Section 9.4).

7.    When all solder paste printing is complete, wash the screen with a solvent such as 1,1,1-trichloroethane. The cleaning of the screens and stencils is generally a manual operation done in a batch solvent cleaner. Now solvent cleaners especially designed for screens and stencils are commercially available to make the job easier for the operators.

8.    Discard any unused paste (unless the paste viscosity has not been modified more than twice by using thinners, in which case the unused portion may be recycled for reuse).

The discussion above applies to solder application by both screening and stenciling. Now let us focus in more detail on the important process variables in each method of solder paste application and their pros and cons.

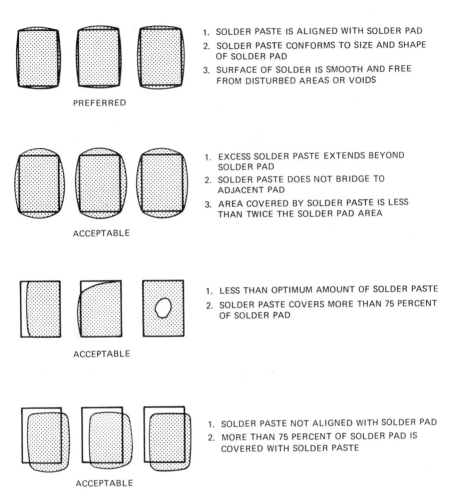

1. SOLDER PASTE IS ALIGNED WITH SOLDER PAD
2. SOLDER PASTE CONFORMS TO SIZE AND SHAPE OF SOLDER PAD
3. SURFACE OF SOLDER IS SMOOTH AND FREE FROM DISTURBED AREAS OR VOIDS

PREFERRED

1. EXCESS SOLDER PASTE EXTENDS BEYOND SOLDER PAD
2. SOLDER PASTE DOES NOT BRIDGE TO ADJACENT PAD
3. AREA COVERED BY SOLDER PASTE IS LESS THAN TWICE THE SOLDER PAD AREA

ACCEPTABLE

1. LESS THAN OPTIMUM AMOUNT OF SOLDER PASTE
2. SOLDER PASTE COVERS MORE THAN 75 PERCENT OF SOLDER PAD

ACCEPTABLE

1. SOLDER PASTE NOT ALIGNED WITH SOLDER PAD
2. MORE THAN 75 PERCENT OF SOLDER PAD IS COVERED WITH SOLDER PASTE

ACCEPTABLE

**Figure 9.14 Acceptable solder prints, showing coverage of at least 75% area on the lands.**

## 9.3.1 Screen Printing

The basic principle of applying paste is essentially the same whether screens or stencils are used. As illustrated in Figure 9.8, a screen made of a woven wire mesh is stretched over a frame with a glued-on photosensitive emulsion. The wire mesh supports the emulsion, which is etched where solder paste is needed. The screen is attached to an aluminum frame, which imparts rigidity and tension, hence a flat but flexible screen surface.

Screen construction was illustrated in Figure 9.9a. A screen contains open wire mesh around which solder paste must flow to reach the substrate surface. The wire mesh is usually made of monofilament polyester or stainless steel. Polyester materials are more resilient than stainless steel and last longer. The diameter of the wire and the size of the opening determine the mesh number, that is, the number of holes per square inch. The typical mesh number of screens used in surface mount assembly is 80. The diameter of the wire and the thickness of the emulsion primarily determine the solder paste thickness that can be deposited, but the snapoff distance (the distance between top of the substrate and the bottom of the screen) also plays a role.

Screen printing has been widely used in the hybrid industry for printing solder paste and in the printed circuit industry for printing wet film solder mask. It has been also adopted for surface mounting because of its many advantages: chiefly lower cost and faster delivery. For quick-turn around prototyping applications, screens are ideal. Screens have some disadvantages, too. For example, aligning them with the land patterns of the substrate is difficult and time-consuming.

The lands on the board cannot be readily seen through the wire mesh. The screens are restricted to off-contact printing because some snapoff distance is necessary to deposit a sufficient thickness of paste. The variables that control print quality and print thickness include snapoff height, squeegee pressure, wire diameter and mesh number, and emulsion thickness. When using screens, the following guidelines are helpful.

1.  Usually screen printing requires a slower speed and a greater snapoff distance than stencil work. Also, lower viscosity pastes should be used for easy printing.
2.  A feeler gauge or other suitable means can be used to adjust the Z-stops so that screen and work holder are parallel within 0.002 inch.
3.  The snapoff distance is adjusted to 0.030 ($+0.020/-0.005$) inch, while maintaining the screen-to-board parallelism within 0.002 inch.

## 9.3.2 Stencil Printing

Screens and stencils are functionally the same but there is a major difference, as mentioned earlier. In stencils, instead of screen mesh, the desired opening is chemically etched out either in metal sheets or in wire mesh covered with emulsion, such that the stencil opening does not obstruct solder paste flow. Thus, a stencil provides 100% open area for the paste to be printed through, whereas the screen provides only about 50% (the actual percentage depends on the mesh wire diameter and mesh number used for the screen).

Depending on the material in which the opening is etched, as shown in Figure 9.9b, 9.9c, and 9.9d (and the corresponding photographs in Figure 9.10b, 9.10c, and 9.10d), stencils can be classified in three major categories.

1. The selectively etched mesh/emulsion screen (Figures 9.9b and 9.10b)
2. The metal (brass, stainless steel, or beryllium-copper) mask (Figures 9.9c and 9.10c)
3. The flexible metal mask, a combination of 1 and 2 above (Figures 9.9d and 9.10d)

Each type of stencil has its advantages and disadvantages. The selectively etched mesh/emulsion type (Figure 9.9b) is made the same way as the screen, but the openings are completely etched out. The main disadvantage of this type of stencil is high cost. Also, because the wire mesh is etched, the screen is less stable.

All-metal mask stencils (Figure 9.9c) are not under constant tension and must be printed on contact only. All-metal mask stencils last longer but are more expensive and take longer for delivery. The premium for shorter deliveries can be high, and the vendor base is limited, especially for very large frame sizes.

The flexible metal mask (Figure 9.9d) combines the advantages of the all metal mask and the flexibility of screens. The metal portion of the mask is held in tension by the border of the flexible mesh to allow off-contact printing. This is the most widely used type of stencil.

The following guidelines should prove useful when using flexible metal mask stencils. First, polyester mesh is preferred over stainless steel mesh as a border for mounting because it allows better tension. However, polyester does not clean as well as a stainless mesh border. Second, the border area should be about 3 inches all around. A smaller border area will force the polyester fibers to stretch beyond the elastic point, and they will become damaged during printing [11]. Finally, the squeegee should stay on the metal mask; it should not be allowed to print across the polyester border.

## 9.3.3  Screen Printing Versus Stencil Printing

Stencils and screens are functionally the same, but they are used differently on the machine. For example, printing in both directions (print/print mode) is used for stencils to avoid the seeping of paste during the flood mode, discussed earlier (Section 9.2). The print/print mode requires the squeegee to jump over the paste at the end of each print stroke.

By contrast, in screen printing, both the flood and print modes are used. Also, stencils are easier than screens to align on the substrate surface because the openings provide clear visibility. Moreover, the holes generally do not become plugged, and it is easy to get consistently good print. Stencils are much easier to clean than screens. They are also sturdier than screens, hence last much longer. The viscosities required can be about the same as used for screening but should be at the higher end of the viscosity range.

Stencils can be used for selective printing but screens cannot. Selective printing can be very beneficial if two different paste thicknesses are required on the same board. For example, if fine pitch packages, larger tantalum capacitors, and other PLCCs and SOICs are to be mounted on the same board, each with a different paste thickness requirement, selective printing can be very desirable. This practice is specially becoming common with the increasing usage of fine pitch devices which require lower paste thickness.

Selective printing can be accomplished in two ways: by selectively etching emulsion/stencil or multilevel etching in metal stencils. Selective etching (i.e. some openings to be etched like screens and others like stencils), is performed in mesh/emulsion type of stencil (Figure 9.9b).

Multilevel etching, on the other hand, accomplishes the same objective but is performed in metal stencils (all metal mask, Figure 9.9c or flexible metal mask, Figure 9.9d). Multilevel etching in a metal stencil is illustrated in Figure 9.15. First, a larger area (larger than the land pattern area of the component to prevent solder skipping and damage to the squeegee) is etched to a desired thickness for the components that require lower paste thickness. For example, if the metal thickness for most of the board is 8 mils, and only 6 mils for a fine pitch package, the metal area for the fine pitch device and additional 100 mils around the package will be first etched down to 6 mils. Then the entire stencil is processed in normal fashion.

In screens, off-contact printing is always used to ensure deposition of an adequate thickness of paste and to prevent smearing during the flood mode (screens use only flood/print mode). In stencil printing, both- off-contact and on-contact printing are used; smearing is not a problem because the flood mode is not used for stencils, only the print/print mode.

Off-contact printing entails less risk of the substrate sticking to the screen. Since stencils are very easy to align with the substrates, hand printing (on contact printing) can be used for quick prototyping jobs. Hand printing is almost impossible with screens because they are so difficult to align.

The metal mask and flexible metal mask stencils are etched by chemical milling from both sides using two positive images. However sometimes in

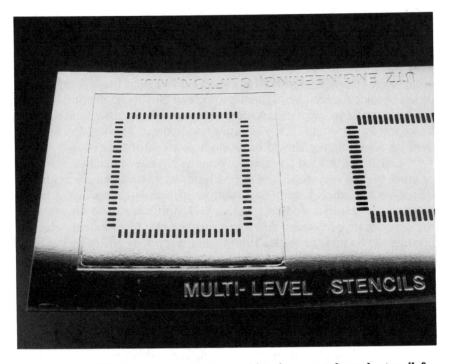

**Figure 9.15 Multilevel or selective etching in a metal mask stencil for depositing different solder paste thicknesses on the same board. (Photograph Courtesy of Intel Corporation.)**

the chemical milling process the openings are larger than desired, causing extra solder deposit. This problem is due to undercutting, which in turn is related to the etch factor. A function of the thickness of the metal being etched, the etch factor is around 25% of the metal thickness. By providing the artwork with reduced pad size at the photographic stage during chemical milling, the openings will be larger than shown in the artwork but will match the land patterns on the board. This process is called compensation, since it compensates for the etch factor. Some vendors can correct for the etch factor at additional cost by modifying the artwork. This is a more desirable option also when one is experimenting with different pad sizes and stencil thicknesses.

Table 9.5 lists the pros and cons of screens and stencils. Stencils have more advantages for solder paste printing. They are used where thicker deposition is required. A print at least 8 mils thick is required for acceptable solder joints in most cases, especially given the general need to compensate for some board warpage and lead coplanarity and still produce acceptable solder joints.

**Table 9.5    Advantages and disadvantages of screens and stencils**

| FLEXIBLE METAL MASKS (STENCILS) | SCREENS |
|---|---|
| ADVANTAGES | DISADVANTAGES |
| • Easy to set up | • Harder to set up |
| • On-contact and off-contact printing | • Off-contact printing only |
| • Can print by hand | • Cannot print by hand |
| • Wider usable viscosity range | • Narrow viscosity range |
| • More durable | • Less durable |
| • Do not plug | • Plug easily |
| • Easy to clean | • Hard to clean |
| • Allows multi-level printing | • Does not allow multi-leel printing |
| DISADVANTAGES | ADVANTAGES |
| • Higher cost | • Lower cost |
| • Longer turnaround time | • Quick turnaround |
| • Not suitable for quick turnaround prototype applications | • Suitable for quick turnaround prototype applications |

## 9.3.4    Dispensing

Solder paste is dispensed by being squeezed through the needle of a syringe. Since paste is dispensed on one land at a time, generally, in a manual operation, it is a much slower process than screen printing. However, dispensing can also be done automatically or semiautomatically, and in such cases the speed can be increased substantially. Most syringes rely on pneumatic dispensing systems, and the problems most commonly encountered entail clogging of needles. Syringing requires a paste of lower viscosity than is used for printing.

Dispensing is not considered for speed of paste application but for its versatility. For example, the dispensing method can be used to print different shapes (squares, rectangulars, circles, or dots) and different materials (paste, adhesive, or temporary solder mask), permitting the same tool to be used for several applications. Also, there is no vendor turnaround time involved, as is the case with screens and stencils.

Dispensing is appropriate when screening cannot be used—for example, in repair, since there are other components on the board. After the defective component has been removed, the solder paste is dispensed on the appropriate land pattern. The replacement component is then placed on the paste-covered land pattern and soldered.

Dispensing may also have special application for fine pitch packages that require lower paste thickness than other components, such as PLCCs and SOICs on the same board. However, as noted earlier, dispensing is not the only way to get different paste thicknesses on the same board. This can also be accomplished by selectively etching mesh/emulsion screens (Figure 9.9b) or multilevel metal stencils (Figure 9.15) as already discussed. Screening and stenciling methods will continue to be the dominant methods for paste application. However, dispensing is needed for specialized applications for which screening is not viable. Refer to chapter 8, Figure 8.12 for a semi-automated dispenser that can also be used for dispensing solder paste.

## 9.4 PASTE PRINTING DEFECTS

Solder paste print defects can be classified in various categories as shown in Figure 9.16, namely smearing, skipping, ragged edges, and mis-

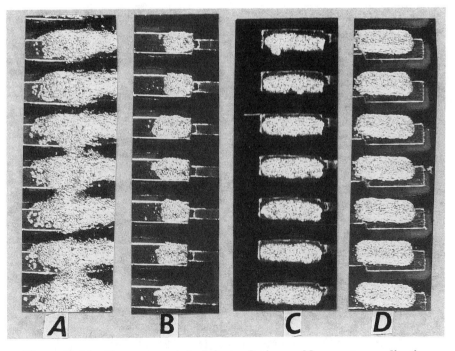

**Figure 9.16 Common print defects during solder paste application: (a) smeared print, (b) skipped print, (c) print with ragged edges, and (d) misaligned print. (Photograph courtesy of Intel Corporation.)**

alignment. Smearing occurs when solder paste is printed on areas on which it is not required (e.g., in between surface mount lands). Smearing is unacceptable because it causes solder bridges and/or solder balls. Figure 9.16a shows a smeared print. Skipping (Figure 9.16b) occurs when there is insufficient deposition of the solder paste on the surface mount lands. Skipping is unacceptable because it will cause insufficient solder joints. Figure 9.16c shows a print with ragged edges, defect that will result in nonuniform solder joints. A misaligned solder paste print (Figure 9.16d) occurs when the land patterns and the openings of screens or stencils are not properly aligned before printing.

The incidence of print defects can be minimized by continuously monitoring the paste quality and by properly controlling the equipment- and non-equipment-related variables discussed next. Also, as discussed earlier in connection with Figure 9.14, solder prints showing land coverage of less than 75% are not acceptable.

## 9.5 PASTE PRINTING VARIABLES

At the beginning of a typical production run of surface mount boards, various solder paste equipment and printing parameters are set to maximize the print quality. (Refer again to the equipment variables summarized in Table 9.3 and the recommended equipment settings shown in Table 9.4.)

Some of these parameters change over the course of a production run. For example, a stray vacuum leak will cause a drop in print quality. If vacuum is used for hold-down, the fixture plate should not allow any leakage. To consistently obtain prints of acceptable quality, various parameters must be maintained within a narrow range of values.

### 9.5.1 Solder Paste Viscosity

Earlier, we discussed viscosity and some variables that affect it, such as particle size, shear force, and temperature. We noted that the instrument commonly used for measuring viscosity is the Brookfield viscometer (Figure 9.5). In this section we discuss the impact of viscosity on the printability of paste. Solder paste viscosity is very critical for obtaining an acceptable print: smearing will occur if it is too low, and there will be skips if it is too high.

Depending on the method of application as measured at 25°C using a Brookfield viscometer (model RTV with a TF spindle at 5 rpm) solder paste should have the following values:

Dispensing: 350,000 ($\pm$50,000) cP
Screening: 500,000 ($\pm$50,000) cP
Stenciling: 550,000 to 750,000 cP

Typically, solder paste viscosity should be in the 550,000 to 750,000 centipoise range for obtaining acceptable prints using stencils. However, some people use as high as 850,000 cP (at 5 rpm, using a TF spindle) for best results for 10 mil thick flexible metal mask stencils [11].

We discussed the impact of temperature on viscosity in connection with Figure 9.4. Solder paste viscosity also increases with the number of boards printed. Generally, after only a 5°C change in temperature or three prints, the viscosity of the paste remaining on the stencil/screen may be degraded substantially enough to affect print quality. Thus to minimize viscosity changes induced by print quantity fresh paste may have to be added to the stencil/screen after three prints. Not all pastes degrade so quickly, however, and this may be one of the factors to consider when selecting a paste.

To minimize temperature-induced viscosity changes, the paste should be printed in a temperature-controlled area. In any event, since the viscosity of the paste will increase as the rheological modifiers it contains evaporate, the paste should be removed from the stencil/screen at the end of each production run.

## 9.5.2   Print Thickness

Print thickness determines the volume of solder in the joints. Too thick a print will result in excessive solder joints or even solder bridges, and too thin a print will result in insufficient solder joints. The thickness of the paste print is determined by the thickness of the metal mask of the stencil (or the emulsion thickness and mesh number for a screen). Generally, the print obtained from a stencil 8 to 10 mils thick will be thick enough to generate acceptable solder joints on 50 mil lead pitch packages using standard land dimensions.

For finer pitch (25 mil or less) packages, paste only about 6 mils thick is required. This requires use of multilevel or selectively etched screens/stencils discussed earlier (Section 9.3.3 and 9.3.4). The openings in the stencils for the fine pitch packages can also be modified (staggered partial printing on fine pitch lands) to allow the same paste thickness for all packages.

## 9.5.3   Squeegee Wear, Pressure, and Hardness

Squeegee wear, pressure, and hardness determines the quality of print and should be monitored carefully. For acceptable print quality, the squeegee edges should be sharp and straight. A low squeegee pressure results in skips and ragged edges. A high squeegee pressure or a soft squeegee will cause smeared prints and may even damage the squeegee and stencil or screen. Generally, a squeegee pressure of about 25 and 50 psi for stencils and screens, respectively, and squeegees with 60 to 80 durometer hardness (70 preferred for stencils) will give prints of acceptable quality. Higher viscosity pastes (above 550,000 centipoise) require harder squeegees (above 70 durometer).

### 9.5.4    Print Speed

Print speed is also critical for obtaining good print quality. A fast print speed will cause planing of the squeegee, resulting in skips. Though a slow speed is generally preferred for paste printing, too slow a speed will cause ragged edges or smearing in the print. A print speed in the range 0.5 to 1.0 inch/second yields prints with acceptable quality. With proper process control on certain screen printers, even speeds of 8 inches/second have been successfully used for stencils [11].

### 9.5.5    Mesh Tension

The mesh should be taut enough to maintain good registration between the land pattern on the substrate and stencil/screen opening, adequately releasing the paste onto the lands by snapping back the stencil/screen and not smearing of the print. However, the tension should not be high enough to elongate the screen/stencil mesh beyond its yield point.

### 9.5.6    Board Warpage

Excessive board warpage will result in skips on areas in which the top of the board does not touch the bottom of the stencil during printing. To minimize board warpage, the boards should be held flat, with vacuum pressure supplied from the underside of the base plate. Vacuum leaks should be minimized. Any stray vacuum leak will result in a poor quality print. Tool plates with tooling pins are also necessary to hold boards in place for accurate paste printing.

### 9.6    SUMMARY

The variables that control print quality are paste, screen printer, method of paste application, and printing process variables. Solder paste or cream is simply a suspension of fine solder particles in a flux vehicle. The composition of the particles can be tailored to produce a paste of the desired melting range. Additional metals can be added to change paste compositions for specialized applications. Particle size and shape, metal content, and flux type can be varied to produce pastes of varying viscosities.

Some of the major features to consider when selecting screen printers are the maximum board size handling capability, controls for accurate screen alignment and repeatability of print, and board hold-down mechanism. Considerable progress has been made in the application methods, such as screening and stenciling for pastes on substrates. Each method of application has its pros and cons, but stenciling appears to be more widely used for both low and high volume production. Although the supplier is essentially responsible for providing the desired solder paste and screens or stencils, the user must control process and equipment variables to achieve good print quality. Even the best of pastes, equipment, and application methods are not sufficient by themselves to ensure acceptable results.

## REFERENCES

1. IPC-SP-819. General Requirements and Test Methods for Electronic Grade Solder Paste, October 1988, IPC, Lincolnwood, IL.

2. Stein, A. M. "How to choose solder paste for surface mounting. Part 1." *Electronics,* June 1985, pp. 57–59.

3. Rooz-Kozel, B. Solder Paste, Surface Mount Technology. International Society for Hybrid Microelectronics Technical Monograph Series 6984-002, Reston, Va 1984.

4. Peterson, G. G. "A practical guide for specifying printing equipment for electronic applications. SMT-III-21, *Proceedings of the SMART III Conference* (IPC/EIA), January 1987.

5. Rosen, R. L. *Fundamental Properties of Polymeric Materials.* New York: Wiley, 1982, p. 202.

6. Mackay, C. A. "Solder creams and how to use them." *Electronic Packaging and Production,* February 1981, TR-1071.

7. Morris, J. "Solder paste tackiness measurement." *IPC Technical Review,* September–October 1987, pp. 18–24.

8. MacKay, C. A. "What you don't know about solder creams." *Circuits Manufacturing,* May 1987, pp. 43–52.

9. Heimsch, R. D. "Framework for the evaluation of precision screen printing equipment." *Hybrid Circuit Technology,* June 1986, pp. 19–24.

10. Atkinson, R. W. "An automated screen printer." *Circuits Manufacturing,* January 1983, pp. 26–29.

11. Borneman, J. D., and Rennaker, R. L. "Paste printing from a pro." *Circuits Manufacturing,* February 1987, p. 38.

# Chapter 10
# Metallurgy of Soldering and Solderability

## 10.0  INTRODUCTION

The solder used in electronic assembly serves to provide electrical and mechanical connections. In through-hole mount assemblies we had to worry primarily about obtaining sound electrical connections, because the plated through holes imparted sufficient mechanical strength. With the advent of SMT, the role of the surface mount solder joint has become very critical, because it must provide both mechanical and electrical connections. The solder joint strength is controlled by the land pattern design and a good metallurgical bond between component and board.

Surface mount land pattern design for providing adequate mechanical strength was covered in Chapter 6. In this chapter we concentrate on the metallurgical aspects of a reliable solder connection, as determined by the solder and the solderability of components and boards. A reliable solder connection must have a solderable surface to form a good metallurgical bond between the solder and the components being joined. An under-standing of metallurgical bonding entails knowledge of phase diagrams, the concept of leaching, surface finish, wetting, oxidation of metallic surfaces.

Metallurgical phase diagrams are used to display the solubility limits of one metal into another and the melting temperatures for metals and their alloys, for a better understanding of intermetallic bonds. Phase diagrams can also be used for better understanding of leaching or dissolution phenomenan (of one metal into another). Finally, to produce solder joints in a cost-effective way, we need to know about the methods, requirements, and economics of solderability testing.

The focus of this chapter is on the practical aspects of metallurgical issues related to solderability of surface mount assemblies. The subject of soldering for surface mount assemblies is covered in Chapter 12. For an expanded coverage of basic metallurgical issues in electronics, refer to Wassink [1] and Manko [2].

## 10.1 PHASE DIAGRAMS

Phase diagrams are used in the electronics industry to determine the existence of intermetallic compounds and the melting points of metals and their alloys. These charts can indicate the solubility of one metal into other at different temperatures, as well as various phases of alloy compositions. Phase diagrams are graphic representations of the equilibrium interrelationships between elements and compounds. A multitude of phase diagrams have been established.

The most commonly used phase diagram in the electronics industry is the tin-lead phase diagram, because tin-lead is used for soldering, but nickel-tin and silver-tin phase diagrams are also used. Some phase diagram systems are as follows.

1. *Unary systems*, which contain only a single element or compound—for example, tin, copper, water. Unary systems show only the melting and boiling points of elements or compounds as a material transforms from the solid to the liquid phase.
2. *Binary systems*, which contain two elements or compounds and are the most widely used. Examples of binary systems in the electronics industry include lead-tin, gold-tin, silver-tin, and nickel-tin.
3. *Ternary systems*, which contain three elements or compounds, such as tin-lead-silver, and used in some solder pastes. These complex systems are more difficult to interpret.

The construction of a phase diagram, by plotting melting point versus compositions for the alloy under evaluation, is a laborious and time-consuming process. Fortunately, others have done the difficult job and we can enjoy the benefits as long as we can learn how to "read" a phase diagram. We begin by stating that as the composition varies, the microstructure and melting points vary. For most alloys, there is a range of melting points. This will become more clear if we examine Figure 10.1, a phase diagram for a binary eutectic system consisting of elements A and B. The Y axis represents temperature and the X axis represents composition. At point A, the composition is as follows: element A 100%, element B 0%. At point B, we have 100% element B and 0% element A.

The solid lines in a binary phase diagram represent an equilibrium between the two phases of the system, namely, the liquid phase and the solid phase. The line above which the alloy is entirely liquid is called the liquidus line, and the line below which the alloy is entirely solid is called the solidus line. The solid phase, lying below the solidus line, has two

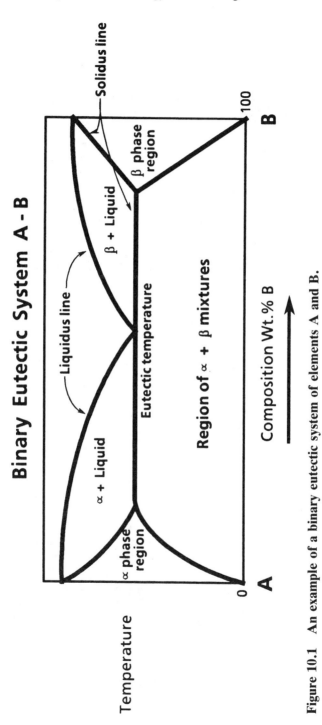

**Figure 10.1** An example of a binary eutectic system of elements A and B.

different phases: α and β. The α phase region shows the maximum solubility of element B in A at different temperature, similarly, the maximum solubility of element A in element B at different temperatures is shown in the β region.

One should not confuse the individual elements with the different phases. For example, the α phase is not element A saturated with element B. It is very different from both elements A and B and has entirely different properties and microstructure. By the same token, the β phase also differs from each of its components. Between the two regions, however, is a mixture of the α and β phases.

The eutectic temperature is the lowest temperature at which melting is observed for the eutectic composition (α and β phases) of elements A and B. At temperatures above the eutectic melting point, the eutectic composition is liquid but any other composition is pasty. To the left of the eutectic composition (above the eutectic temperature), the pasty region contains the liquid and α phase; to the right, the pasty region contains the β and liquid phases.

At temperatures below the eutectic temperature, the alloy compositions (of elements A and B) to the left and right of the eutectic composition are called the hypoeutectic AB and hypereutectic AB compositions, respectively. For determining the percentages of different microstructural constituents (such as liquid, α, β, element A, element B) at different locations on the phase diagram, a rule known as the lever rule is used. These and other details of phase diagrams are not necessary for the points I want to make in this chapter. Readers interested in such details should refer to Wassink [1] and Manko [2].

If we say that element A is lead (Pb) and element B is tin (Sn), we have an actual Pb-Sn phase diagram, as shown in Figure 10.2, which can be used to obtain some useful information on solder alloys. The terminology and the principles are the same as those just discussed. Thus from Figure 10.2 we see that the α phase is the solid solution of tin in lead. The maximum solubility of tin in lead (or lead in tin) occurs at the eutectic temperature. The eutectic solder composition of 63% tin and 37% lead, which melts at 183°C (361°F), has a lower melting point than either lead (327°C) or tin (231°C), as shown in Figure 10.2.

The Sn-Pb phase diagram also shows the limits of variations in compositions that can be easily soldered. The horizontal line superimposed as the reflow soldering temperature clearly shows that the solder composition should not vary too much from the eutectic composition. Otherwise the reflow soldering temperatures, normally around 220°C, would have to be considerably higher.

The rule of thumb is that the soldering temperature should be about 30 to 50°C above the melting point of the alloy being used for soldering.

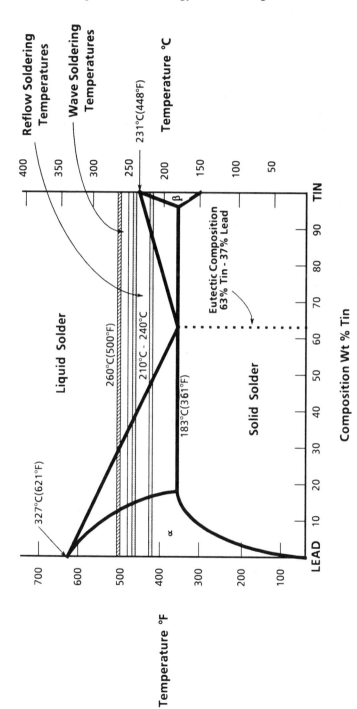

**Figure 10.2 The tin-lead phase diagram.**

Figure 10.2 shows the reflow (vapor phase) soldering temperature to be 215°C. For infrared, the temperature may be higher (230°C). Wave soldering temperatures are even higher (generally 240-260°C) because the solder not only should be molten but must have very low viscosity and suitable hydraulic flow characteristics to permit recirculation in the solder pot. Of course soldering can be accomplished at lower temperatures (above the melting point of the alloy), but, depending on temperature, it may take longer. Thus a lower temperature may be technically acceptable but not practical in a manufacturing environment.

The noneutectic solder has a pasty range as defined by the temperature range between liquidus and solidus lines. The pasty range is also illustrated by two peak temperatures shown in Figure 10.3 by differential scanning calorimetry (discussed in Chapter 8). DSC is a quick way to determine the solidus and the liquidus points of a noneutectic solder. The first peak in Figure 10.3 (179°C) indicates the solidus line. According to the Sn-Pb phase diagram (Figure 10.2), however, this temperature should be 183°C for any Sn-Pb composition. What is the explanation for the difference?

It should be noted that phase diagrams represent equilibrium conditions and assume that sufficient time has been allowed for the alloy to melt completely. A DSC curve, on the other hand, shows only the first occurrence of melting; thus at this point the entire sample may not have melted. The second peak in Figure 10.3 indicates the peak liquidus temperature. Between these two peaks, depending on its composition, the solder exists both in the liquid phase and in either the $\alpha$ or $\beta$ phase.

Can we distinguish between $\alpha$ and $\beta$ phases from a DSC curve? Not in general. In the case of Figure 10.3, however, it is possible to tell that the region between the two peaks is the $\beta$ phase. The clue is provided by the maximum peak temperature (276°C) in Figure 10.3. If we refer back to the Sn-Pb phase diagram in Figure 10.2, we see that the maximum temperature for the $\beta$ phase is only 231°C.

However, if the second peak temperature shown by Figure 10.3 were less than 231°C, the region between the two peaks could be either the $\alpha$ or the $\beta$ phase and the liquid phase. Thus a DSC curve may not give the exact composition of an alloy, but it will provide an approximate range. The DSC curve for pure elements or a eutectic compositions will show only one peak, since the melting point is fixed in these cases.

Although eutectic Sn-Pb solder is the most widely used composition, a tin-lead-silver ternary system consisting of 62% Sn, 36% Pb, and 2% Ag, is also used. This is a eutectic composition of the ternary system with a melting point of 179°C (354°F). The equilibrium microstructure at room temperature is $\alpha$, $\beta$, and silver-tin intermetallic ($Ag_3Sn$). The chemical formulas (e.g., $Ag_3Sn$) of compounds are not derived from the phase diagram but require other analytical tools. Refer to Wassink [1] for details.

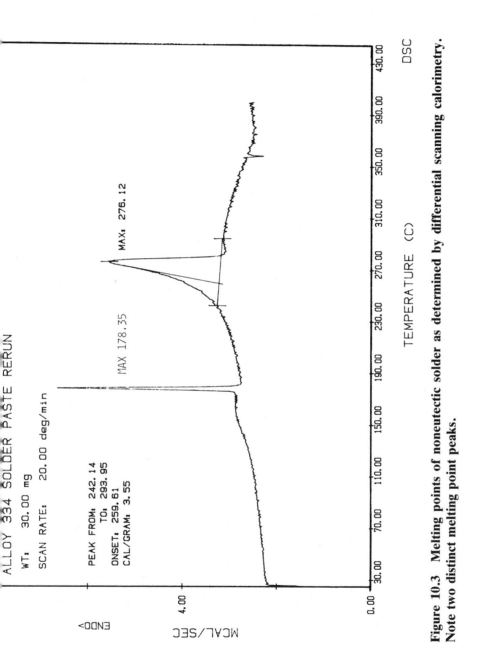

**Figure 10.3** **Melting points of noneutectic solder as determined by differential scanning calorimetry. Note two distinct melting point peaks.**

## 10.2 METALLIZATION LEACHING IN PASSIVE SURFACE MOUNT COMPONENTS

Leaching is a kinetic metallurgical phenomenon in which one element dissolves into another at soldering temperatures. In surface mounting, the term "leaching" designates the dissolution of silver metallization in surface mount resistors and capacitors during soldering.

Why is leaching a problem? To provide electrical connections to the internal electrodes or terminations, surface mount ceramic chip capacitors and resistors require an adhesion layer of precious metal such as gold or silver. A silver adhesion layer is more common. After a silver adhesion layer has been applied by a thick film process (printing of paste containing silver), the terminations are either dipped in solder or plated with tin-lead solder. During soldering, the silver is prone to dissolution in the tin. This causes the underlying ceramic surface to be exposed and results in a poor solder fillet or no fillet at all.

How is leaching avoided? By using a barrier layer under the terminations or by using silver in the solder paste. Let us discuss the barrier layer approach first. The rate of dissolution of metals in other metals is determined by the metal used for plating component metallization. For example, gold, silver, palladium, copper, nickel, and platinum dissolve at decreasing rates in solder, as indicated in Figure 10.4.

Bader conducted an experiment by dissolving pure metal wires, 20 mils in diameter, in solder; different temperatures and times were used, and the rate of dissolution was measured [3]. Figure 10.4, adapted from this study, shows relative rates of dissolution only; for actual rates of dissolution for various metals in molten solder at various temperatures, refer to Table 10.1.

From Figure 10.4 we see that platinum and nickel have the lowest dissolution rates in solder and gold and silver have the highest rates. If platinum or nickel is used between silver and solder on surface mount resistors and capacitors, it will act as a barrier and prevent the dissolution of the silver metallization. Because of its high cost, platinum is not used as a barrier layer. Nickel is much cheaper and nearly as effective and therefore is commonly used.

When nickel is used to prevent leaching, the nickel-tin intermetallic compound formed are $Ni_3Sn_4$ (liquid-solid reaction) and $Ni_3Sn$ (solid state reaction). The formation of $Ni_3Sn$ intermetallic compounds continues even when the solder has solidified. Once the solder joint has solidified, however, the reaction rate is much slower.

How thick should the nickel barrier layer be? This question can be answered by knowing the growth rate of nickel in tin as a function of time

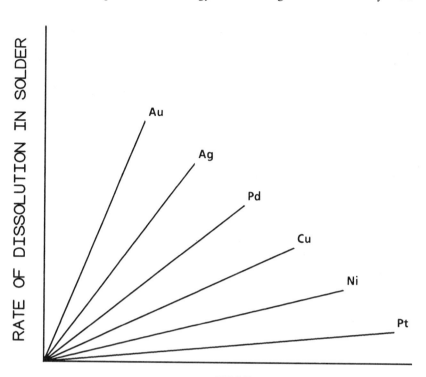

**Figure 10.4    The rate of dissolution of various metals in solder [3].**

and temperature. This means that we need to take into account not only the soldering time and temperatures but also the time and temperatures in service. As noted above, solid state reactions (diffusion) cause leaching to continue even after solidification, although at a much slower rate.

It is known that the rate of dissolution of silver at 525°F (near soldering temperatures) is more than 50 times the rate for nickel. It is difficult to precisely compare dissolution rates for the silver and nickel, since Table 10.1 does not show data at common temperatures for all metals. This is because the rate of nickel dissolution at temperatures under 700°F is so low that it is hard to measure. Even at 700°F, nickel dissolves at only 1.7 microinches/second, whereas the dissolution rate for silver at a lower temperature (600°F) is 190.7 microinches/second. The dissolution rates are higher at higher temperatures, where reaction rates increase.

Surface mount components spend a maximum of 5 seconds in the wave and about 30 seconds or more in reflow. Since the dissolution rate of nickel in solder is almost negligible at soldering temperatures and is 1.7 microinches/second at 700°C, a nickel barrier layer of about 25 to 30 mi-

**Table 10.1 Dissolution rates of various metals in solder at different temperatures [3].**

| METAL | TEMPERATURE (°F) | RADIAL DISSOLUTION RATE (microinches/ second) |
|---|---|---|
| Gold | 390 | 35.0 |
| | 420 | 68.5 |
| | 450 | 117.9 |
| | 486 | 167.5 |
| Silver | 390 | 20.7 |
| | 450 | 43.6 |
| | 525 | 97.0 |
| | 600 | 190.7 |
| Palladium | 450 | 1.4 |
| | 525 | 6.2 |
| | 600 | 3.6 |
| | 700 | 14.0 |
| | 800 | 40.5 |
| | 900 | 103.1 |
| Platinum | 700 | 0.83 |
| | 800 | 5.0 |
| | 900 | 16.9 |
| Copper | 450 | 4.1 |
| | 525 | 7.0 |
| | 600 | 21.2 |
| | 700 | 61.5 |
| | 800 | 143.0 |
| | 900 | 248.0 |
| Nickel | 700 | 1.7 |
| | 800 | 4.4 |
| | 900 | 11.3 |

croinches should be sufficient to prevent leaching of silver during wave and reflow soldering.

A nickel barrier of about 50 microinches is commonly used on all chip components to prevent leaching during soldering. This provides some margin for storage and for two soldering steps on double-sided assemblies. It should be noted that once an assembly has been soldered, the leaching phenomenon becomes less important.

If a component lacks a nickel barrier, a second approach to prevent leaching is to use a solder paste containing 2% silver, in 36% lead- 62% tin base. Leaching still occurs, but the sacrificial source for silver is the solder paste on the land, not the essential silver adhesion layer under the component termination. This approach can be used only for reflow soldering (vapor phase or infrared), not for wave soldering.

Leaching is a more serious problem in wave soldering than in reflow soldering because of the higher temperatures in wave soldering. Can users add silver to the wave solder pot to prevent leaching? No. This would be wasteful if the same wave soldering machine were also used for through-hole assemblies, which are not affected by leaching, and even for dedicated wave soldering equipment for SMT, the price is too high.

The most appropriate and commonly used approach is to buy surface mount resistors and capacitors with about 50 microinches of nickel barrier from a component supplier. This allows flexibility in using the same components for either wave or reflow soldering. The nickel barrier approach is becoming an industry standard and should be named on the procurement specifications.

# 10.3   SOLDER ALLOYS AND THEIR PROPERTIES

Solder is most often thought of as an alloy of tin and lead. This is partly true. Tin and lead alloys are the most common, but many other compositions of solder are gaining acceptance. Two important selection criteria for solder are melting point and strength; these properties for familiar alloys are shown in Table 10.2. Refer to Chapter 9 for commonly used solder compositions in a solder paste.

The properties of solder vary widely depending on composition and microstructure. Fo all the properties, fatigue resistance is most important. Relative fatigue resistance values for commonly used solders are shown in Table 10.3 as developed by DeVore [4], who used eutectic solder as the control. DeVore also characterized the fatigue resistance of various solders based on their microstructure, as shown in Table 10.4 [4].

It should be noted from Table 10.3 that the fatigue resistance of 63/37 tin-lead solder is three times less than the rarely used 96/4 tin-silver solder. Table 10.2 shows very little difference between the tensile strengths of these solders: 7700 and 8900 psi, respectively. This shows that strength plays the least important role in the selection of solder, since the commonly used solders (63/37 and 60/40 tin-lead solders in Table 10.3 and 10.4) have the poorest fatigue resistance.

For a given solder, the fatigue resistance of a solder joint depends on shear strain, frequency, and temperature, as shown in Figure 10.5. This

**Table 10.2 Properties of electronic grade solders.**

| ALLOY DESIGNATION | | MELTING POINT OR RANGE (°C) | MECHANICAL STRENGTH (PSI) | |
|---|---|---|---|---|
| | | | IN TENSION | IN SHEAR |
| 62 Sn/36 Pb/2 Ag | Sn 62 | 179 | | |
| 63 Sn/37 Pb | Sn 63 | | | |
| | ASTM 63A | 183 | 7700 | 5400 |
| | ASTM 63B | | | |
| 60 Sn/40 Pb | Sn 60 | | | |
| | ASTM 60A | 183-189 | 7600 | 5600 |
| | ASTM 60B | | | |
| 96.5 Sn/3.5 Ag | Sn 96 | | | |
| 96 Sn/4 Ag | ASTM 96 TS | 221 | 8900 | 4600 |
| 96 Sn/5 Ag | | 221-245 | 8000 | |
| 95 Sn/5 Sb | Sb 5 | | | |
| | ASTM 95 TA | 232-240 | 5300 | 5360 |
| 10 Sn/88 Pb/2 Ag | Sn 10 | 268-299 | | |
| 10 Sn/90 Pb | ASTM 10B | 268-302 | | |
| 5 Sn/92.5 Pb/2.5 Ag | | 280 | | |
| 5 Sn/93.5 Pb/1.5 Ag | | 296-301 | | |
| 5 Sn/92.5 Pb/2.5 Ag | | 300 | | |
| 5 Sn/95 Pb | Sn 5 | | | |
| | ASTM 5A | 301-314 | | |
| | ASTM 5B | | | |
| 1 Sn/97.5 Pb/ | Ag 1.5 | | | |
| 1.5 Ag | ASTM 1.5 S | 309 | | |
| 60 Sn/40 Pb | | 174-185 | 4150 | |
| 50 Sn/50 Pb | | 180-209 | 4670 | 2680 |

**Table 10.3 Fatigue Resistance of Solders (Life relative to Sn 63) [4].**

| ALLOY | | FATIGUE LIFE (relative to Sn 63) |
|---|---|---|
| 96/4 | Sn/Ag | 3.03 |
| 95/5 | Sn/Sb | 2.10 |
| 62/36/2 | Sn/Pb/Ag | 1.14 |
| 63/37 | Sn/Pb | 1.00 |
| 59/37/4 | Sn/Pb/Sb[a] | 0.66 |
| 40/60 | Sn/Pb | 0.61 |

[a]Approximate composition.

**Table 10.4    Fatigue characteristics of solders [4].**

| PERFORM-ANCE | COMPOSITION (wt %) | MICROSTRUC-TURE |
|---|---|---|
| Poor | 63 Sn, 37 Pb | 2 phase, eutectic |
| | 60 Sn, 40 Pb | 2 phase, near eutectic |
| | 62 Sn, 36 Pb, 2 Ag | 2 phase, Ag hardened B |
| | 65 Sn, 35 In | 2 phase, variable compo- |
| | 42 Sn, 58 Bi | sition |
| | 50 Sn, 50 In | 2 phase, eutectic |
| | 50 Sn, 50 Pb | 2 phase, variable compo- |
| | | sition |
| | | 2 phase, eutectic plus |
| | | proeutectic |
| Fair | 15 Pb, 80 In, 5 Ag | — |
| | 99 In, 1 Cu | 2 phase, eutectic |
| | 90 Sn, 10 Pb | 2 phase, mostly proeu- |
| | 99.25 Sn, 0.75 Cu | tectic B |
| | | 2 phase, eutectic |
| Good | 99 Sn, 1 Sb | 1 phase, solid solution |
| | 50 Pb, 50 In | 1 phase, solid solution |
| | 100 Sn | 1 phase |
| Excellent | 48 Sn, 52 In | 2 phase, eutectic |
| | 96 Sn, 4 Ag | 1 phase, particle hard- |
| | 95 Sn, 5 Sb | ened |
| | 80 Au, 20 Sn | 1 phase, particle hard- |
| | | ened |
| | | 2 phase eutectic AuSn + |
| | | hexagonal close packed |
| | | solid state |

graph, developed by Wild [5], shows that in eutectic solder, the fatigue resistance (mean cycles to failure) at a given shear strain is much lower at lower frequencies (cycles/day) at a given shear strain. The difference can be 2 orders of magnitude when the frequency is changed from 1 cycle per day to 300 cycles per day. Slower frequencies mean longer dwell times at test extremes (thermal or mechanical cycles), leading to complete stress relaxation, loss of ductility, and premature failure. See Chapter 5 (Section 5.8.1).

The most commonly used solder is eutectic solder with 63% tin and 37% lead; its melting point is 183°C. Since tin is more expensive than lead,

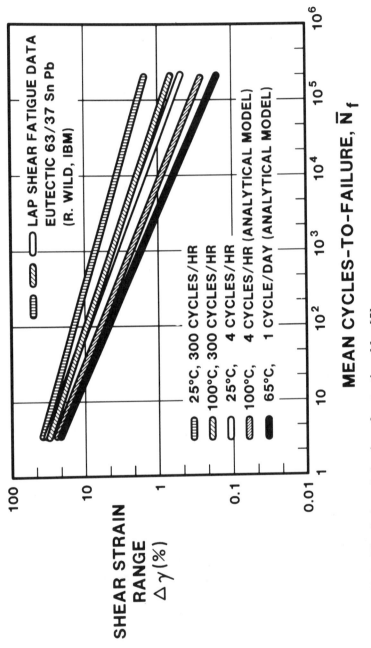

**Figure 10.5** The fatigue behavior of eutectic solder [5].

the eutectic (63 SN/37 Pb) solder costs slightly more than the 60 Sn/40 Pb solder. For wave soldering applications, the two forms are equally good. For surface mounting, eutectic solder is more common because of its slightly lower melting point. However, the noneutectic solder alloys with a pasty range are more effective than a eutectic solder in minimizing solder wicking (the traveling of solder up the lead, leaving with no solder on the pad and causing an open solder joint) in reflow soldering.

The noneutectic solders melt over a range of temperatures, and some of the solder paste remains on the pad by the time the lands have reached the melting point of the paste. Thus the incidence of wicking may be reduced considerably by using a noneutectic solder alloy. Wicking is process dependent, being more common in vapor phase than in infrared reflow. (See Chapter 12 for details on the mechanism of solder wicking.) Wicking of solder joints also can be minimized by adding 2% silver to produce a 62 Sn/36 Pb/2 Ag solder, but it is slightly more expensive. Selecting devices that meet coplanarity requirements and using infrared reflow soldering are other measures to reduce solder wicking (see Chapter 12 for details).

Solders can be divided in two broad categories: high temperature solders (having a melting point above 183°C) and low temperature solders (solder alloys melting below 183°C) [1]. Use of solders of different melting points is also helpful in step soldering, in which one side of the board is soldered with high temperature solder first, then the secondary side is soldered with low temperature solder, thus preventing the melting of the high temperature solder.

Step soldering may be used if it is necessary to keep larger devices from falling off the board. Surface tension may not be sufficient to hold onto PLCCs larger than 84 pins [6]. Step soldering is rarely used, however. Instead, boards requiring very large devices should be designed such that these devices are on the side that is reflowed last.

Special flux is necessary when low temperature solders are used, because the standard flux may not be active at lower temperatures. The flux used with low temperature solders must have an activation temperature lower than is needed for eutectic tin-lead solders. Another problem associated with low temperature solders is the reduction in wetting properties caused by the lower fluidity at subeutectic temperatures.

For low temperature applications, solders containing the semiprecious metal indium (In) are gaining some acceptance. One indium alloy being used by some companies because it provide better rework/repair characteristics has 52% In and 48% Sn. Since the alloy melts at 244°F (117°C), rework can be performed at lower temperatures many times without causing thermal damage. If the printed circuit board must be plated with gold as an antioxidant, indium solder can be used to prevent leaching of gold [7].

Despite their better properties, indium-based alloys or other specialty solders are rarely used because of electromigration problems. In addition, the specialty solders cost more. Also, the specialty solders, including indium-based solders, are not included in federal specification QQ-S-571 or in the American Society for Testing and Materials standard ASTM B32, which control the compositions of solder used in the United States.

In a discussion about solder composition, it is necessary to briefly consider the effects of solder contamination. The acceptable contamination levels (weight percent) in 63 Sn/37 Pb solder in QQ-S-571 are: aluminum, 0.0006; antimony, 0.5 maximum; arsenic, 0.03; bismuth, 0.25; cadmium, 0.005; copper, 0.30; gold, 0.2; iron, 0.02; silver, 0.1; zinc, 0.005; and nickel, 0.01. The tin composition can vary from 59.5 to 63.5%, and the remainder, about 37% is lead.

Contamination levels of aluminum above 0.004%, (the specification allows 0.006%) and copper above 0.2% (specification allows 0.3%) cause gritty-looking solder joints. Entrapped dross in solder may also cause gritty solder joints. Noneutectic solder, because of its pasty melting characteristics, also contributes to a gritty solder joint appearance. These factors may act alone or together to cause grittiness.

Gritty appearance by itself is not a matter of concern. It is to be expected with noneutectic solders. If it is due to dross or copper or aluminum, however, the solder joint properties may be degraded. Gritty solder joints are not common in reflow soldering. They are generally seen in wave soldering because the solder is prone to contamination as a result of repeated use of the same solder.

Gritty solder joints should not be confused with dull solder joints. Reflow solder joints are often dull, not bright and shiny like the wave soldered joints, because they cool at a slower rate than the wave soldered joints. A dull appearance is simply a reflow process characteristic and is not a matter for concern. See the thermal profiles of different processes in Chapter 12.

# 10.4  SOLDERABILITY

The commonly accepted definition of "solderability" is the ability to solder easily. This applies to components and boards alike. If both do not solder easily, rework costs increase and the reliability of the product is compromised. There is complete agreement in the industry that to provide good solder joints, boards and components should have good solderability.

Components require good solder joints because they are subjected to mechanical and thermal stresses during handling, shipping, and field service. This is especially critical in surface mount, since the solder joint provides both electrical and mechanical connections. The clinched leads and the plated through holes, which impart added mechanical strength in conventional assemblies, are absent in surface mount.

Surface mounting has made the solderability of components and board even more critical by changing the soldering parameters (time and temperature of soldering) and using less active fluxes in the solder paste. In the rest of this chapter we discuss various aspects of such solderability issues as mechanisms of solderability, solderability test methods and requirements, and approaches to ensure solderability. There are three basic mechanisms of solderability: wetting, nonwetting, and dewetting. We shall deal with these in turn.

## 10.4.1    Wetting

Wetting of the surfaces to be soldered is important for the formation of a metallurgical bond. Just because a metallic surface is covered with solder does not mean that it is wetted. If there is an oxide layer between the solder and the metallic surface, the solder will not adhere to the surface and the joint will fail in service when subjected to mechanical stress.

In the soldering of electronic assemblies, the oxide is removed by the application of flux in the presence of heat, which not only activates the flux but also reduces the surface tension of the solder. The higher the temperature, the lower the surface tension of the solder, and the better the wetting action. In the absence of good wetting action, either a nonwetted or a dewetted solder joint will result.

In electronic assemblies the wetting action takes place during soldering, in the interface between the copper and the tin. The reaction products of this wetting action are the copper-tin intermetallic compounds $Cu_3Sn$ (near copper) and $Cu_6Sn_5$ (near solder) [8]. The reaction must take place rapidly to ensure the formation of a good solder joint.

Degrees of wetting also can be characterized by the angle of contact of the solder on the base metal. A smaller contact angle between two surfaces is an indication of better attraction, hence a stronger bond. On the other hand, a larger contact angle, as between water and a waxed surface, indicates very superficial attraction between the surfaces.

The concept of the contact angle is illustrated in Figure 10.6, where the circles surround solder joints and the vertical angle in each case can be assumed to be the lead. In the preferred condition, the solder contact angle is less than 90°; an angle of 90° exactly is acceptable, but a solder contact angle that exceeds 90° should be classified as an unacceptable, nonwetted solder joint.

Figure 10.7 schematically illustrates good solder wetting, with the char-

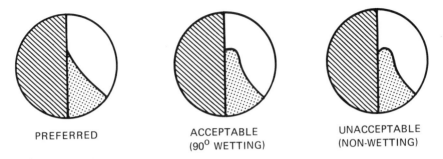

PREFERRED       ACCEPTABLE       UNACCEPTABLE
                  (90° WETTING)       (NON-WETTING)

**Figure 10.6 Acceptable and unacceptable contact angles for solder joints.**

acteristic concave fillet; the contact or wetting angle is less than 90° with good feathering at the solder lead or termination interface. Also, in good wetting a uniform and smooth layer of solder stays on the surface. In figure 10.8, a photomicrograph of a wetted solder joint, the surface is smooth and an intermetallic layer clearly indicates that a good metallurgical bond

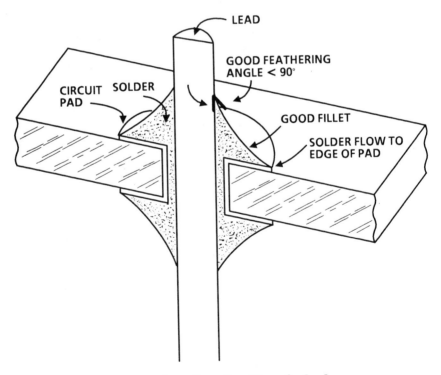

**Figure 10.7 Schematic view of good wetting of a lead.**

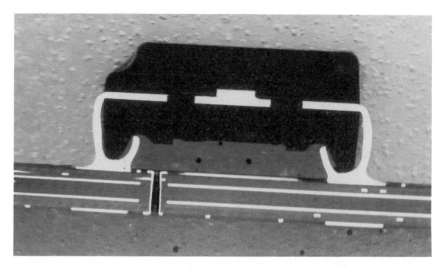

**Figure 10.8   A photograph of a well wetted surface. Clearly visible are good inner and out solder fillets along with the cross-section of the substrate and via hole (bottom of the photograph. The photograph also shows the silicon die over the lead frame paddle or die pad (top center of the photo). (Photograph courtesy of Intel Corporation).**

has been formed. There is generally no problem in identifying a wetted surface in concave fillets, but an excess of solder can make it impossible to tell whether a surface is well or poorly wetted.

## 10.4.2   Nonwetting

In nonwetting the solder simply does not adhere to the surface. This condition, the opposite of wetting, is caused by a physical barrier (intermetallic or oxide) between the solder and the base metal. The predominant cause is the total consumption of the layer of solder, which turns into an intermetallic compound of the base metal. In other words, if the solder coating is thin, the intermetallic layer will be exposed. Intermetallic compounds have poor wetting characteristics. The second mechanism of nonwetting is the formation of a thick oxidized layer that may be too thick for the flux to remove within the time allotted for soldering. Oxidation of the surface is generally not the primary cause, since oxidation is a very slow process. Nonwetting is a very easy condition to identify.

### 10.4.3 Dewetting

A dewetted surface is characterized by irregular mounds of solder caused by pulling back of solder, as if it had changed its mind about wetting. Dewetting is difficult to identify, since solder can be wetted at some locations while the base metal is exposed in other places. It may appear as a partial wetting condition in which some regions of the surface are wetted and others are nonwetted.

All dewetting is the result of gas evolution during soldering. The source of gas can be either thermal breakdown or hydration of organic materials [8] generated by the processes used for the plating of leads and boards. The gas thus released leads to the formation of voids, which often remain dewetted, interfering with the integrity of the joint.

In severe cases, if enough organics are deposited during the plating process, the extent of gas evolution can be sufficient to passivate the surface, and a nonwetting condition will result. The problem is not corrected by increasing the time and temperature of soldering or by using a more active flux. Instead the problem may be compounded, with evolution of more gas, hence more dewetting and nonwetting. Pretinning such a lead may not prevent dewetting either. The problem can be corrected only at the source, where parts are plated. An example of a dewetted condition is shown schematically in Figure 10.9. Figure 10.10 shows several solderability defects, including a dewetted surface with gas voids. Voids and pinholes in a joint can lead to stress concentration and eventual fatigue failures. Good substrate and component solderability not only provide good solder fillets, but also minimize voids and pinhole problems and reduce the level of overall touch-up and repair. Poor solderability is not the only cause of voids and pinholes; it can, however, compound these problems.

## 10.5 VARIOUS APPROACHES FOR ENSURING SOLDERABILITY

Good soldering results, free of dewetted or nonwetted joints, call for the use of substrates and components with good solderability. Achieving good solderability, however, is far from a simple technical matter; rather, it entails a complex mixture of technical (test method and requirement), quality and reliability (rework/repair, flux), financial (inventory, pretinning,) standardization, and vendor/management (getting the users and suppliers to agree to a common requirement) issues. When parts are purchased from distributors, dealing with the original manufacturer becomes

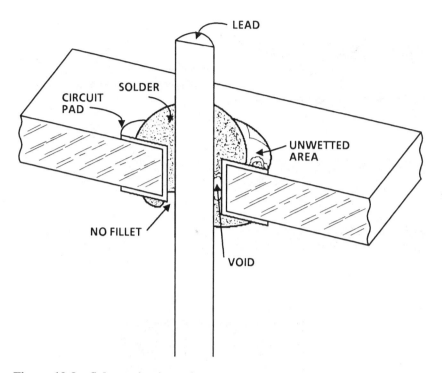

**Figure 10.9    Schematic view of a dewetted solder joint.**

even more complicated. For these reasons, solderability is a controversial and complex subject.

The solderability problem has been with us since the advent of the electronics industry and it will remain at least for the foreseeable future. However, when assemblies are soldered using components of unknown or variable solderability, it is almost impossible to obtain good manufacturing yield and consistent quality. This leads to excessive touch-up, which not only costs money but may degrade product reliability. Not surprisingly, there are many different approaches to solving solderability problems, each with its advantages and disadvantages. No matter how the problem is approached, parts must have good solderability at the time of soldering. Let us briefly discuss some of the approaches commonly being practiced in the industry.

Some companies focus on the storage environment and the time of storage. A part that is solderable when shipped by the vendor may lose its solderability after a certain period of time because of the slow growth of intermetallics at storage temperatures. Thus some users maintain a minimum inventory and do not store components for long periods. Others

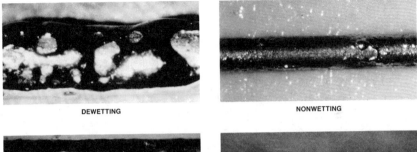

DEWETTING                    NONWETTING

PIN HOLES                    POROSITY

FOREIGN MATERIAL

**Figure 10.10    Types of solderability defect. (IPC-S-805, Solderability Test Specification.)**

may store components for more than a year. The storage conditions selected vary as well, from high humidity and temperature to a very controlled environment.

The current approaches for solving the solderability problems fall into six major categories: (a) using a very active flux, (b) doing in-house solderability testing and either pretinning the failed parts or returning the shipment to the vendor (i.e., forcing the vendor to supply solderable parts), (c) pretinning all incoming parts, (d) controlling solderability at the vendor's facility to ensure receipt of solderable parts, (e) establishing a usable industry-wide, consistent solderability specification, and (f) requiring a surface finish on leads and substrates to ensure solderability. These approaches are interdependent and cannot be adopted in isolation.

In this section we discuss the first four approaches; these offer only short-term fixes, but they are the most popular. Approaches (e) and (f), which are discussed in Sections 10.7 and 10.9, respectively, have the most direct effect on solderability and truly provide a long-term solution.

Theoretically speaking, almost any metallic surface can be soldered if we use a flux that is active enough at the highest possible temperature and for the longest possible time. From a practical standpoint, we do not have the luxury of selecting these variables because higher soldering temperatures and longer times can increase the intermetallic thickness so greatly that the resulting joint will be unsolderable, brittle, or both. Also, as mentioned in connection with dewetting, if the underplating of the components and the board contains organics, a higher temperature and a longer time simply compound the problem of gas evolution [8].

A very active flux is inherently able to cause failures by corrosion, as discussed in Chapter 13. Thus if the components and boards cannot be soldered within the time, temperature, and flux requirements of the process being used, active fluxes are out of the question. Under schedule pressure, for example, many users either pretin parts or use poorly solderable parts and then touch up defective joints. One may not agree with this approach, but it is not uncommon. Other users rely on suppliers to provide solderable parts and do not do any in-house solderability testing or pretinning.

The companies that simply pretin all incoming parts consider this to be the cost-effective alternative. They tend to be users of parts of many different types, none in any significant quantity, and the solderability of components and boards varies from excellent to nonexistent. A fairly common practice among firms supplying the military sector is to pretin all incoming parts, store for a predetermined time, and then tin again, generally after 2 years of storage. This approach is limited in that it is difficult to retin an unsolderable part. For through-hole parts either the pretin or the touch-up option can be used, depending on the company's philosophy, as well as cost, volume, and schedule constraints.

For surface mounting, however, it is not practical to pretin components, especially small devices. For example, how can one pretin resistors and capacitors and package them back on tape and reel? For all practical purposes, controlling solderability at the vendor's facility is the only cost-effective solution for surface mounting.

Now there is a trend in the industry to require vendors not only to meet the solderability requirement but also to process parts with a specific finish to ensure solderability. This would seem to be a simple solution but it is not. The suppliers are not unwilling to perform the solderability tests or to provide the lead finishes requested. The problem lies in the considerable variations in requirements for both solderability test methods and lead finishes that exist in the industry.

## 10.6 SOLDERABILITY TEST METHODS AND REQUIREMENTS

The lack of consistent solderability requirements and test parameters in the industry compounds the solderability problem. A typical supplier is faced with a multitude of idiosyncratic requirements from different users. Two major tests for solderability are used by the industry: the dip and look test, which is the most common, and the wetting balance test.

The specifications that govern these tests and another one, called the globule test, are shown in Table 10.5. These specifications are established by industry standards organizations such as the Institute of Interconnecting and Packaging Electronic Circuits, the Electronic Industries Association,

**Table 10.5  Different solderability test specifications for Dip and Look and Wetting Balance.**

| TYPE OF TEST | SPECIFICATIONS |
| --- | --- |
| Dip and look for board and leads | Mil-Std-202F, method 208F |
| | Mil-Std-750, method 2006.3 |
| | Mil-Std-883C, method 2022 |
| | ● JEDEC-Std 22A |
| | - B-102 part 1 |
| | - B-102 part 2 |
| | ● ASTM-B67B-86 |
| | ● RS-186-9F, method 9 |
| | ● IEC Technical Committee 50 |
| | - Test and guide |
| | ● IEC-68-2-20 |
| Dip and look for leads | ANSI/IPC - 805 |
| Dip and look for boards | ANSI/IPC - 804 |
| Wetting balance for leads[a] | Mil-Std-883C, method 2022 |
| | IEC-68-2-20 test T |
| Globule test for leads[a] | IEC-68-2-20 |

[a]Can also be used for flux selection/characterization if lead solderability is kept constant.

the International Electrotechnical Commission (IEC), and the military. These specifications are inconsistent (see Table 10.6) and cause considerable confusion in the industry. We discuss the inconsistencies and the current trend toward their resolution in the next section. But let us first present the pros and cons of these two test methods.

In the dip and look test, which also applies to the solderability testing of substrates, the component or substrate is dipped in a pot of molten solder and solderability is evaluated by determining the percentage of wetted area. This method is subjective, and it is generally difficult to estimate the nonwetted/dewetted area, especially if solderability is marginal. Estimating excessive or minimal dewetting is generally not a problem. Refer to Figure 10.11 as a guide for estimation of the percentage of defective area in the field of view. One of the IPC-S-805 or IPC-S-804 specifications shown in Table 10.5 will give details on equipment and test procedures. For the component or the board to be considered acceptable, no nonwetted or dewetted area should be present. Since, however, the suppliers would never agree to a requirement for 100% wetting, the industry has settled on an allowable maximum of 5 to 10% noncoverage, depending on the specification.

In the second commonly used solderability test, the wetting balance method, the lead is dipped in a solder pot and the time required to reach the maximum wetting force is measured. For the details of the test equipment and procedures, refer to reference IEC document-68-2-20 or Mil-Std-883C, method 2022.

The wetting balance method, by virtue of its interaction with the liquid solder, allows a quantitative measurement of the forces acting on the sample. The forces acting vertically on the specimen are measured as a function of time. The shorter the time to reach the maximum wetting force, the more solderable the component. Figure 10.12 gives a typical curve generated by the wetting balance test method. As indicated in the insert to Figure 10.12, when the specimen is first lowered in the solder bath, the upward or buoyancy force is recorded as the negative wetting force on the specimen. The specimen heats up in the bath and the flux becomes active, and then the wetting begins. The slope of the curve moving up, as the wetting continues, is very important. The time it takes to reach the maximum force or maximum wetting is defined as the wetting time.

The third type of test, the globule test, is widely used in Europe. Instead of dipping into a solder pot, a solder globule of prespecified size is placed on a pretinned aluminum block and heated. The globule test generally is used to measure the wetting time for through-hole mount components but can also be used for surface mount devices.

The wetting balance test is difficult to use for small chip resistors and capacitors because they do not have the long terminations necessary for

**Table 10.6 Inconsistent solderability test parameters or various dip and look test specifications in Table 10.5.**

## TEST PARAMETER REQUIREMENTS IN VARIOUS TEST SPECIFICATIONS

| | |
|---|---|
| Aging | None |
| | 1 hour |
| | 4 hours |
| | 16 hours |
| | 16-24 hours |
| | 24 hours |
| Flux type | R |
| | R, 25% solids |
| | R, 35% solids |
| | R, 40% solids |
| | RMA |
| Solder composition | Sn 60/Pb 40 |
| | Sn 63/Pb 37 |
| | Sn 62/Pb 36/Ag 2 |
| Solder pot temperature ($\pm 5°C$) | 215°C |
| | 230°C |
| | 235°C |
| | 245°C |
| | 260°C |
| Dwell time in solder | 5 ± 0.05 seconds |
| | 3 ± 0.03 seconds |
| | 4.5 ± 0.5 seconds |
| | 20 ± 2 seconds (vapor phase) |
| Magnification for inspection of noncoverage | 2 to 10X |
| | 8 to 12X |
| | 10 to 20X |
| | 15 to 25X |
| | Not specified |
| Accept/reject criteria | 95% wetted, 5% dewetted/ nonwetted, nonlocalized |
| | 90% wetted, 10% total nonwetted/dewetted, 5% spot |

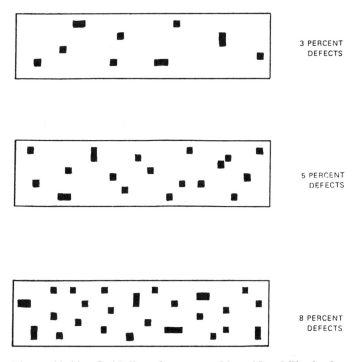

**Figure 10.11    Guidelines for acceptable solderability in the dip test. (IPC-S-805, Solderability Test Specifications.)**

this test. Fixturing of surface mount components is a problem for the globule test as well, however.

The basic difference between the wetting balance and the dip and look tests is that the latter is subjective whereas the former is considered to be quantitative. Actually, however, the time requirement for definition of solderability in the wetting balance test is subjective even though the test method itself is quantitative. For example, as shown in Figure 10.12, the military requires that two-thirds of the maximum wetting force be reached in less than 1 second. Also, the wetting force must reach zero within 0.6 second. The military requirements, which are not consistent with those established by other specifications such as IPC-S-805 and IEC-68-2-20, are considered too unrealistic and have not been accepted by the civilian industry.

Despite some problems, many companies use wetting balance testing for process control, to determine whether to accept parts. Parts that fail the wetting balance test are rechecked with the dip and look method before being rejected.

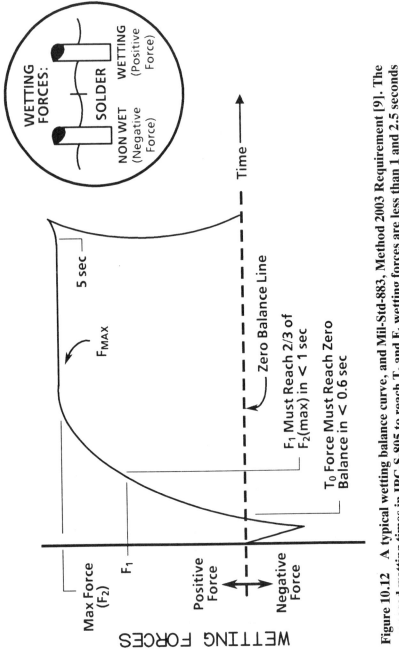

**Figure 10.12** A typical wetting balance curve, and Mil-Std-883, Method 2003 Requirement [9]. The proposed wetting times in IPC-S-805 to reach $T_0$ and $F_1$ wetting forces are less than 1 and 2.5 seconds respectively.

Different countries have adopted different tests for solderability, although both types discussed here should be used. For example, there has been something of a divergence in the approach to solderability testing found in Europe and in the United States. Whereas Europeans decided to concentrate on the wetting balance test because of its quantitative features, American companies continued to improve upon the dip test because it is simple and cheap.

Now, however, the wetting balance test is being adopted slowly by the United States and has been included in Mil-Std-883 and IPC-S-805. But as mentioned earlier, the definition of what is solderable as determined by the wetting balance method remains to be resolved. Where do we go from here? In the next section we discuss some recommendations for ensuring solderability.

## 10.7   RECOMMENDATIONS FOR SOLDERABILITY TEST METHODS AND REQUIREMENTS

The confusion in the industry with regard to adopting an adequate solderability test method has two main causes. First, there is a lack of consistent solderability requirements; lacking, too, are hard data correlating test methods used and solder defects reported. The various specifications, too, are inconsistent, as illustrated in Table 10.6.

In addition to inconsistencies, some of the current test parameters are really not applicable for surface mounting. For example, although reflow soldering of surface-mounted components is performed at 220°C, most of the dip and look solderability tests are performed at 245 to 260°C. There is also inconsistency between the flux used for the test and the flux used in soldering processes. For example, a mildly activated RMA flux in the solder paste is used for reflow soldering, but in wave soldering it is common to use activated rosin. In commercial applications even organic acid fluxes are used for wave soldering.

Because of the process differences discussed above, the test methods and requirements used for through hole may not be applicable for surface mounting. Thus it might be argued that in determining solderability, the components should be soldered using a production process to simulate production conditions. Such an approach is not workable, however. Given the many reflow processes used by different companies, no supplier could possibly duplicate all the different manufacturing setups used by its customers. This absence of replicability could cause a real problem in the correlation of solderability results obtained by the users and suppliers. It would be hard, for example, to determine whether a solderability problem was caused by component, board, paste, or reflow equipment. Using a manufacturing process may be a good development tool for users; it could not feasibly be adopted as an industry standard.

A basic understanding of the issues should help in deciding the right approach for solving solderability problems for surface mounting. If the industry were to agree on a consistent set of standards, many solderability problems would disappear, since the suppliers would know the requirements and could gear up to meet them, as they do now for large customers.

Our discussion in this section is focused on dip and look test parameters applicable for surface mounting because the wetting balance and the globule tests are awkward for surface mounting. The important test parameters in dip and look solderability tests are aging requirement, solder temperature, dwell time, and type of flux (see Table 10.6). In discussing these test parameters, I first highlight the inconsistencies as summarized in Table 10.6 and then discuss the test parameters that should be used for dip and look tests.

Except for the solder bath and dwell time, all the test parameters discussed below apply to both through-hole mount and surface mount components and boards. We should keep in mind that a mixed assembly board contains both through-hole and surface mount components. Thus a discussion on solderability of surface mount must include through-hole components. Let us start with the aging or pretreatment requirement.

Specified aging requirements vary from none to 24 hours and everything in between. There is some question about whether aging is needed because it may not simulate actual storage conditions. Also, if a company is run with very little inventory, aging tests may distort the results by causing failures that might not have occurred in manufacturing. There is some substance to these arguments. Nevertheless the aging portion of solderability testing provides conservative results and should be required.

Aging alone is rarely the cause of poor solderability. As far as the duration of aging is concerned, after 8 hours of aging, no appreciable deterioration in surface condition takes place (major deterioration stops after 16 hours) [9]. The industry is basically debating between 8 and 16 hours of aging. The IPC is leaning toward 16 hours of aging, but the military has changed its requirement from 16 to 8 hours. However, the 16 hour aging requirement may be more realistic [9].

The container for steam aging should be noncorrosive. Distilled water should be used, and the part should be suspended in a manner that prevents condensation on the leads. Boards that are to be aged for solderability testing should be baked before dipping in the solder bath to ensure that moisture on the board surface does not influence the solderability results.

The solder temperature and dwell times are the most controversial parameters. Solder bath temperatures vary greatly. For example, different specifications call for temperatures of 215, 230, 235, 245, and 260°C. There is confusion in establishing a temperature requirement because a given component (e.g., chip resistor or capacitor) may be used in both wave and reflow soldering. Similarly, a printed circuit board for a mixed assembly will go through both reflow and wave soldering. Which temperature should be used?

Should one be conservative and use the lower temperature? If this route is selected, for surface mount components, the industry should use a 215°C solder bath with 10 seconds dwell time. The solder bath temperature is the only important solderability test parameter that needs to be different for surface mount to more realistically simulate the reflow soldering environment, and with surface mount, 215°C is more realistic than the higher temperatures.

All the military specifications, however, have changed to 245°C for through-hole mount components with a 5 second dwell time. The IPC is leaning toward 235°C with a 10 second dwell time for all components and boards. The dwell time is not as important a variable as temperature. If a part is indeed solderable, it can meet the short dwell time (5 seconds) requirement.

The type of flux used for solderability testing are R (rosin) and RMA (rosin mildly activated) with solids contents of 25, 35, and 40%. It is important to use the R flux to get conservative test results. The solids content is not very important as long as a fixed percentage is used, for better correlation of results. In this regard, 30 to 40% solids content is widely used and should be acceptable. Only the areas under test should be coated (by dipping the part in flux for a few seconds. The excess flux should be removed and the part then dipped into the solder pot within a minute after fluxing.

Different specifications state various solder composition requirements: Sn 60/Pb 40, Sn 63/Pb 37, and Sn 62/Pb 36/Ag 2. To keep the test results consistent, a tin content of 59 to 64% should be chosen, since this range is commonly used. If the composition is other than eutectic (Sn 63/Pb 37), the melting point of the solder will be slightly higher, hence the results slightly conservative. Solder composition may make a slight difference in results for marginally solderable components, but it is not a major issue.

The type of solder pot may affect the results significantly, however, and a static solder pot is recommended. The solderability results obtained with a static pot are more conservative than those with an agitated solder pot.

After testing, the flux residue should be cleaned off using a suitable solvent such as isopropyl alcohol. The magnification to be used for inspection after testing can vary from 2X to 25X. Generally a 10X magnification is sufficient for initial evaluation, but higher magnification (30-40X) can be used for verification of dewetted or nonwetted areas.

Finally, what is the acceptable solderability requirement? An argument can be made that since the user cannot ship a few percent defective assemblies, no dewetting or nonwetting should be allowed. All the components and boards should be perfect and should show 100% wetted surface. First of all, this is not realistic, and second, it is not necessary.

We are talking about a test to simulate good solderability, not an actual soldering process. After all, in a manufacturing environment, the components and boards are not aged. Also, some activators are used in the flux. It has been found that a part that shows less than 5% dewetted or nonwetted area during testing will provide an easily solderable surface. The next section discusses the impact on solderability of the surface finish of substrates and leads.

## 10.8 EFFECT OF SUBSTRATE SURFACE FINISH ON SOLDERABILITY

In soldering surface mount or through-hole mount devices, the surface finishes of substrates and leads or terminations play the key role in determining the solderability of components and substrates. Generally, a high percentage of solderability problems are caused by boards. One problem area is found near the "knee" of plated through holes; weak knees are due to exposed intermetallics and their passivation.

For boards, the critical surface finish requirement is to ensure sufficient solder thickness. The general recommendation in the industry about maintaining at least 0.3 mil (300 microinches) is essential for long-term solderability, as found by Mather [10]. Solder coatings thicker than 0.3 mil have good long-term solderability and meet solderability requirements without difficulty even after aging for 24 hours.

On the other hand, solder coatings less than 0.3 mil thick dewet after as little as 4 hours of aging. Similar conclusions were reached by Mackay [11], who found that 5 μm (200 microinches) of tin coating and 7.5 μm (300 microinches) of 60 Sn/40 Pb provide good soldering even after storage in some artificial aggressive environments.

Surprisingly different results were found in an industry-wide round robin test program conducted by the IPC [12]. These boards retained solderability after 2 years of storage even when they had very little solder coating. Thus solder coating thickness alone is not important. The quality of coating is also very important. The plating must be free of deleterious organic materials under the plated surfaces, and the plating or coating thickness should be uniform.

The hot air leveling process, in which the copper surface is dipped in hot solder and the excess blown away with hot air knife, shows great variation in solder thickness. A 80 to 1400 microinches (1.4 mils) is common in hot air leveling. There are problems with both thin and thick surface coatings, especially in surface mounting.

Thinner surfaces (< 100 microinches) are not desirable because they are almost entirely consumed during reflow, especially if the assemblies are reflow soldered twice (primary and then secondary side, such as double-sided boards). Since intermetallic compounds passivate easily if exposed, they have poor solderability.

Thicker coatings, especially with wide variations within the same land pattern of the same package, are not good because they compound the coplanarity problem for active devices. Thicker coatings are good for solderability, but anything exceeding 300 microinches is really not necessary. Hot air leveling has many advantages, such as acting as an instant solderability tester in a production process (solder will not adhere to a poor surface in HAL). Uneven solder thickness is its main disadvantage.

Difficulties occur in plating, too, if the current density is not properly controlled. Uneven composition may result, for example, if the tin composition is decreased as the current density is lowered. This would cause formation of a lead-rich solder, which is less solderable. Even with good control, however, weak "knees" are encountered in the plated through holes.

## 10.9 EFFECT OF COMPONENT LEAD OR TERMINATION FINISH ON SOLDERABILITY

Surface finishes are applied to component leads and terminations by plating or by hot dipping. The requirements placed on active devices for military applications favor hot dipping, but for commercial applications a plating process may be better; there are pros and cons for both approaches. For all applications an appropriate finish is necessary to achieve a good solder joint.

The hot dipping process provides sufficient solder thickness but the coating is generally uneven, displaying a "dog bone" appearance with thicker terminations and a narrower body. Such a shape tends to cause problems during placement because it is difficult to accurately pick and place the "dog bone" parts.

Plating of terminations is preferred to achieve symmetrical shapes and uniform coating. However, if a plated surface is reflowed before shipment, most of the tin may be consumed in intermetallic formation, causing solderability problems. Hence plated terminations should be reflowed only at the time of soldering.

Let us briefly describe the process sequence in lead finishing for commercial and military applications and then discuss the pros and cons of plating and hot dipping processes. (For active devices, both processes are used, the plating and dipping requirement being related primarily to the burn-in sequence.)

In commercial applications, the lead frames are molded in the plastic compound, and then the leads are plated. The packages may or may not be burned in before shipment. The burn-in time is relatively short (48-68 hours). Both plating and hot dipping processes are used.

Hot dipped leads show solderability problems at the edges and on the bottom. Not only does the solder tend to pull away from the edges of the leads, resulting in insufficient solder thickness at the edges and too much solder in the center, but at the lower bends of the "J" in PLCCs, the solder thickness is minimal. Hot dipping may also compound the coplanarity problem.

Plating, on the other hand, produces uniform thickness, but the coating tends to be somewhat porous. Porosity can be reduced by controlling the process to produce a fine grain size. Also, the lead ends of plated leads, such as SOIC and gull wing fine pitch packages, have exposed copper because the tie bars holding the leads are cut after plating. The exposed copper may cause poor/dewetted outer fillets. They should be acceptable as long as a good inner fillet at the heels of the leads exists. In general, however, the solderability of plated leads has been found to be superior to that of hot dipped leads by Belani et al. [13]. Their dip and look and wetting balance results appear in Tables 10.7 and 10.8, respectively. The tables also show the test conditions.

Table 10.7 shows that no solderability test failures were found in the plated leads, but the hot dipped leads showed 14 to 27% failures for zero and 24 hour aging, respectively. These dip and look results were confirmed by the wetting balance tests shown in Table 10.8: the wetting times for the hot dipped leads are considerably higher.

For military applications, the sequence is different. After die attach and glass sealing, the leads are tin plated. After that they are burned in for about 168 hours. They are hot dipped either before or after burn-in.

**Table 10.7  Solderability test result (Failures in sample size of 22) comparison of plated and hot dipped copper leads using Dip and Look tests [13].**

| STEAM AGE TIME (HOURS) | SOLDER PLATED | SOLDER DIPPED |
|---|---|---|
| 0 | 0/22 | 3/22 |
| 24 | 0.22 | 6/22 |

**Table 10.8**  Solderability test result comparison of plated and hot
dipped copper leads using Wetting balance tests [13].

| STEAM AGE TIME IN (HOURS) | WETTING TIME (SECONDS) | |
|---|---|---|
| | SOLDER PLATED | SOLDER DIPPED |
| 1 | 0.30 | 0.30 |
| 16 | 0.38 | 0.89 |
| 24 | 0.41 | 0.63 |

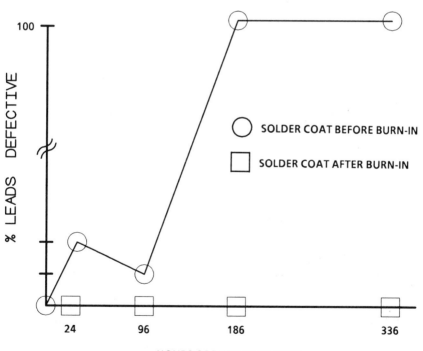

**Figure 10.13**   Impact of burn-in on solderability of alloy 42 leads before
and after hot solder dip. (Illustration courtesy of Intel Corporation.)

Serious problems can be caused if leads are burned in after hot dipping, however. As shown in Figure 10.13, burn-in has a serious impact on lead solderability. The alloy 42 (58% copper, 42% nickel) leads, which were burned in after hot dip coating failed the solderability test miserably; but when tin-plated leads were burned in and then hot dipped in solder, no solderability defects were found. The hot dipping process which produces thinner coating at the edges of the leads and causes poor solderability. The burn-in process uses burn-in sockets, which can also cause mechanical damage to the lead finish. Because such mechanical and metallurgical phenomena degrade the solderability of an otherwise good lead finish, it is better to perform the final lead finishing operation after burn-in.

In many cases, plating and hot dipping complement each other. In applications that call for a long burn, for example, both plating (before burn-in) and hot dipping (after burn-in) may be required to provide the needed solderability. This is true for both commercial and military applications.

How thick should the plated or hot dipped surface be? The requirements for leads and for boards should not differ. Thus whether the leads are plated or hot dipped, 300 microinches of solder is preferred. If the base material is pure copper (as opposed alloy 42, for example) a thickness less than 150 microinches is insufficient because most of the solder is consumed as a copper-tin intermetallic compound.

In addition to solder thickness, the grain size influences solderability [14]. The difference in wetting times for fine and coarse grained microstructures becomes even more significant with the steam aging of leads before solderability testing. Grain size variation is seen essentially in the plating process with finer grain sizes resulting in denser finishes which are less susceptible to oxidation.

## 10.10  SUMMARY

The metallurgy of soldering is the key to the understanding of the mechanisms of intermetallic formation and solderability testing. A basic knowledge of phase diagrams of the metals used in soldering permits one to determine the solubility of one metal in another. The solubility, in turn, controls such important variables as alloy composition, melting point for soldering, and occurrence of leaching.

There is a general agreement that for successful soldering results, components and boards must have good solderability, but the industry has not yet resolved the issues of methods and requirements for either the dip test or the wetting balance test.

Component suppliers must conform to many different specifications related to solderability requirements. One of the underlying causes of confusion has been the insonsistent requirements found in different specifications. In addition to pointing out the inconsistencies in this chapter, we made recommendations for achieving good solderability.

The lead finish not only plays the key role in solderability but also has an impact on component placement in passive devices and coplanarity in active devices. To control the solderability of printed circuit boards or components, one must have adequate solder thickness and finer grain size.

## REFERENCES

1. Wassink, R. J. *Soldering in Electronics*. Ayr, Scotland: Electrochemical Publications, 1984.

2. Manko, H. H. *Solders and Soldering*, 2nd ed. New York: McGraw-Hill, 1984.

3. Bader, W. G. "Dissolution of Au, Ag, Pd, Pt, Cu, and Ni in molten tin-lead solder." *Welding Research Supplement*, December 1969, pp. 551-557.

4. DeVore, J. "Fatigue resistance of solder." *NEPCON Proceedings*, February 1982, pp. 409-414.

5. Wild, R. N. Some Fatigue Properties of Solders and Solder Joints. IBM Technical Report 73Z000421, January 1973.

6. Hart, C. "Double-sided attach using reflow solder." *NEPCON Proceedings*, February 1985, pp. 46-53.

7. Keeler, R. "Specialty solders outshine tin/lead in problem areas". *EP & P*, July 1987, pp. 45-47.

8. DeVore, J. A. "Solderability." *Journal of Metals*, July 1984, pp. 51-53.

9. Wild, R. N. "Component lead solderability versus artificial steam aging, II." *Proceedings of the Naval Weapon Center Soldering Conference*, January 1987, NWC TP 6789 pp. 299-320.

10. Mather, J. C. A Need for Tighter Requirements on Circuit Board Solderability. IPC-TP-484, September, 1983.

11. Mackay, C. A. "Surface finishes and their solderability." *Brazing and Soldering Supplement, Metal Fabrication*, January/February 1979.

12. IPC-TR-462 Printed/board protective coating solderability evaluation over long term storage, Round Robin Test Program. October 1987, Lincolnwood.

13. Belani, J., Sajja, V., Mathew, R., and Patel, A. "Plated lead finishing for reliable solder joints." *NEPCON Proceedings*, February 1986, pp. 467-470.

14. Geiger, A. L. "Solderability of capacitor lead wire." *Proceedings of the Naval Weapon Center*, Soldering Conference, February 1986, NWC TP 6707, pp. 111-129.

# Chapter 11
# Component Placement

## 11.0 COMPONENT PLACEMENT

Surface mount components are placed on a printed circuit board after deposition of adhesive or solder paste. Generally solder paste and adhesive are deposited at a separate work station by a screen printer and by the placement equipment, respectively. Thus the main function of most placement equipment is adhesive deposition, and of course component placement. Placement equipment is commonly referred to as "pick-and-place" equipment. Sometimes it is called "onsertion" equipment, to differentiate it from the insertion equipment used for through-hole components. The term "onsertion" has never become popular, however.

Several factors make it almost mandatory to use placement equipment for surface mounting. Seldom, for example, are surface mount components marked, especially passive components. Moreover, packages with finer and finer lead spacings or "pitches" are being used. Finally, manual placement of surface mount components, which is neither reliable nor economical, is used for production.

The pick-and-place machine is the most important piece of manufacturing tooling for placing components reliably and accurately enough to meet throughput requirements cost effectively. Typically, surface mount pick-and-place equipment constitutes about 50% of the total capital investment required for a medium volume surface mount manufacturing line (Type I SMT).

Also, the throughput of a manufacturing line is primarily determined by the pick-and-place machine. The majority of manufacturing defects that require rework stem from placement problems. The type of feeder system used on the pick-and-place equipment also plays an important role in determining both throughput and the reliability of placement. Since no placement equipment is best for all applications, the effort required in selecting an equipment with an appropriate feeder system should constitute a major part of the effort spent in selecting SMT capital equipment.

In this chapter, after briefly discussing manual placement, we concentrate on automated-placement equipment and its selection criteria. Using the selection criteria as a guide, we will survey some of the popular placement equipment on the market. The reader can use the information in choosing placement equipment for individual applications. A detailed questionnaire is provided in Appendix B for this purpose.

## 11.1 MANUAL PLACEMENT OF PARTS

As mentioned earlier, manual placement of surface mount components is neither reliable nor economical, but it can be used for prototyping. Also, if the placement equipment requires considerable programming time or if the appropriate feeders are not on hand, manual placement may be used for quick-turnaround prototypes.

Since most passive components do not have any part markings, one of the main problems in manual placement is preventing part mixup, and a procedure must be put in place to handle this problem. One common method is by using bins or containers marked with part numbers or values. If the parts so classified get mixed up, they must be positively identified or thrown away.

Another problem is manual placement—the increased potential for placing even marked components in the wrong orientation—and this typically crops up when tantalum or other polarized passive devices such as SOICs or PLCCs are to be placed. The operator must know how to identify pin 1 of active devices (look for a dot over pin 1) and the polarity of tantalum capacitors (the positive terminal has either a beveled edge, a notch in the termination, or a welded stub). Most important of all, manual placement is operator dependent, and not everyone is well suited for the job. Only a person with a steady hand and good dexterity can consistently place parts accurately. Some operators can drop a fine pitch package exactly on location almost every time.

Placement accuracy is more critical when placing components onto solder paste than onto adhesive because of potential paste smudging across the pads, which may cause bridging. It is not a good idea to rely on self-centering during reflow of a misplaced component on solder paste. If components do get misaligned, it is better to effect realignment and take a chance on solder paste smudging.

In component placement, there are two main functions, pickup and placement (hence the name "pick-and-place" for this type of equipment). In manual placement, the components parts are picked up either by tweezers or by a vacuum pipette. For passive components tweezers are adequate, but for multileaded active devices, a vacuum pipette is very helpful in dealing with component rotation. An example of vacuum pipette used for placement of surface mount components is shown in Figure 11.1. When placing components manually, the following additional guidelines should be kept in mind.

1.  Care must be taken to avoid mixing parts that look identical but may have different values. Parts that have been dropped and recovered should be positively identified or discarded. We have mentioned this point earlier, but it is worth repeating.
2.  Undue tension or compression on the components should be avoided.
3.  Tweezers or other tools that may damage the part should not be used to pick up components.
4.  Parts should be gripped by their bodies, not by their leads or terminations.
5.  Care should be exercised to keep tweezers free of adhesive or solder paste. Since some contact with adhesive or solder paste

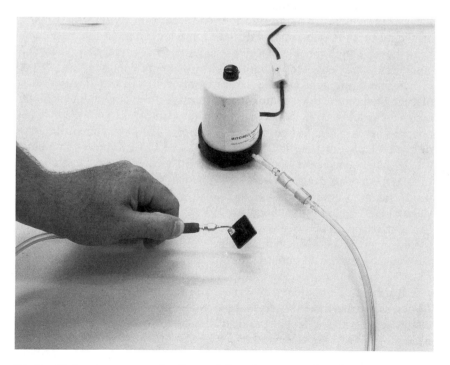

**Figure 11.1    A vacuum pipette used for placement of surface mount components.**

may occur during part placement, tweezer tips should be cleaned regularly with a solvent, such as isopropyl alcohol or Freon.

6. Incorrectly placed parts should be discarded or properly cleaned before reuse. During cleaning, damage to terminations or leads is a real concern.

7. Programmable devices such as programmable logic arrays (PLAs) are handled manually for programming before placement. If this is not done properly, especially when the devices are pried out of the programming sockets, lead damage can cause significant lead coplanarity problems. This will result in solder opens during soldering.

## 11.2 AUTOMATED PLACEMENT OF PARTS

Accuracy requirements almost mandate the use of autoplacement machines for placing surface mount components on the board. The placement machine is the most important piece of equipment required for surface mounting. It absorbs the highest capital investment, and it also determines the overall economy of manufacturing. The placement equipment available on the market today basically falls into four categories: in-line placement equipment, simultaneous placement equipment, sequential placement equipment, and sequential/simultaneous placement equipment.

In-line equipment (Figure 11.2a) employs a series of fixed-position placement stations. Each station places its respective component as the PC board moves down the line. Cycle times vary from 1.8 to 4.5 seconds per board. The simultaneous equipment (Figure 11.2b) places an entire array of components onto the PC board at the same time. Typical cycle times are 7 to 10 seconds per board. Sequential placement equipment (Figure 11.2c) typically utilizes a software-controlled X-Y moving table or a moving head system. Components are individually placed on the PC board in succession. Typical cycle times are 0.3 to 1.8 seconds per component. Sequential/simultaneous placement equipment (Figure 11.2d) features a software-controlled X-Y moving table system. Components are individually placed on the PC board from multiple heads in succession, with a typical cycle time of approximately 0.2 second per component.

Placement equipment can also be classified based on its flexibility and throughput. The flexibility to place components of many different types comes at a price: the higher the flexibility, the lower the throughput. For example, robots are placement machines that provide the ultimate in flexibility. They can be used for placing surface mount components, for solder mask or paste dispensing, for soldering, and for lead tinning. Their hardware cost is relatively low, but software and hardware development can

In-Line
Placement

**b**

Simultaneous
Placement

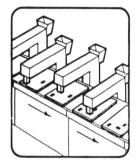
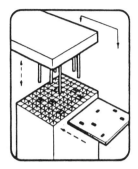

Moving board/fixed head

Each head places one component

1.8 to 4.5 seconds/board

- Fixed table/head
- All components placed simultaneously
- Seven to ten seconds/board

Sequential
Placement

**d**

Sequential/
Simultaneous Placement

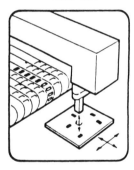
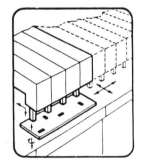

X-Y movement of table/head

Components placed in succession
individually

0.3 to 1.8 seconds/component

- X-Y table/fixed head
- Sequential/simultaneous firing of
  heads
- 0.2 seconds/component

**Figure 11.2** (a) In-line placement equipment, with stationary head and moving board. (b) Simultaneous placement equipment, with simultaneous placement of components on a stationary board. (c) Sequential placement equipment, with X-Y movement of head and table. Components are placed individually, one at a time. (d) Sequential/simultaneous equipment, with sequential or simultaneous firing of heads.

be quite expensive. Robots are very flexible, but they are extremely slow and require considerable development for each type of application.

## 11.3 SELECTION CRITERIA FOR PLACEMENT EQUIPMENT

In each of the categories discussed in Section 11.2, there are numerous pieces of equipment available, and new models are constantly being introduced in the marketplace. The cost can vary from $20,000 for a benchtop model to millions of dollars for a very high speed placer. The cheaper models ($20,000-$100,000) are generally used for prototyping but serve in production by many companies. They have limited capacity and may not be very accurate.

The production equipment can be either flexible, able to place surface mount components of all types, or dedicated, placing only components of certain types. Evaluation and selection of a pick-and-place system is a very complex process, which is made more difficult by the general lack of packaging standards on components and feeders. Where the standards do exist, the tolerances are poor. Also, various pieces of equipment used in a surface mount line may not be compatible with each other because there are no hardware and software standards for integration. The surface mount equipment manufacturers association (SMEMA) is addressing the equipment standardization issues, however.

The selection of the appropriate autoplacement machine is dictated by many factors, such as the complexity of the device, the applicable packaging and equipment standards, the type and number of parts to be placed, and current and future needs for volume and flexibility. Thus one must establish some guidelines for selection of a machine. A matrix detailing the desired features versus features of the available equipment will help reduce the number of choices for a given application. Here recommendations by existing users can be the most important selection criteria. Refer to Table 11.1 for a summary of selection criteria. Appendix B offers a detailed questionnaire that will help in the final selection of a pick-and-place machine. The questionnaire should be used to narrow the list, which can be further evaluated with some form of weighted points for features important for the given application. (What kind of parts are to be handled? Will they come in bulk, in a magazine, or on tape? Can this machine accommodate future changes in tape sizes? How will the board be handled? Does the equipment come with an automated board handling system?) Some of the major selection criteria for pick-and-place system are discussed next.

**Table 11.1 Summary of selection criteria for a pick-and place equipment for surface mount components. (Refer to Appendix B for details.)**

Model number _____ Price range _____
Maximum board size/placement area _____
Maximum 8 mm tape input _____

Acceptable types of feeder input:
8 mm tape \_\_\_\_\_ 12 mm tape \_\_\_\_\_ over 12 mm tape \_\_\_\_\_
7 inch reels \_\_\_\_\_ 13 inch Reels Tube (stick) \_\_\_\_\_ Bulk \_\_\_\_\_
Waffle pack \_\_\_\_\_

Adhesive application: Yes \_\_\_\_\_ No \_\_\_\_\_
Adhesive application speed _____

Acceptable component types:
Rectangular chips \_\_\_\_\_ Cylindrical chips \_\_\_\_\_ SOT \_\_\_\_\_
Fine pitch \_\_\_\_\_ SOIC/PLCC up to 20 leads \_\_\_\_\_
SOIC/PLCC 20 leads or more \_\_\_\_\_

Placement per hour \_\_\_\_\_ Placement accuracy: X-Y \_\_\_\_\_ rotation \_\_\_\_\_
Programming: on-line \_\_\_\_\_ off-line \_\_\_\_\_

Special features:
Component on-line testing: Yes \_\_\_\_\_ No \_\_\_\_\_
Component-missing verifier: Yes \_\_\_\_\_ No \_\_\_\_\_
Statistical quality control: Yes \_\_\_\_\_ No \_\_\_\_\_
Software for optimization of feeder placement: Yes \_\_\_\_\_ No \_\_\_\_\_

Description of the equipment:

# 11.3.1 Maximum Substrate Size Handling Capacity

Although many placement machines developed for the hybrid industry can handle only small substrates, for surface mounting it is not uncommon to use 12 inch by 18 inch or even 18 inch by 18 inch substrates. Thus when selecting placement equipment, the maximum size of the substrate or panel that the apparatus can handle is probably the place to start.

Even when the substrate size is small, it is generally economical to use large panels containing multiple breakaway substrates. If fixtures are to be used for handling, their maximum size should be used instead of substrate size as the selection criterion for a pick-and-place machine.

Various types of fixtures are available to hold the printed circuit board in place while components are being placed. Figure 11.3 shows an example of an unique PCB holder with movable magnetic supports. Depending on the board, the magnetic supports can be moved quickly to desired locations within the fixture to accommodate different configurations and sizes. Plastic edge clamps (on magnetic supports), tooling pins in the fixture frames are used to hold the PCB in place over the magnetic supports. To prevent board buckling in large boards, a few pin supports on the base of the fixture plate can also be used if necessary. (See Figure 11.3.) The fixture holding the board can be transported either on a conveyor belt or on rails and presented to the placement head.

## 11.3.2 Maximum Feeder Input or Slot Capacity

Maximum number of feeder input positions is another measure of placement equipment capacity. Since this number provides a measure of capacity to process an assembly with different part types (or different values of the same part type) in a single pass through the machine, it is a very important consideration.

To determine the slot or input capacity needed, one must first analyze the product requirements and ascertain the maximum number of part types that will be used. (By "part type" we mean not only parts of different mechanical outlines but also parts of different electrical values in a given outline.) Future expansion needs that can be anticipated to arise when almost all components become available in surface mount should be included in the study.

A standard measure of feeder input capacity is the 8 mm tape feeder. The more 8 mm slots a machine can accommodate, the higher its input capacity. However, only a limited number of parts come on an 8 mm tape. Larger parts supplied in a tube take up multiple 8 mm slots, and the same components on a 13 inch reel require two to three 8 mm slots. Components supplied in bulks or in waffle packs affect input slot requirements differently.

Another option for compensating for insufficient input capacity is to complete the placement of components in two or three passes or to sequence many parts of the same size but differing values in the same tube feeder. This error-prone approach is definitely not recommended for production, where its adverse effect on placement rate could not be tolerated.

While conducting the in-house study on slot requirements, side benefits may arise. Perhaps, for example, some part numbers can be totally eliminated (e.g., a 1% tolerance resistor if all you need is a 5% tolerance part), or some part numbers can be consolidated. Using a minimum number

MAGNETIC BOARD SUPPORTS

PIN SUPPORT

TOOLING PIN

**Figure 11.3** A printed circuit board holder with movable magnetic supports to accommodate different sizes of boards. (Photo courtesy of Intel Corporation.)

of part types not only conserves the input slots but also can reduce inventory costs and provide increased leverage with vendors due to higher volumes.

### 11.3.3 Placement Rate and Flexibility

In selecting a placement machine or set of machines to meet production requirements, one must determine the product mix, the number and types of components per board, and the annual production volume. Knowing the current requirement is not sufficient. Future needs and manufacturing plans should be factored in. Thus the placement rate is very important.

Actual throughput will depend not only on the placement rate of the machine but also on the component mix and the types of feeder used. The placement rate, in turn, depends on the location of the feeders with the most widely used components. If the product mix changes significantly, it may be difficult to place feeders of the most widely used components to provide the shortest possible distance between the pick location and the board. Some machines come with software that optimizes placement head movement for the shortest overall travel time for the entire board.

Another factor that slows placement rate is component testing before placement, especially if testing is done on the fly or if a test nest is used. If vision is used to locate components, the placement rate is further reduced.

Placement rate is also affected by board size and by component and feeder types used. For example, a tube or waffle pack feeder that requires constant loading and unloading may cause interruption of operation of the machine. Larger boards with an increased number of component types on the machine will require greater travels, which in turn will slow the throughput of the machine. Most important, a machine that is constantly down or in need of repair, even though it runs very fast when working, will have an adverse impact on throughput.

It becomes very obvious that determining the actual placement rate is not easy. As a general guide, the rated placement rate quoted by the vendor should be derated by 30-40% to arrive at a conservative number. The experience of other users who have tried a similar product mix with the machine under consideration can also give an idea about the placement rate.

Thus we see that placement rate cannot be considered in a vacuum. Dedicated machines with limited types of parts placement capability are going to be much faster than machines that can place components of all types and sizes, test them before placement, and ensure accurate placement using a vision system.

## 11.3.4　Placement Accuracy/Repeatability

As components with larger sizes and finer pitches come into common use, the need for accurate placement stands out even more. There are many ways to define accuracy, but one useful characterization is as follows: accuracy is the greatest tolerable deviation of the component lead from the center of its corresponding land after placement [1].

Many times suppliers specify the machine resolution, which is defined as the smallest discrete resolution the machine can discern. This does not mean that what the machine can discern can be repeated time after time, however. Also, machine resolution generally assumes the use of CAD data for programming. Thus if the machine is programmed by "teach mode" or by using a digitizer instead of the CAD-supplied coordinates, additional errors can be introduced during programming and "teaching" of the machine.

Repeatability instead of accuracy is a more useful guideline, then, that is, the consistent ability of the placement head to place a part at the specified target within a specified limit. Accuracy, for all practical purposes, simply means the placement of components on the land pattern within the acceptable deviation or shift. Depending on the application, the maximum shift of leads or terminations from their pad generally varies from 25% to 50% of pad width [2]. A 50% shift is excessive and may degrade solder joint reliability; a 25% shift is the maximum acceptable, as long as the component does not shift any further during reflow soldering.

For 50 mil center packages, the pad width generally used is 25 mils, and for fine pitch packages only 12 mils. Thus a 6 mil accuracy requirement should be good enough for 50 mil center packages, and 3 mil placement accuracy should be sufficient for fine pitch packages. This is the X-Y accuracy requirement. However, the X-Y requirement alone is not sufficient because it does not take into account rotational accuracy and the additional deviations caused by manufacturing processes such as handling (between placement and soldering) and reflow soldering.

Rotational accuracy is not the same as X-Y accuracy. The same degree of $\Theta$ or rotational deviation will produce a larger offset in some pads for bigger devices than it will in smaller devices. For example, a one-degree rotation error will displace a lead near the corner of an 84 pin, 50 mil center PLCC by more than 0.010 inch [1]. If we add to this the deviation caused by the artwork and PCB manufacturing processes, the problem becomes serious. The total component shift may exceed 50%, which is totally unacceptable.

The accuracy requirement will vary for different applications, but a 0.002 to 0.004 inch in X-Y repeatable accuracy from the target on a 14 inch by 18 inch substrate, irrespective of programming method, and 0.2 to 0.5 degree rotational accuracy should be sufficient for most production applications. The situation is improved by the usage of a vision system, as discussed in the next section.

## 11.3.5    Vision Capability

One way to offset inaccuracy in placement is to use a vision system, to tell how far a component lead is from its corresponding land and to instruct the machine to correct for the discrepancy. To implement the instructions of a vision system, the hardware design of the placement machine must afford the needed resolution or repeatability.

The vision system is a good way to compensate for deviations in land pattern locations due to PCB manufacturing processes and poor tolerances of component manufacturing. It also can compensate for the relatively larger deviations from the reference location in larger boards.

When using a vision system, a set of alignment targets should be designed on the board. Location of alignment targets on a board are shown in Figure 11.4, which shows the location of tooling holes along with the alignment target. The dimensions A, B, C, and D shown in Figure 11.4 will depend on the form factor for a given product. The basic intent of Figure 11.4 is to show the location of 3 alignment target on the board. The alignment target shapes are either diamond shape or round. And the sizes of the alignment targets vary from 20 mil diameters to 80 mils. No matter what the shape, the alignment target must be free of solder mask and should be flat within one mil. Some equipment manufactures require that the target be flat within 0.4 mil. If hot air solder leveling process is used, it becomes difficult to achieve flatness within 0.4 mil. One must check with the placement equipment manufacturer about the alignment target requirements. By referencing part placement to these targets instead of to tooling holes or board edges, one can avoid such tolerances as tooling hole/ tooling pin fit, tooling pin size, tooling hole size, and tooling hole location with respect to the land patterns.

There are two types of vision systems: single-camera and two-camera systems. In a one-camera system, the land patterns are viewed and the placement coordinates adjusted appropriately. In a two-camera system, both the component and the land pattern are viewed and compensated for. For finer pitch packages with lead pitches of 25 mils or less, it is almost mandatory to use a two-camera system that can match the land pattern with the corresponding component leads.

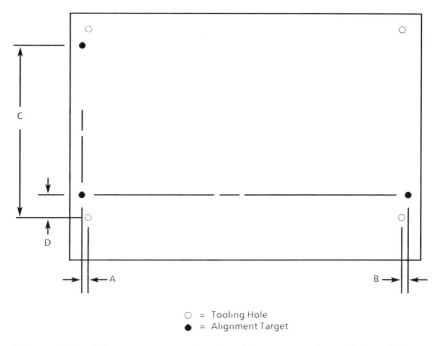

○ = Tooling Hole
● = Alignment Target

**Figure 11.4   Alignment target for the vision system in a pick-and-place equipment.**

Currently, very few fine pitch devices are used on a single board. For these hard-to-align fine pitch parts, another set of alignment targets like that shown in Figure 11.4 should be placed inside each fine pitch (lead pitch 25 mils and under) land pattern. This is in addition to the set of alignment targets already on the board. The second set of targets compensates differently in various regions of the same board.

## 11.3.6    Adhesive Dispensing Capability

Nonconductive adhesive is used to hold components in place temporarily before wave soldering. The adhesive is dispensed, the component is placed, and the board is heated to cure the adhesive, thus ensuring that the components will withstand the rough action of the wave during wave soldering.

There are various ways to dispense adhesive as discussed in Chapter

8, but the most commonly used method is syringing. An adhesive dispensing system using a syringe is generally an integral part of the pick-and-place equipment. Either dedicated adhesive dispensing heads with their own X-Y table can be used, or the component placement X-Y table can be shared for adhesive dispensing.

Many equipment suppliers offer both options—a dedicated X-Y table for adhesive dispensing or a common X-Y table for both adhesive dispensing and component placement. The latter option is cheaper but slower because the placement head will either be dispensing adhesive or placing component. Thus selection should be based on production volume requirements.

Adhesive dispensing parameters were described in detail in Chapter 8. It is simply worth noting here that having the desired adhesive dispensing capability may be one of the decisive factors in the selection of a particular placement system if the machine is intended for Type II or III SMT assembly. Figure 11.5 shows an adhesive dispenser as an integral part of the pick-and-place system.

## 11.3.7    Other Important Selection Criteria

Many other features of a pick-and-place machine are very important for most applications and thus should be considered. Their significance may be more important in some applications than in others.

1.    The pickup mechanism can have a vacuum nozzle or mechanical jaws or both. Most equipment picks up parts with a vacuum nozzle but uses jaws to center components before placement.

2.    In some placement equipment, both the placement head and the X-Y table (where the board is held) move during placement. In other cases the X-Y station remains stationary during placement and only the head moves. For example, in the Zevatech PPM the head moves and the X-Y table is stationary. In the Fuji and TDK machines, the head is stationary and the X-Y table moves. In the Dynapert MPS 500, both the head and the X-Y table move. This is an important feature to keep in mind, because if the acceleration and deceleration of the X-Y table is not properly controlled, the tackiness of the adhesive or solder paste may not be sufficient to hold components in place. In general, the incidence of component misalignment is less in machines of stationary X-Y table design. This feature becomes even more critical in the placement of the larger fine pitch devices, in which case a moving table may cause component shift, hence repairs that otherwise would not have been necessary.

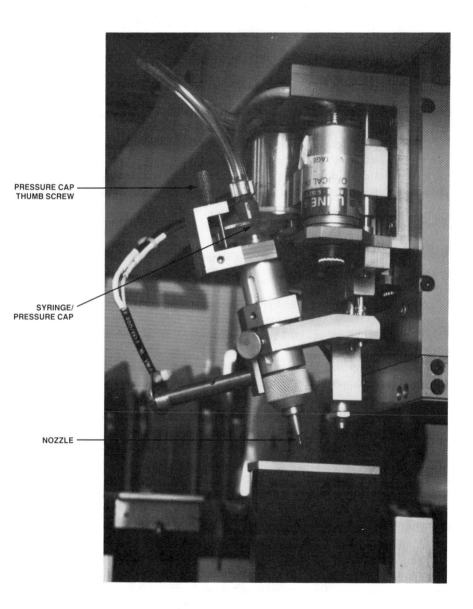

PRESSURE CAP
THUMB SCREW

SYRINGE/
PRESSURE CAP

NOZZLE

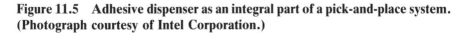

**Figure 11.5    Adhesive dispenser as an integral part of a pick-and-place system. (Photograph courtesy of Intel Corporation.)**

3.  The pickup head should be able to sense that the part has been picked up. If this fails to happen, machine operation should halt. This is an important feature because components can be held too tightly or too loosely if tolerances of either the components or the tape cavity are poor.

4.  The control system for the placement head should provide CAD downloading of component locations but should also allow for manual editing of adhesive placement locations and part pickup and placement locations. It should also provide management information reports, such as number of poor part picks and total components used.

5.  The placement head should have programmable Z axis travel. The pickup head should be able to place the parts onto the adhesive or the solder paste within the preprogrammed Z axis displacement. The displacement force should be minimal, to avoid damage to the parts or the substrate.

6.  Component test before placement is a nice feature. A test that can be performed in flight without much loss of placement speed is preferable. Testing in flight is generally accomplished on passive devices only. Active devices require a separate test nest, which can considerably slow the placement rate.

Testing will prevent placement of wrong components if, as quite commonly happens, tapes are mounted at incorrect feed locations. To have a minimum effect on placement rate, the testing of passive devices may be programmed for only the few first devices to be placed. Thus the test "programming" effort for passives is trivial. Instruments are readily available to test resistance, capacitance, and inductance, but testing active devices requires a considerable test development effort. Also, the equipment cost for testing actives is at least an order of magnitude greater than that for testing passives.

7.  Flexible placement systems such as sequential placement machines should be able to accept such commonly used feeders as 7 and 13 inch reels, tubes, and bulk and waffle packs.

8.  The advance mechanism for the reels should be adjustable in 4 mm increments because in a given feeder width, the feeder pitches vary in 4 mm increments. This allows the use of the same feeder for a given tape width but with varying tape pitches.

9.  An important feature that may be very desirable is on-line programming capability for PLA devices, to allow the use of tape and reel packaging. Currently these devices are programmed off-line and then put in a tubes for placement. Since PLA devices generally have high fallout rates, the pick-and-place system must be able to test them before placement. Some work is being done in the industry in this area, but commercial on-line PLA programming units as an attachment to pick-and-place systems are not available at this time.

It should be noted that PLA programming is a labor-intensive operation, and the potential for lead damage is significantly increased during manual handling, especially when the programmed PLAs are extracted from their programming sockets. Buying already programmed and tested devices should be considered, so that the parts can be used in tape and reel.

## 11.4    SELECTION OF FEEDERS FOR PLACEMENT EQUIPMENT

Generally much attention is given to the selection of placement equipment and not enough to the selection of feeders. The importance of feeders, however, cannot be overemphasized. In the absence of reliable feeders that can feed components consistently and with minimum operator intervention, the placement rate suffers because the operator must constantly interrupt machine operation to correct the feeder problems.

The feeder system for the pick-and-place machine essentially determines reliability and throughput in placement. We use the term "feeder system" advisedly, because simple chip resistors, capacitors, and transistors do not use the feeders of same type required by the complex, multileaded devices such as PLCCs, SOICs, and fine pitch packages. In many cases there are overlaps, though, allowing the same type of feeder to be used for components of different types.

The determining factors in the selection of a shipping and handling medium should be desired quantity, availability, part identification, capability of the pick-and-place machine, cost of components, inventory issues, and potential for damage during shipping and handling. There are four types of component feeders: tape, tube, bulk, and waffle pack.

In general, tape and reel feeders are used for all surface mount components, but the bulk feeders are generally used only for very limited, simple components such as resistors and capacitors. The tube feeders are used for SOICs and PLCCs, but the waffle packs are limited to fine pitch devices. There are exceptions of course. Each method has its inherent advantages and disadvantages. The details are discussed next.

## 11.4.1    Tape and Reel Feeders

Tape and reel is used for chip resistors, capacitors, MELFs, SOTs, SOICs, PLCCs, and SOJs. It is by far the most popular and desirable feeding method, especially if large volumes of components are to be placed. As shown in Figure 11.6, the components are nestled in individual pockets

**Figure 11.6** Tape and reel feeder: the components are nestled in individual pockets in the

in the tape and are covered with a plastic cover, which is peeled back at the time of placement. Tape widths vary from 8 mm to 56 mm for different components. Every tape width requires a different feeder. Figure 11.7 shows a row of tape feeders mounted on a pick-and-place system.

The reels protect the devices from damage during shipping and handling, save on loading and unloading time on the pick-and-place equipment, and prevent both part mix-up (by compensating for the lack of part markings on many surface mount components) and placement of components in the wrong orientation. Placing components in the wrong orientation is not uncommon when using tube feeders, which require frequent loading and unloading.

Why aren't the tape feeders used exclusively, then? First, not all component are available on tape and reel, and all the pick-and-place machines do not have tape and reel handling capability. The pick-and-place machines that do have such capability require more slot space than the tube feeders. Also, components such as PLAs that require programming before placement cannot be handled on a tape and reel because few if any pick-and-place machines have on-line programming capability.

**Figure 11.7    A row of tape feeders mounted on a pick-and-place machine. (Photograph courtesy of Intel Corporation.)**

Another factor preventing the universal use of tape and reel becomes important especially if smaller quantities are desired, since tape and reel is best for large quantities. This factor is related to the standard that controls tape and reel construction, namely EIA-RS-481. Because the tolerances in this specification are not very tight, reel quality varies greatly from one supplier to another. Thus tapes often have poor tolerances, especially in center-to-center distance of sprocket holes. This can cause variation in pick location, which may even prevent part pickup.

Also, there may be problems in cover tape removal. The EIA specification allows ±30 pounds in peel strength. Thus one of the most common problems in tape feeding is the wide variation in pull force required to peel off the cover tape to expose components for placement. Either all pick-and-place machines should be able to handle these differences or the peel strength tolerance should be tightened. The EIA-RS-481 committee is considering a new peel strength criterion.

Still another problem involves the very broad reel size requirement. The EIA specification (EIA-RS-481) establishes the maximum reel size at 13 inches but gives no minimum. The de facto standards for passive and active components have become 7 and 13 inch reels, respectively. Since active components generally are available only in the 13 inch reel size, inventory cost may be a real problem for small users. The minimum quantities that come on a 13 inch reel are shown in Table 11.2. The user must be willing to purchase such a minimum quantity to use tape and reel at all. Table 11.2 also shows the quantity of components that could be handled if 7 and 10 inch reel sizes became available. Thus the desired quantity generally is of prime importance in the selection of feeder equipment. If the desired quantity is large, tape and reel is the natural choice.

If small quantities are needed, selection becomes more difficult. Suppliers are generally not interested in offering smaller or partial reels because they would rather sell more on a larger reel than less on a smaller or partial reels. Users who need small quantities end up buying in tubes. This may add another mil or two in lead coplanarity during shipping and handling, but suppliers get blamed nevertheless.

Tape and reel is a good medium for shipping containers for suppliers and users alike, and at the request of the Surface Mount Council, the EIA RS-481 committee once considered a change in the specification to allow both 7 and 13 inch reels for both active and passive devices, to meet the needs of small and large users. (Tape and reel has not been a major issue for passive components, which are fairly inexpensive, such that inventory costs generally are not a problem.)

Even with the contemplated change, if it does ever materialize, the supply of components for active devices on smaller reels will depend on market forces. It should also be noted that the IEC (International Elec-

**Table 11.2   Parts count in various sizes of reels for tape and reel feeders. (If active components became available on 7 or 10 inch reels, the number of components per reel may also be standardized)**

| TYPES OF PACKAGES | TOTAL COMPONENTS PER REEL | | |
|---|---|---|---|
| | 7 INCH[a] | 10 INCH[a] | 13 INCH |
| SOIC 8, 14, 16L | 1057 | 2323 | 4158 |
| SOIC 20, 24, 28L PLCC 18, 20L SOJ 20L | 341 | 851 | 1468 |
| PLCC 28 | 168 | 422 | 769 |
| PLCC 44, 52 | 112 | 281 | 513 |
| PLCC 68 | 52 | 140 | 259 |
| PLCC 84 | 46 | 124 | 230 |

tronics Council) allows reels of 7, 10, and 13 inches and other sizes. A change by EIA-RS-481 would promote compatibility with other standards, since components are purchased beyond the national boundaries.

Every attempt should be made to use tape and reel whenever possible. The larger reel size will not be an issue when SMT takes a firm hold and most users are handling larger quantities of surface mount components. However, there will always be some need for smaller reel sizes, even for large users. For example, even a high volume product takes from 6 to 12 months for design and prototyping, and smaller quantities are needed during this phase.

Changing an industry specification, is a slow process but the placement equipment suppliers could minimize at least one problem by designing the feeder transport mechanism to prevent sudden accelerations and decelerations, which can dislodge components from their pockets. The user, in turn, may be able to reduce the variation from reel to reel by limiting dealings to as few vendors as possible.

## 11.4.2   Bulk Feeders

Bulk feeders may be acceptable in prototyping and in very low volume applications, but they are very inconvenient for feeding components in automated assembly. Generally only passive devices are fed by this method—that is, from a vibrating bowl into a track. The track width limits the width of component that can be fed.

There are many problems associated with bulk vibrating feeders. For example, if the intensity of vibration is not properly controlled, the components sometimes jump out of the feeder; since one or more components may not be placed when this happens, rework is likely to increase. In addition, termination solderability may be lost as a result of constant rubbing of components against each other and against the bowl feeder. Bulk feeding should be avoided because of potential damage to terminations and leads, as well as misplacement and part mixup problems.

### 11.4.3    Tube or Stick Feeders

Although high volume SMT production almost mandates the use of tape and reel feeders, currently there are few companies so heavily committed to this technology. Thus despite the advantages of tape and reel, tube feeders are the most commonly used, because of the large inventory cost of tape and reel for small volume users. Also, many pick-and-place machines, especially the benchtop models, will not accommodate tape and reels feeders.

Tube feeders can be divided into three types: horizontal, stick-slope, and ski-slope. The horizontal tube feeders have multiple tracks that are machined to accommodate components of different widths. The number of tracks per feeder depends on the width of the component body to be used. Figures 11.8 and 11.9 show a horizontal tube feeder and a row of tube feeders mounted on a pick-and-place machine, respectively.

The stick-slope feeder is simply a modification of the horizontal feeder in which the plastic tube can be directly loaded into the individual tracks of the horizontal feeders. Since the components do not have to be taken out of their plastic shipping tubes, the stick-slope feeders have all the advantages of horizontal feeders without any of the disadvantages. Both the horizontal and the stick-slope feeders use some form of electromagnetic vibration to move components along to the pick position.

The ski-slope feeders are like the stick-slope feeders, but the components are moved along by spring action, not by vibration. Also, unlike the horizontal tube feeders and the stick-slope feeders, the ski-slope feeders do not use machined tracks for different components; rather, the tracks are an integral part of the equipment. The ski-slope feeders require tighter component tolerances and will jam up if the tolerances vary as a result of plastic flashes left on the component body.

The ski-slope feeders are more expensive and require additional space on the machine to accommodate their length. Greater length can be an advantage, however, because ski-slope feeders can take more components than either the horizontal or the stick-slope feeders, hence they do not

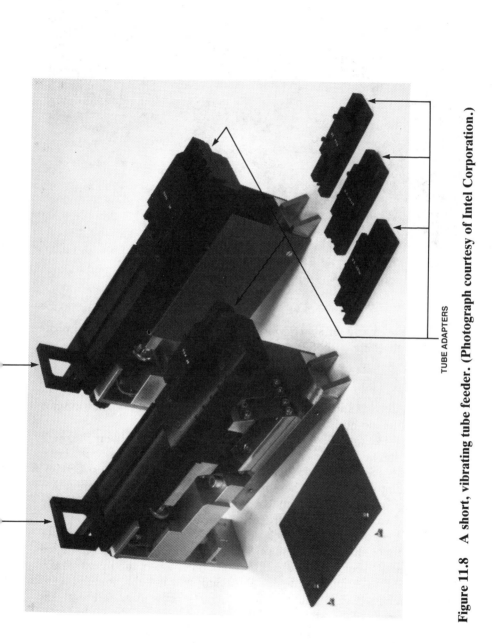

TUBE ADAPTERS

**Figure 11.8    A short, vibrating tube feeder. (Photograph courtesy of Intel Corporation.)**

**Figure 11.9   A row of short, vibrating tube feeders mounted on a pick-and-place machine. (Photograph courtesy of Intel Corporation.)**

need to be loaded and unloaded as frequently. Nevertheless, they are not as popular as the horizontal or the stick-slope feeders.

The horizontal tube feeders are fairly reliable and relatively inexpensive; they can be used on almost all pick-and-place machines, both large and small. Since the tube feeders are short, however, parts must be constantly fed into the tracks. These feeders do not cause lead or termination damage because the components are not dumped in a bowl. However, the intensity of vibration must be properly controlled.

Both the horizontal and stick-slope feeders should use tracks that are machined to accommodate some variation in component dimensions. Since the tolerances in component dimensions are fairly loose, the tracks should be machined to accommodate the maximum and not the nominal component dimension. This will allow use of components from different vendors that meet JEDEC standards. However, the track should not be wider than necessary, since excess track width may compound inaccuracies in placement by shifting the lateral pick positions.

One common problem experienced when using tube feeders is feeding components with the wrong orientation. It is very important to keep this possibility for error in mind, and operators should be trained to identify the location of pin 1.

## 11.4.4 Waffle Packs

Waffle packs, as the name implies, use flat machined plates with pockets or plastic carriers resembling waffles. Because waffle packs are large, they tend to consume a lot of input slot space. Also, the X-Y movement range of the pick-and-place equipment is generally limited. Since the pickup position is different for each individual component when using waffle pack, a very limited number of components can be picked, depending on the X-Y movement range of the placement head. In comparison, for a tube or for tape and reel, the pickup position is always the same for all components in a given tube or tape.

In addition, waffle packs have very small capacity and must be constantly replenished. Limited capacity may not be a serious problem, however, if only one or two large packages are used per board. To alleviate the capacity limitation imposed by waffle packs, waffle pack handlers can be used. They pick up parts from the waffle packs and bring them to a fixed location and thus maintain the same pickup position for pickup by the placement head, as in tubes or tape and reels. However, because the waffle pack handlers transport part to an intermediate fixed location for pick up by the placement head, the over all placement rate is decreased (instead of one pickup, two pick up steps are involved for the same part and those steps can be performed only sequentially). The need for waffle pack handlers also increases equipment cost since they cost around $100,000 dollars.

Despite the adverse effect of waffle packs on placement rate and equipment cost, there are instances when use of waffle pack is the only viable option. For example, as discussed in Chapter 3, the Japanese fine pitch packages generally do not have corner bumpers to protect the fragile leads (fine pitch devices made in the United States have corner bumpers (See Figure 3.23) to allow packaging in tubes). The waffle packs are the only option for handling these bumperless fine pitch devices, where part movement within the tube cannot be tolerated because of the danger of lead damage.

## 11.5 AVAILABLE PLACEMENT EQUIPMENT

The commercially available placement machines can be classified into three categories: low, medium, and high flexibility, based on the types of components they can place and the types of feeder they can accommodate. Low and high flexibility machines have been available for some time, but the machines in the middle category are recent developments. A fourth category also exists—namely low throughput and high flexibility but low cost. Such equipment is used mostly in laboratories, but some companies use it in production.

No particular category of machine is the best for every application. Some companies require one machine from each category to meet various product needs. For example, one manufacturing line may need a flexible machine for product development and medium production, whereas another line may call for one machine with low or medium flexibility and one with high flexibility to provide a good balance between throughput and flexibility requirements. In addition, most companies need a low cost laboratory model for prototyping and low volume production.

Table 11.3 summarizes the major features of six selected systems from each category. The questionnaire in Appendix B should be used for a detailed evaluation of these or other machines, before a final selection is made.

## 11.5.1 Equipment with Low Flexibility and High Throughput

A low flexibility machine is one that accommodates only passive components such as resistors and capacitors. These are generally in-line machines, referred to as chip placers. They can have one or multiple heads. The Japanese manufacturers such as Fuji, Panasonic, and TDK have dominated this market, and for good reason. The analog circuit designs of their products require passive devices (resistors and capacitors) in large quantities. It is not surprising that the Japanese companies dominate the market for both passive placement equipment and wave soldering equipment since their machines in this category place passive devices at a rate exceeding 10,000 per hour. The U.S. manufacturers such as Universal (Onserter) have attempted to capture some market share in this category of machines but without much success.

One European manufacturer has done extremely well in this category, however. Various MCM models made by Philips are widely used by companies in the automotive industry, such as Delco. The automotive industry is another user of analog circuit designs that rely extensively on passive components. MCM machines are designed for very high throughput. For some salient features of the MCM VI (Figure 11.10), refer to Table 11.3.

**Table 11.3**  A summary of major features of selected pick-and-place systems with various levels of throughput and flexibility.

| MANUFACTURER: MODEL NUMBER: | PHILIPS MCM VI | SIEMENS MS-72 | ZEVATECH PPM | FUJI CP-II | QUAD STAR 150 | CELMAC 85G |
|---|---|---|---|---|---|---|
| Placement area (inches): | 12 × 17 | 13 × 12 | 16 × 20 | 14 × 18 | 18 × 22 | 13 × 17 |
| 8 mm tape input slot space: | 256 | 72 | 80 | 100 | 60 | 36 |
| 8 mm tape | Yes | Yes | Yes | Yes | Yes | Yes |
| 12 mm tape | Yes | Yes | Yes | Yes | Yes | No |
| >12 mm tape | Yes | Yes | Yes | Yes | Yes | No |
| 7 inch reels | Yes | Yes | Yes | Yes | Yes | Yes |
| 13 inch reels | Yes | Yes | Yes | Yes | Yes | No |
| Tube | No | Yes | Yes | No | Yes | Yes |
| Bulk | No | Yes | Yes | No | Yes | No |
| Waffle pack | No | Yes | Yes | No | Yes | No |
| Rectangular chips | Yes | Yes | Yes | Yes | Yes | Yes |
| Cylindrical chips | Yes | Yes | Yes | Yes | Yes | No |
| SOT | Yes | Yes | Yes | Yes | Yes | Yes |
| SOIC/PLCC, >20 L | No | Yes | Yes | No | Yes | Yes |

**Table 11.3** (*Continued*)

| MANUFACTURER: MODEL NUMBER: | PHILIPS MCM VI | SIEMENS MS-72 | ZEVATECH PPM | FUJI CP-II | QUAD STAR 150 | CELMAC 85G |
|---|---|---|---|---|---|---|
| SOIC/PLCC, <20 L | No | Yes | Yes | Yes | Yes | Yes |
| Fine pitch | No | Yes | Yes | No | Yes | No |
| Placements/hour | 10-80k | 4k | 1-2k | 7-14k | 0.5-4k | 0.5-1k |
| Adhesive application | Yes | Yes | Yes | No | Yes | Yes |
| On-line programming | No | No | Yes | No | No | No |
| Vision | Yes | Yes | Yes | Yes | Yes | No |
| Component on-line testing | No | Yes | Yes | No | Yes | No |
| Component-missing verifier | Yes | Yes | Yes | Yes | Yes | No |
| Software for placement optimization | Yes | Yes | Yes | Yes | Yes | No |

**Figure 11.10    A pick-and-place machine (Philips MCM VI) with low flexibility and high throughput. (Photograph courtesy of Philips.)**

## 11.5.2    Equipment with High Flexibility and Low Throughput

Equipment with high flexibility has naturally lower throughput. Thus the equipment in this category is almost the opposite of the equipment described in the preceding section. High flexibility is needed in applications using digital circuit designs, as are common in telecommunications and in the computer industry. Whereas the Japanese have dominated the consumer products market, the digital market has been largely controlled by European and American companies for both reflow soldering and placement equipment.

The high flexibility machines are of the sequential type (Figure 11.2c), in which components are placed one at a time. The equipment can place almost all types of component, from a chip to a large PLCC, including fine pitch devices. These machines are very accurate for placement; they are generally modular and can be used alone or in conjunction with equipment in other categories to meet differing manufacturing requirements.

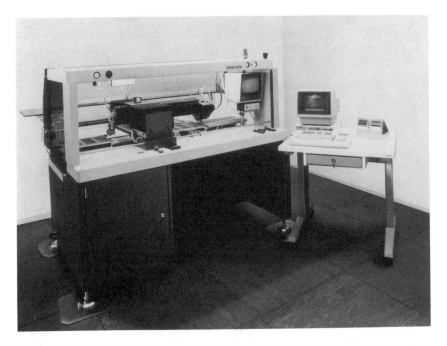

**Figure 11.11   A pick-and-place machine (Siemens MS-90) with high flexibility and low throughput. (Photograph courtesy of Siemens.)**

Among the many machines and manufacturers in this category are the Dynapert MPS-500, the Siemens MS-72 and MS-90, and the Universal Omniplace, and Quad Systems, and Zevatech. The Siemens MS-90 and the Zevatech PPM are shown in Figures 11.11 and 11.12 respectively. Some of the machines (e.g., the Zevatech PPM: see Table 11.3) have unique features such as on-line testing of passive components and vision capability without much loss of placement speed for passive components. This machine, as mentioned earlier has fixed X-Y table which prevents component shift during placement.

## 11.5.3   Equipment with Medium Flexibility and Throughput

Despite a long-standing need for equipment that provides a good balance between flexibility and throughput, such machines did not exist until very recently. In most cases the machines in this category are simply high speed chip placers, modified to be able to place small active devices (<20 leads) such as SOICs and PLCCs and 64k and 256k DRAM (memory)

**Figure 11.12  A pick-and-place machine (Zevatech PPM) with high flexibility and low throughput. (Photograph courtesy of Zevatech.)**

devices. If the pin count increases beyond 20 leads, this category of machine cannot be used.

Since chip placers are made mostly by Japanese companies, the modified equipment is also essentially from Japan. The Fuji CP-II (Figure 11.13) is a good example of a machine in this category. A manufacturing line with a machine like this, in addition to a highly flexible machine like the Zevatech PPM (Figure 11.12), can meet the requirements of high throughput and high flexibility.

## 11.5.4  Equipment with Low Cost and Throughput But High Flexibility

So far we have been discussing machines that cost upward of $200,000. There is a need for low cost machines as well, for use in the laboratory for quick-turnaround prototype work and also for low volume production. The cost of such machines varies widely and indeed is so insignificant compared with the cost of other machines discussed above that this category of machine should be at least be considered for the lab.

Equipment in this category may offer an inexpensive way to get into the surface mount business, and when production volume picks up, it can

**Figure 11.13   A pick-and-place machine (Fuji CP-II) with medium flexibility and throughput. (Photograph courtesy of Fuji.)**

be used for prototyping or as a short-term backup for the production machines. In many applications it can be used in production if volume requirements are low. Some companies use these machines in larger numbers as a substitute for an actual production machine [3]; if one of them is down, the others can still get products out the door. The low cost machines should not considered to be tinker toys or entry level machines [3].

The Quad Star 150 and the Celmac 85G (Figures 11.14 and 11.15, respectively) are good examples of machines in this category. In addition to placing chip devices, they can place "all sizes of" PLCCs. See Table 11.3 for some major features of these machines. Other machines in this category are the Excellon MC-30, the Dynapert MPS-118, the Celmac SMT-8400 and many others.

## 11.6   SUMMARY

The pick-and-place machine is one of the most important pieces of equipment for surface mounting. There are many types of placement equip-

**Figure 11.14  A pick-and-place machine (Quad Star 150) with low cost and throughput but high flexibility. (Photograph courtesy of Quad Systems.)**

ment available on the market today with different placement rates and flexibility. Some machines can place all types of components, accurately but slowly; others are intended for speed or throughput but not flexibility, and some combine flexibility and throughput. Features such as off-line programming, teach mode, edit capability, and CAD/CAM compatibility may be very desirable. Special features such as vision capability, adhesive application, component testing, board and waffle pack handling, and reserve capability for further expansion may be of interest for many applications. Vision capability will be especially helpful for accurate placement of fine pitch packages. Reliability, accuracy of placement, and easy maintenance are important to all users.

The user must define current and future needs before embarking on the evaluation and selection of a placement system. A detailed questionnaire should be used to evaluate all candidates before making a final decision, since no single piece of equipment can meet all needs. In some companies, a combination of machines in various categories may be needed to create a placement system, as opposed to a single placement machine to meet both flexibility and throughput needs.

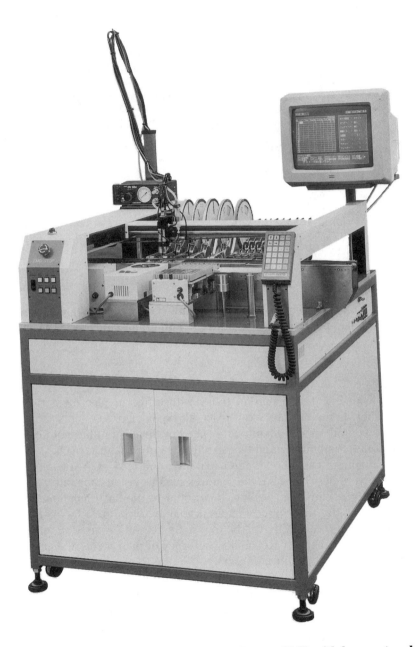

**Figure 11.15    A pick-and-place machine (Celmac 85G) with low cost and throughput but high flexibility. (Photograph courtesy of Celmac.)**

# REFERENCES

1. Amick, C. G. "Close does not count." *Circuits Manufacturing*, September 1986, pp. 35–43.

2. ANSI/IPC-S-815B. General Requirements for Soldering Electronic Interconnections. IPC, Lincolnwood, IL.

3. Tuck, J. "Low volume placers stir high volume interest." *Circuits Manufacturing*, May 1987, pp. 54–60.

# Chapter 12
# Soldering of Surface Mounted Components

## 12.0 SOLDERING

Welding, brazing, and soldering processes for joining metals together differ basically with respect to the temperature at which the joining takes place. Welding is generally used for joining ferrous metals and is accomplished at high temperatures (1500–2000°F). Brazing is used for joining nonferrous metals at relatively lower temperatures (800–1000°F). Soldering, which is used mostly for electronic products, occurs at the lowest temperatures (400–500°F).

In all cases, either two similar metals are joined, or certain dissimilar metals or alloys. The joining mechanism is governed by the formation of intermetallic compounds between the metals to be joined. A basic knowledge of metallurgical properties and of phase diagrams is necessary for an understanding of the principles involved in joining. The metallurgy of soldering is discussed in Chapter 10. Now we focus on the processes and equipment used in this form of joining.

The earliest electronic products were hand soldered. In the 1950s hand soldering was replaced by the wave soldering process for mass soldering of through-hole components. Then in the 1970s, reflow soldering came into widespread usage for surface mount components. However, for surface mounting, depending on the component mix, both wave and reflow soldering processes are used.

The basic differences between wave and reflow soldering lie in the source of heat and the solder. For example, in wave soldering, the solder wave serves the dual purpose of supplying heat and solder. The source for the supply of solder is unlimited because the wave pot holds more than 500 pounds. In reflow soldering, however, solder paste is applied first in a predetermined quantity, as discussed in Chapter 9, and during reflow, heat is applied to melt (i.e., reflow) the solder paste. Thus a more appropriate name for reflow soldering might be reflow heating. Reflow soldering takes place at lower temperatures than are used for wave soldering, but for a longer time.

In the beginning most of the surface mount wave soldering equipment came from Japan. This is not surprising, since the market for consumer products which use surface mount discrete components extensively in their analog circuit designs, is dominated by Japanese companies, and wave soldering is the most commonly used process for soldering discrete components glued to the bottom side of a through-hole assembly. Now surface mount wave soldering equipment is also available from suppliers in the United States and in Europe.

The reflow soldering equipment market is essentially controlled by the United States, where widely used reflow soldering equipment, especially vapor phase, was developed. The digital circuit designs that are common in the telecommunications and computer industries feature active devices, which generally use reflow soldering, and this market is dominated by American companies.

In addition to wave and reflow soldering processes, hand soldering is also used for repair and prototype work. However, the hand soldering process requirements (temperature, time, and pressure on pad) are essentially the same as in through-hole assemblies. No special consideration is required for hand soldering of surface mount components. Some discussion on hand soldering is included in Chapter 14; here we concentrate on the details of commonly used production volume soldering processes for surface mount assemblies of various types.

## 12.1 WAVE SOLDERING

Wave soldering is the main process used for soldering component terminations en masse in conventional through-hole mount assemblies. It is also the most widely used process for soldering surface mount discrete components (resistors, capacitors, diodes, etc.) glued to the bottom of Type III and Type II SMT assemblies. (See Chapter 1 for definitions of assembly types.)

Other surface mount components amenable to wave soldering are small outline transistors (SOTs). Four-sided leaded packages, such as PLCCs, are difficult to wave solder. Small outline integrated circuit (SOIC) packages can be wave soldered but it is generally not desirable to wave solder either the PLCCs or SOICs because of *potential* problems with reliability (active fluxes seeping inside the package through the lead frames) and manufacturability (shadowing in PLCCs and bridging in SOICs).

Problems may be encountered in the wave soldering of surface mount components if appropriate actions in design and manufacturing are not taken. Various issues in wave soldering and some approaches for over-coming them are summarized in Table 12.1. Let us elaborate on the details of the potential problems and then suggest some solutions.

## 12.1.1 Design and Process Variables in Wave Soldering

In through-hole mount assemblies the lead provides an easy path for the wicking of solder through the plated through holes. As soon as the

**Table 12.1 Technical issues and their resolution in wave soldering process.**

| ISSUE | APPROACH |
|---|---|
| • Accommodation of different PWA sizes and compatibility with in-house material handling system | • Evaluate tradeoffs: Automated finger conveyor width adjustment Standard PWA pallets Robotic loader and unloader |
| • Maintenance of specific gravity, activity, and required flux wave and foam height | • Evaluate automatic monitoring and control systems Flux density Dump inactive flux Flux wave and foam height sensor and controller |
| • Even preheat of boards | • Evaluate automatic monitor and feedback temperature control: Preheaters Conveyor speed |
| • Critical wave features for accommodating various PWA configurations | • Investigate wave features best suited to various PWA configurations: Smooth, rough, single, double Dry, oil, gas, intermixed Wave-height adjustment Wave-pot adjustment Solder bath temperature control Solder chemistry control Solder level and feed control |

lead end touches the solder wave, the wetting force helps the solder climb up the lead. Inside the plated through hole the capillary action pulls up solder, and again on the top the wetting force of the pad spreads the solder onto the pad. Advancing solder pushes out flux vapors and air, thus filling the hole.

In surface mounting there are no holes and leads. Instead, flat or round components sit on flat pads, making soldering more difficult because of sharp corners formed by the component and the board surfaces. The classic solder wave with its laminar solder flow hits the underside of chip resistors and capacitors tangentially and may not always get to the corners formed by the rectangular components and the flat board surface.

Because of sharp corners, outgassing and solder skips are two main concerns during the wave soldering of resistors and capacitors. Outgassing (Figure 12.1), which is believed to be caused by insufficient drying of flux, can be corrected by raising the preheat temperature. Providing escape holes in surface mount pads can also alleviate outgassing problems by the same mechanism as in through-hole mount soldering. Solder skips are due to the shadow effect of the part body on the trailing terminations (Figure 12.2). The shadowing effect also may be encountered if the smaller components are just behind larger components when going over the wave. Shadowing of trailing component terminations also will occur when staggered components are not placed far enough apart.

Bridging and solder skips caused by shadowing in multileaded devices such as SOICs and SOTs are very common. To minimize solder skips, the board should be designed with the land patterns in the recommended orientation with respect to the board travel direction, as shown in Figure 12.3. Running the board over the wave in the recommended direction will

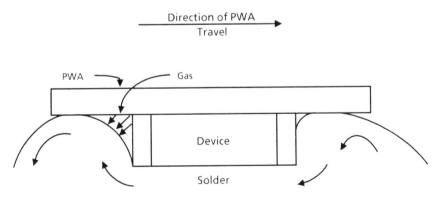

**Figure 12.1 Outgassing in surface mount chip components during wave soldering.**

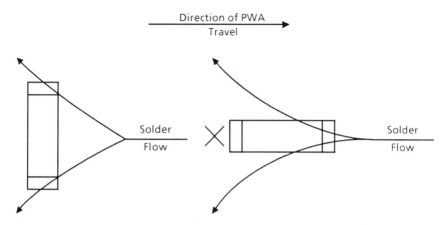

**Figure 12.2  Solder skip ("x" in right figure) in surface mount components during wave soldering.**

prevent shadowing in chip components, SOICs, and SOTs. Failure to observe this convention can lead to solder skips.

If, the layout of the components on the board makes it impossible to solder all terminations as recommended, the board should be oriented in such a way that the majority of terminations comply with the recommended orientation shown in Figure 12.3. Wrong orientation of components should not be allowed during the design phase. Even with the correct orientation

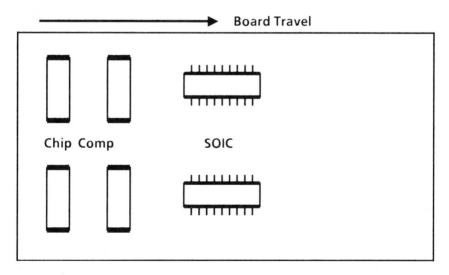

**Figure 12.3  Recommended component orientation for wave soldering.**

of SOICs, bridging on trailing leads is fairly common. Adding one extra dummy pad on each side takes care of this problem. Bridging, if it then occurs, does not disrupt circuit operation because the bridged pad is not connected to a functional circuit.

Providing adequate interpackage spacing and orienting components in such a way that both terminations are soldered simultaneously solves most shadow effect problems. Refer to Chapter 7 for information on interpackage spacing and rules for staggered components to prevent shadowing of terminations during wave soldering.

The other question that is generally asked is whether immersion of components in the wave poses reliability problems. It worth noting that in through-hole assembly, only the leads experience the soldering temperatures. Component bodies stay relatively cooler. There is some question about the reliability of plastic surface mount packages when subjected to reflow soldering conditions. (See Chapter 5, Section 5.7, for details on moisture-sensitive packages, which may be prone to cracking unless baked before reflow soldering.)

A similar question about component reliability can be asked with respect to components that go through the wave. That is, can the surface mount resistors and capacitors withstand the wave soldering temperature when they are dipped in the solder wave at 500°F (260°C) during wave soldering? I have found the maximum shift in tolerance of resistors and capacitors after wave soldering to be 0.2% [1]. This is a negligible amount, considering that the part tolerance of commonly used components is 5 to 20%. The components generally spend about 3 seconds in the wave, but they are designed to withstand soldering temperatures (500°F) for up to 10 seconds.

The impact of the soldering wave on reliability may be different for active plastic packages such as SOICs and PLCCs. In addition to potential shadowing and bridging problems, these packages may be adversely affected if flux seeps in through the lead frame during wave soldering. This can happen because the lead frame and the molding compound have different coefficients of thermal expansion. Seepage of flux inside the package is not a concern in reflow soldering bacause the packages are not predipped in flux. The solder paste flux is generally confined to the leads of the packages.

Passive components may undergo leaching of terminations during wave soldering; that is, there may be dissolution of the precious metal (gold or silver) adhesion layer on the ceramic body. If the precious metal adhesion layer is dissolved, the unsolderable ceramic body is exposed, preventing the formation of solder fillets or causing poor ones. Chapter 10 presents a detailed discussion of leaching and the rates of dissolution of various metals in molten solder.

Leaching can be prevented by a 50 microinch (minimum) nickel barrier underplating between the precious metal adhesion layer and the solder coating. This should be required and should be part of the component procurement specification as discussed in detail in Chapter 3. However, a nickel barrier is not absolutely guaranteed to prevent leaching if the dwell time in the solder wave is too long. Dwell time in the solder wave thus should be kept to a minimum (3-4 seconds).

No matter which soldering process is used, cleanability is a concern because of its impact on product reliability. This is addressed in the next chapter. The cleanability of wave soldered assemblies is especially critical if an active flux is used.

It is repeated many times in this book, and the point cannot be overemphasized, that to ensure process compatibility, all components must be qualified for the processes to which they will be subjected. This must be done before each component is designed into the product.

## 12.1.2    Process and Equipment Variables in Wave Soldering

A variety of equipment is available to combat solder skips, outgassing, and bridging problems in the wave soldering of surface mounted components. Using an effective wave soldering machine is not by itself sufficient to achieve a good yield, however; many equipment- and non-equipment-dependent variables play a role. Before we discuss the equipment itself, let's look at the main generic equipment variables.

The important variables in wave soldering, as suggested in Figure 12.4, are board handling, fluxing, solder profile, and solder wave geometry. These must be characterized and then controlled to achieve good soldering yield. In board handling, allowance in the conveyor system must be made for the thermal expansion of the board during preheating and soldering to prevent board warpage.

In fluxing, flux density, activity, and ratio of flux foam to wave height must be closely monitored. A system must be in place to determine flux density, flux activity, and the timing for dumping the old flux. Machines that come with a flux density controller automatically add flux or thinner to maintain flux density within the specified limits; there are many such machines.

Flux activity monitoring must be done manually, either chemically or by solderability tests on known good terminations or leads. When the flux gets old, the solderability of the known good lead will deteriorate and the flux should be discarded. The soldering equipment presently available lacks

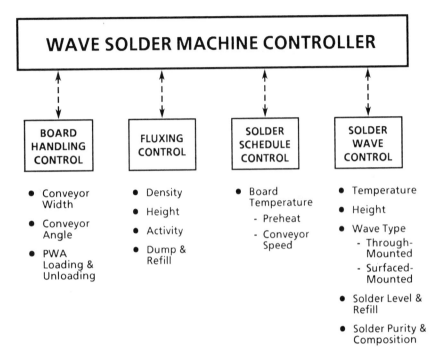

**Figure 12.4   Common equipment variables in a wave soldering machine.**

the capacity to monitor flux activity and give a signal for dumping inactive flux.

Monitoring of flux activity is not an issue if a foam fluxer is used, because the quantity of flux required is small. It may be more economical to dump the flux daily than to determine its activity. However, if a wave fluxer containing many gallons of flux is used, monitoring of flux activity may be desirable to control solder defects.

The role of the solder profile in achieving good soldering yield cannot be overemphasized. A good solder profile helps not only to reduce solder defects but also to prevent cracking of surface mount capacitors, which are prone to thermal shock damage. To understand why this is true, let's examine a bad solder profile (Figure 12.5). Here the rate of heating is too rapid: the board has seen a maximum preheat of only 176°F (80°C) before hitting the first solder wave (first peak in Figure 12.5). The board should attain a top side preheat temperature of 220 to 240°F (104-116°C) before it enters the solder wave. Surface mount capacitors must have good thermal shock characteristics and should be gradually heated, and the temperature

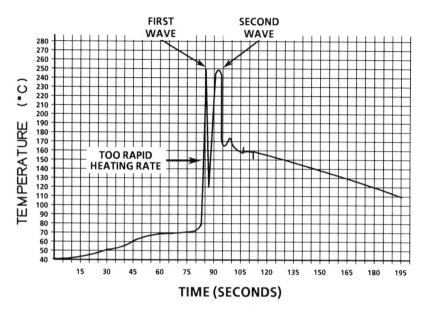

**TIME (SECONDS)**

**Figure 12.5 A bad example of solder profile for dual-wave soldering.**

differential between the preheat zone and the soldering zone should be minimized. This not only prevents thermal shock in surface mount capacitors but also reduces solder skips caused by outgassing of undried flux (see Chapter 14: Section 14.3.2 [Component-Related Defects]).

In controlling the solder profile, preheating a printed wiring assembly in two or three stages mimimizes thermal shock damage and improves the service life of the assembly. Uniform preheating is achieved by developing a solder profile that specifies preheat settings and conveyor speed for each type of board.

Solder wave geometry is the most important variable of all; it must permit the effective soldering of both surface mount and through-hole mount components, both of which are present on the board in a mixed assembly. Wave geometry is important for preventing icicles and bridges in through-hole mount components and for preventing outgassing and solder skips (shadowing) in surface-mounted components.

No matter which wave geometry is used, it must promote the pushing of solder between the tight spaces and into the corners. There are various approaches to accomplish this, as discussed in the next section. Among all the variables mentioned above, wave geometry is the most critical and must be evaluated extensively for its effectiveness in different applications.

## 12.2 TYPES OF WAVE SOLDERING FOR SURFACE MOUNTING

Solder waves designed for the wave soldering of surface-mounted components are intended to solder both tightly packed and incorrectly oriented, difficult-to-solder trailing terminations. Does this mean that the components should not be oriented in the recommended direction as discussed previously? No. If the components are incorrectly oriented, the terminations may be soldered by using specialized wave geometries, but oversized fillets will be formed on the trailing terminations. Such oversized fillets may stress and crack chip components.

The issue of wave geometry was introduced in Chapter 7 to emphasize the need to ensure correct orientation during the design cycle. We bring it up again here because even if the design rule is followed as discussed in Chapter 7, the process engineer should not count on dual-wave or any other wave geometry to solve a problem of uneven fillets. The board must be oriented correctly over the wave (see Figure 12.3) to reduce solder skips and uneven solder fillets.

There are many types of wave soldering equipment available. The main difference lies in the wave geometries. All such equipment ensure that tight spaces, where a conventional single wave might not be very effective, are filled with solder. This does not mean that the other process and equipment variables discussed earlier are not important. They are, but wave geometry is the most critical.

Wave geometries commonly available on the market include unidirectional and bidirectional; single and double; turbulent or vibrating, smooth and dead zone; oil intermix, dry, and bubbled; and with or without a hot air knife. The wave geometries widely used in the industry can be classified as dual wave (with and without hot air knife) and vibrating wave (with nitrogen bubble or ultrasonic transducer).

### 12.2.1 Dual-Wave Soldering

One of the first fixtures on the market for providing solder in the tight spaces between surface mount components was the dual-wave system. As the name indicates, there are two waves: a turbulent wave and a smooth or laminar wave, shown schematically in Figure 12.6. Figure 12.7 shows a dual wave such as would be produced by the equipment illustrated schematically in Figure 12.6.

In the dual-wave system, the turbulent section of the wave, which

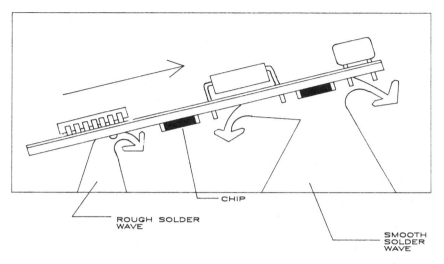

**Figure 12.6    Schematic view of a dual-wave soldering system for SMT.**

**Figure 12.7    Actual view of a dual wave for SMT. (Photograph courtesy of Electrovert.)**

ensures adequate distribution of solder across the board, is the one that prevents solder skips. The solder is pumped through a narrow slit at somewhat high velocity so that the solder can penetrate between the tight spaces. The jet is pointed in the same direction as the board travel. The turbulent wave alone cannot solder components adequately. It leaves uneven and excess solder on joints; hence the need for the second wave.

The second laminar or smooth wave eliminates the icicles (sharp needlelike protrusions of solder on lead ends) and bridges produced by the first or turbulent wave. The laminar wave is actually the same wave that has been used for a long time for conventional through-hole mount assemblies. The laminar wave can either be with or without oil intermix. A thin film of oil reduces oxide formation and loss of tin as tin oxide; but an excessive amount can cause voids in solder joints.

The two pots of the dual-wave system have independent electric motors so that the turbulent wave can be turned off when conventional assemblies are being soldered on the same machine.

In the most common variation of the dual-wave system on the market today, the turbulent wave moves back and forth and the solder jets come out of tiny nozzles instead of one long, narrow slit as discussed above. The moving nozzles are more effective than the narrow slit in preventing solder skips because they both create turbulence and effect washing.

It is not always necessary to buy an entire wave soldering machine to realize the benefits of the dual-wave system. Most conventional wave soldering machines can be retrofitted just by adding the dual-wave pots, at a fraction of the cost of a new machine. The rest of the equipment functions are essentially like conventional soldering equipment.

## 12.2.2    Vibrating Wave Soldering

In the second method for minimizing outgassing and solder skips, nitrogen is injected in the solder wave. The nitrogen bubble wave was introduced by the Koki Company in Japan, but it did not become very popular in the United States. Another variation of the same concept is the ultrasonically vibrating wave known as the Omega Wave in the Electrovert system.

In the Omega Wave a transducer is introduced to generate low frequency vibrations at a controlled amplitude as schematically shown in Figure 12.8. The solder bubbles produced by the vibration of the solder wave (Figure 12.9) penetrate the tight spaces and corners between surface mount components and displace the gas bubbles that cause solder skips. As the board exits, the oscillations gradually decrease, and components

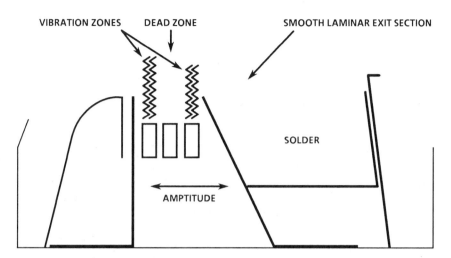

**Figure 12.8    Schematic view of the Omega Wave vibrating single-wave soldering system for SMT.**

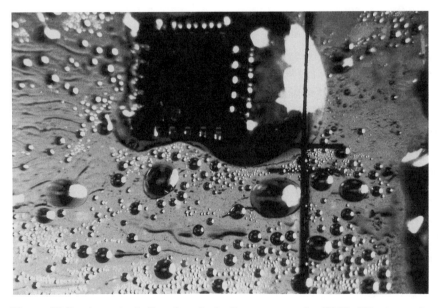

**Figure 12.9    An actual vibrating single Omega wave for SMT. (Photograph courtesy of Electrovert.)**

travel through the conventional laminar wave to eliminate bridging and icicles [2].

The Omega Wave replaces the dual-wave system from Electrovert and may be preferred for economic reasons. For example, the latter requires dual pots and dual controls for temperature and wave height and an excessive amount of solder (1200 pounds for dual wave versus 500 pounds for the Omega Wave).

There will be more loss of tin in a dual-wave pot because of the additional atmospheric exposure of solder. Thus tin will have to be added more frequently to maintain the required percentage of tin in the solder pot. This may have a substantial impact on the initial and subsequent outlays of expenses for solder. According to one study, the dual wave is not as effective in filling holes and eliminating solder skips as the Omega Wave [3].

For reasons of economy and quality, therefore, a single vibrating wave may be preferred over a dual wave. A wave soldering machine providing dual and Omega wave geometry options is shown in Figure 12.10. Some companies prefer system in which a solder wave is combined with a hot air debridging system.

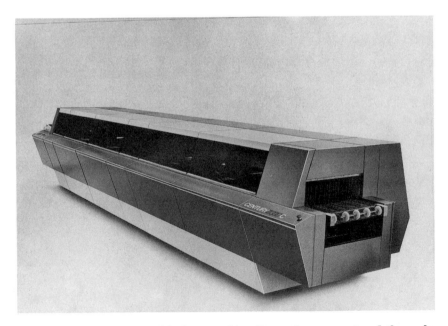

**Figure 12.10  A wave soldering machine for surface mount and through-hole mount assemblies. (Photograph courtesy of Electrovert.)**

## 12.2.3 Modified Wave Soldering

Although the wave solder systems have long been used in industry, they have been plagued with undesirable icicles (protrusions of solder on the lead ends) and bridges (solder deposits between circuits). These defects are not acceptable because icicles can pierce through conformal coating and can also adversely affect printed wiring assembly (PWA) performance especially in high speed circuits, and bridges cause short circuits. The literature chronicles many attempts to reduce icicles and bridges by such means as varying the amount of oil in the wave, preheating the boards, changing conveyor speed, and the hot air knife system just mentioned. But the wave geometry plays the most critical role. To prove this point the author conducted a series of systematic tests designed to determine the basic cause of icicles and bridges at the Boeing Electronics Company Systems Division (now Boeing Electronics Company). The wave solder machines at various commercial and military production sites were producing icicles and bridges on 75-90% of the solder joints of some products. These defects had to be touched up which not only increases cost but also impacts product reliability.

The objective of improving solder joint quality was achieved by modifying the conventional Hollis turbulent wave (Figure 12.11a) into a wave with a static, or "dead," zone in the solder wave (Figure 12.11b). This was done by attaching a welded stainless steel dike (Figure 12.11c) to the far end of the wave where the assembly exits. The dike redirects waves moving upstream back downstream, so that they collide with other waves moving upstream. A static or dead zone is created where the waves collide. See Figure 12.11b. The static waves peel off much more slowly than the conventional waves, which fall rapidly toward the bottom of the solder pool. The modified wave virtually eliminates icicles and bridges.

The dike also improves top side fillets and prevents excess oil entrapment (oil entrapments cause voids in the solder joints). The modified wave creates a solder pool next to the dead zone. The pool postheats the bottom side of the board as they pass over it, preventing premature freezing of solder. This helps in formation of better top side fillets. Oil entrapment is reduced because the modified solder wave prevents the formation of icicles and bridges on the PWA, permitting the amount of oil required to be reduced substantially.

The dike is simple, easy to install and costs under $100, yet the system is highly effective in eliminating icicles and bridges in addition to reducing oil entrapment and improvement in solder fillet quality. The dike was installed on all Boeing Electronics Systems Hollis machines. For reducing the touch-up cost significantly, it was selected as one of the best 10 inven-

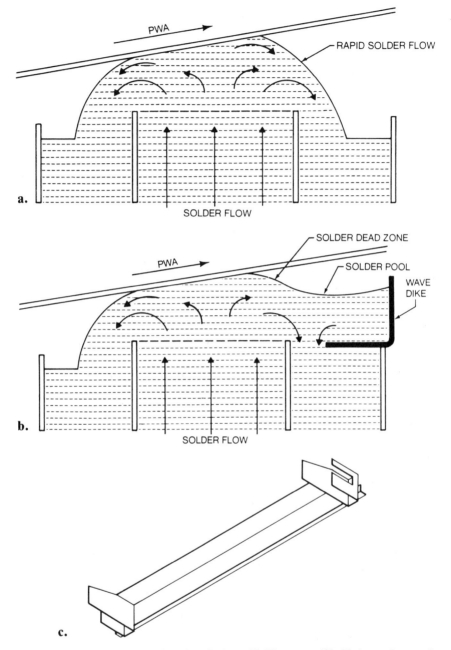

**Figure 12.11    A conventional turbulent Hollis wave (12.11a) can be modified into a static wave with "dead" zone (12.11b) by the wave dike (12.11c). The dike which virtually eliminates icicles and bridges was developed by the author. It proves the importance of wave geometry in reducing solder defects.**

tions by the Boeing Aerospace Company in 1982. So as mentioned earlier, the design of the wave geometry is most cortical in reducing solder defects. This is true not only for conventional assemblies but also for the surface mount assemblies and should be the primary consideration when selecting a new wave soldering system.

## 12.3   WAVE VERSUS REFLOW SOLDERING

The selection of a specific soldering process for surface mounting depends on the mix of components and the form of SMT assembly (Type I, II, or III) to be soldered. In addition, throughput requirements, use of temperature-sensitive components, and the need for specialized rework or soldering also come into the seletion of a soldering process. Each process has its pros and cons, and none is suitable for all applications.

Wave soldering is well suited for through-hole and surface mount component combination (Type III SMT). Actually no other process can compete with the cost effectiveness of the wave soldering process for through-hole and Type III SMT assemblies. However, for Type I SMT, wave soldering is not even considered. Reflow soldering remains the only viable option.

## 12.4   SINGLE-STEP SOLDERING OF MIXED ASSEMBLIES

It is generally accepted that wave soldering and reflow soldering are the most effective means to solder Type III and Type I assemblies, respectively. However, the mixed assemblies (Type II SMT) require both reflow and wave soldering processes. Generally the soldering of Type II SMT assemblies is accomplished in a two-step soldering process. The surface mount components are placed on solder paste on the top side and reflow soldered. Then the through-hole mount components are inserted and the entire assembly is wave soldered.

Equipment is available to solder mixed assemblies in one step. The driving force for single-step soldering is cost savings. There are two approaches to soldering such an assembly in one step. In the first approach, a wave soldering machine is outfitted for reflow capability by the installation of infrared heating panels or hot air systems. Both wave and reflow soldering operations are then performed by the same machine. The surface mount components are placed on screened solder paste and then the through-hole mount components are inserted. The assembly goes through a wave soldering system for through-hole mount parts and immediately passes through a hot air reflow system for soldering of surface mount components. Such a system is available from Hollis Automation.

The Hollis system also uses a hot air knife for removing icicles and bridges from through-hole and mixed assemblies. It can be argued that a hot air debridging system is necessary only if the wave geometry was not adequately designed to prevent formation of icicles and bridges in the first place. On the other hand, such a system can be used as a solderability tester in a production environment because the hot air completely removes solder from otherwise nonwetted solder joints only.

Another approach to accomplish the task of one-step soldering for mixed assemblies is by combining the wave pot in a vapor phase machine. In a system available from the Dynapert Corporation, fluorocarbon liquid (FC-70 or FC-5311) heats the wave solder pot for soldering through-hole mount parts, and the surface mount components mounted on the top side are reflowed with the fluorocarbon vapor.

The driving force for a single-step soldering process is the cost saving realized by eliminating a soldering step and a cleaning step. The cost saving may disappear, however. If the surface mount components sitting on solder paste shift when the through-hole mount components are inserted, touch-up may be required. Besides, soldering time is a small fraction of total assembly time; placement and testing times contribute most to the cost of assembly. Thus one should carefully evaluate the advantages and disadvantages of single-step soldering for mixed assembly before using it in production.

## 12.5   SINGLE-STEP SOLDERING OF DOUBLE-SIDED SMT ASSEMBLIES

Generally, the soldering of a two-sided full surface mount assembly is accomplished in two steps. In this case no adhesive is necessary because when the assembly is turned over after the first side has been reflowed, the surface mount components are held in place by the surface tension of the solder. However, reflow soldering of a double-sided assembly can be accomplished in one step by using adhesive on one side.

In a single-step soldering process, the bottom side surface mount components are placed on the solder paste and also glued in place with adhesive, which is dispensed after screening of the paste. Adhesive is needed because the tackiness of the paste alone would not hold components in place when the assembly is turned over, and the solder cannot hold components in place by surface tension unless it has had a chance to reflow. The top side surface mount components are held in place on the solder paste; no adhesive is used.

When the assembly passes the reflow system (vapor phase or IR), both sides can be soldered simultaneously. The process steps are as follows: screen paste, dispense adhesive, place components, and cure/bake the adhesive/paste; invert the board and repeat the process, except for dispensing the adhesive.

As in the single-step soldering of mixed assemblies, the soldering of double-sided surface mount assemblies saves money by skipping one soldering/cleaning step. However, the time saved in the one-step soldering of double-sided surface mount assemblies may be spent in dispensing and curing glue to hold the bottom side components. In fact, depending on the number of components, even more time may be spent in dispensing adhesive than is saved on the soldering and cleaning steps.

Thus the advantages of single-step soldering over double-step soldering are not quite obvious unless the bottom side surface mount devices must be held with adhesive because of poor lead solderability. Components with poor solderability do not provide enough surface tension to prevent the bottom side components from falling off. However, instead of trying to use adhesive to solve a solderability problem, an attempt should be made to eliminate the problem.

## 12.6   VAPOR PHASE SOLDERING

Vapor phase soldering (VPS), also known as condensation soldering, uses the latent heat of vaporization of a liquid to provide heat for soldering. This latent heat is released as the vapor of the inert liquid condenses on the component leads and PCB lands. In VPS the liquid produces a dense saturated vapor, which displaces air and moisture. The temperature of the saturated vapor zone is the same as the boiling point of the perfluorocarbon liquid. The peak soldering temperature is the boiling temperature of the inert liquid at atmospheric pressure.

VPS does not require control of the heat input to the solder joints or the board. It heats uniformly, and no part on the board, irrespective of its geometry, ever exceeds the fluid boiling temperature. VPS is also suitable for soldering odd-shaped parts, flexible circuits, pins, and connectors, as well as for reflow of tin-lead electroplate on boards and surface mount package leads.

VPS is an easily automated process. Both in-line and batch-type equipment systems are used. Batch systems are designed for research and development or for low volume production. In a batch system, as shown schematically in Figure 12.12, the board is lowered in a basket from the top through the secondary vapor into the primary vapor. After a predefined dwell time in the primary vapor zone, the assembly is lifted into the secondary zone, and kept there for about 30 to 60 seconds to drain off the primary fluid.

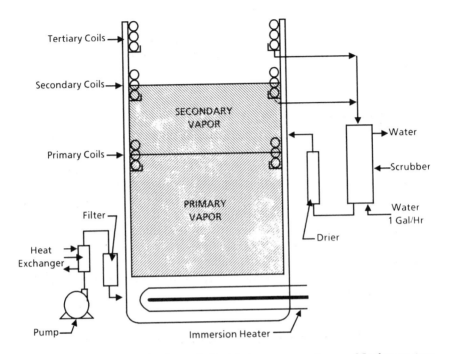

**Figure 12.12   Schematic view of a batch-type vapor phase soldering system for SMT.**

The primary vapor is an inert perfluorocarbon, such as FC-70 or FC-5311 (from 3M Corporation), and the secondary vapor is usually Freon TE. The boiling point for FC-70 and FC-5311 fluids is 419°F (215°C). The boiling point for Freon TE is about 118°F (48°C). However, superheating by the primary vapor causes the temperature in the secondary zone to reach 185 to 225°F (85-107°C).

In-line systems are designed for mass production. In an in-line vapor phase system, the board travels on a conveyor from the left to right; no secondary vapor is used. Generally in the older equipment, the conveyor belt is inclined before and after the reflow zone (i.e., the belt slopes down before entrance to the reflow zone and then slopes up when exiting the reflow zone). If the boards travel in the inclined position, however, the components, especially large ones, tend to slide or move. New equipment is available from some vendors that allows boards to travel in a level position (Figure 12.13), thus eliminating this problem [4].

The major disadvantages for batch and in-line processes are the high cost of the primary liquid (about $600/gallon) and vapor loss during use. Even though the vapor loss rate is similar in both processes, the batch

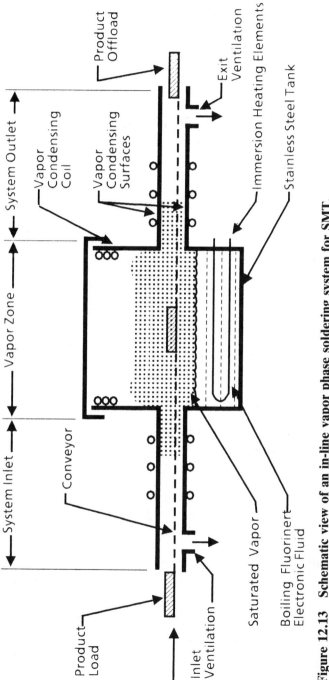

**Figure 12.13 Schematic view of an in-line vapor phase soldering system for SMT.**

process uses a less expensive secondary fluid as a blanket over the primary fluid, minimizing the cost of vapor loss. In both systems, cooling coils help in reducing vapor loss. In addition, some equipment vendors have designed vapor recovery systems that have reduced fluid costs to less than $2/hour or $0.02/board for a 12 inch in-line system [4].

## 12.6.1    The Heat Transfer Mechanism in Vapor Phase Soldering

When the assembly to be soldered enters the primary vapor zone, the vapor condenses on its surfaces, releases its heat of vaporization, and forms a thin, continuous liquid film. The heat transfer rate to the surface of any part at any instant is controlled by the equation

$$Q = hA \ (T_v - T_s) \tag{1}$$

where $Q$ = heat transfer rate from vapor to part (Btu/hour)
$h$ = heat transfer coefficient (Btu/hour·ft$^2$·°F)
$T_v$ = saturated vapor temperature (°F)
$T_s$ = part surface temperature (°F)
$A$ = product surface area (ft$^2$)

and $h$ is determined by the thermal conductivity, viscosity, and density of the liquid condensate, and by whether the surface of a part is vertical, horizontal, or oblique [5].

It is apparent from equation (1) that only the maximum surface heating rate and the maximum surface temperature of any part on a surface mount assembly can be controlled during VPS.

The maximum surface heating rate of a part occurs when $T_s$ is minimum, that is, when the part is initially immersed in the primary vapor zone. The initial $T_s$, hence the maximum heating rate of any part, can be controlled by preheating the part before immersion in the primary vapor zone.

The maximum surface temperature of any part will depend on the dwell time of the assembly in the vapor zone. With increasing part dwell time in the primary vapor, $T_s$ will asymptotically approach $T_v$, and the heating rate will approach zero. Given sufficient dwell time in the primary vapor, the surface temperature of all parts of a surface mount assembly will eventually reach a maximum of $T_v$.

The dwell time of the assembly in the primary vapor zone is fixed by the conveyor speed in an in-line systems and by the elevator speed and dwell time in batch systems. As shown in Figure 12.14, if the dwell time is too short, the surface of lead does not reach the maximum possible temperature. Figure 12.15 shows that when the dwell time is sufficient, the lead temperature reaches the maximum possible (215°C). For successful soldering, it is recommended that all PCB land and component lead surfaces on a surface mount assembly reach a temperature 30 to 50°C above the melting point of the solder.

The total heat requirement from vapor condensation is determined by the thermal mass of the assembly. Assemblies with a larger thermal mass will require a larger or supplementary heating coils. VPS systems with small capacities may be unable to supply enough vapor quickly enough to solder assemblies with large thermal masses because it takes some time for the liquid to boil.

Some VPS systems provide a constant supply of vapor by having large thermal heat sinks above the liquid level, inside the primary zone, in addition to heating coils below the liquid. A heat sink here acts as a thermal flywheel and keeps the vapor available in the quantity necessary for the reflow of larger masses of assemblies.

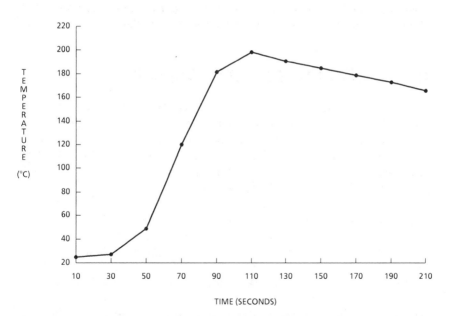

**Figure 12.14** **The temperature-time soldering profile of a PLCC lead, illustrating the impact on maximum lead temperature of a short dwell time (10 seconds) in the primary zone of a batch VPS system.**

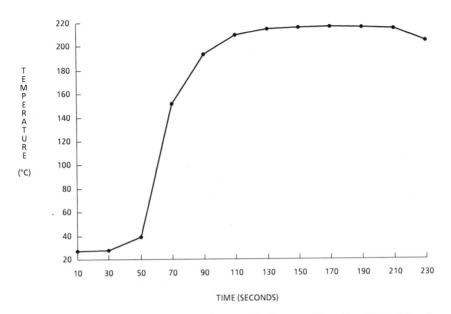

TIME (SECONDS)

**Figure 12.15  The temperature-time soldering profile of a PLCC lead, illustrating the impact on maximum lead temperature of a long dwell time (2 minutes) in the primary zone of a batch VPS system.**

## 12.7  INFRARED REFLOW SOLDERING

The convection/infrared (C/I) and near-IR infrared soldering processes differ in their heat sources and in their heating mechanisms. For C/I, direct infrared radiative heating from panels occurs in the middle infrared wavelength region (2.5-5 μm) and accounts for only 40% of the heat input to the parts being soldered. The remaining 60% of the heat is provided by convection of the hot air in the oven.

For near-IR, almost all the heating is achieved by infrared radiation in the short wavelength range. Convection accounts for less than 5% of the heating. The wavelength range for infrared radiation is 0.72 to 1000 μm. However, the wavelength range of 1 to 5 μm is used for the thermal processing of electronic assemblies; the 1-2.5 μm range is considered to be short or near wave IR and the 2.5-5 μm range medium wave IR [6].

Figure 12.16 shows a schematic diagram of a typical IR oven. Such equipment has 8 to 20 independently controlled heating panels with built-in thermocouples for temperature control. IR panels heat the board traveling on the conveyor from top and/or bottom, as necessary. In the C/I ovens, internal fans circulate hot air or inert gas such as nitrogen. Cooling fans are generally used at the exit.

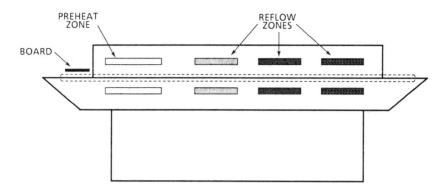

**Figure 12.16 Schematic view of an in-line infrared reflow soldering system for SMT.**

A gradual heating of the assembly is necessary to drive off the volatile constituents of the solder paste. After an appropriate preheat time, the assembly is raised to the reflow temperature for soldering and then cooled.

## 12.7.1 The Heat Transfer Mechanism in Infrared Soldering

Unlike conduction and convection, radiative heat transfer does not require physical contact. Gases such as air and nitrogen are transparent to infrared radiation and do not play any role in pure radiation heat transfer systems. The radiation heat transfer rate between two bodies at different temperatures is controlled by the following equation [6]:

$$Q = F_e F_v A K_B (T_e^4 - T_P^4) \qquad (2)$$

where $Q$ = radiation heat transfer rate (Btu/hour)
$F_e$ = emissivity factor for emitter and product
= 1 for a blackbody (a perfect emitter)
= 0.80 to 0.90 for lamp systems on a printed circuit board
$F_v$ = geometric view factor (amount of incident IR energy on product surface)
= 0.95 to 0.98 due to openings for conveyor belt
$A$ = emitter area (ft²)
$K_B$ = Boltzmann constant ($0.1714 \times 10^{-8}$ Btu/hr-ft²-R⁴), where R is the absolute temperature (Rankine)
$T_e$ = emitter temperature (°R)
$T_p$ = product temperature (°R)

The wavelength of the radiation produced is determined by the emitter temperature as defined by Wien's displacement law given below:

$$W_p = \frac{K_W}{T_e} \tag{3}$$

where $W_p$ = peak emission wavelength ($\mu$m)
$K_W$ = Wien's constant (5215.6 $\mu$m-°R)
$T_e$ = emitter temperature (°R)

From equation (3), shorter peak wavelengths will require higher temperature emitters. Near-IR systems have lamp heaters, which produce shorter wavelength radiation because they operate at higher temperatures. The panel heaters in C/I systems run at lower temperatures, hence produce longer wavelength radiation.

## 12.7.2    Convection/IR Versus Near IR

Because of the nature of the heat source, as discussed above, each system has its pros and cons. In C/I, solder paste may not be completely cured because of a tough "skin" that forms as a result of circulating air. Formation of a skin can lead to "explosion" during reflow [7]. In a near-IR system, the short wave penetration into the board allows faster heating and complete curing of solder paste. This prevents "explosion"; hence solder splatter and solder balls are minimized [7].

The near-IR system has drawbacks too, namely color sensitivity and the so-called shadow effect. Let us take color first. Short wavelength infrared radiation, prevalent in near-IR systems, is not absorbed uniformly across the color spectrum. Hence an assembly consisting of packages having different colors, such as ceramic and plastic component packages, will heat nonuniformly. Furthermore, since convection heating is virtually absent, regions on a board that are not directly exposed to the radiative heat sources [small $F_V$ in equation (2)] will be cooler than others having a direct exposure—again resulting in nonuniform heating. This is the "shadow" effect.

In C/I systems, on the other hand, lower temperature panel heaters produce longer wave-lengths, and heat transfer takes place by natural convection and IR radiation of medium wave-length. The C/I system, by relying on both IR and convection energy, eliminates such once-common IR reflow soldering problems as shadow effects [8]. Color sensitivity is also diminished, since the absorptivity of medium infrared wavelength radiation is uniform across the color spectrum.

Which is better for reflow soldering—near IR or C/I? In any IR system, the product is heated by conduction, convection, and radiation. The convective portion is dominant in panel systems but plays a minor role in lamp systems. Variation in component sizes and colors can be overcome with near-IR systems if a fine-tuned solder profile is developed. Also, if the products in a company's line do not vary significantly, a near-IR system may be entirely acceptable. However, for a company that makes products of different varieties, a near-IR system may require more work in solder profile development than a C/I system. By properly characterizing the process and by developing a detailed profile, either system can give good results. Excellent results have been produced by near-IR systems [9] as well as by C/I systems [10].

## 12.8 VAPOR PHASE VERSUS INFRARED REFLOW SOLDERING

There is considerable debate in the electronics industry over reflow soldering processes: Which should be preferred; VPS or IR? For each process, there are technical issues and ways to resolve them. Tables 12.2

**Table 12.2 Technical issues and their resolution in infrared soldering process.**

| ISSUE | APPROACH |
|---|---|
| • Board discoloration | • Investigate use of nitrogen in the oven<br>• Evaluate new solder profile<br>• Evaluate new solder mask |
| • Uneven heating of board between center and edge, excess heating of some parts, and board warpage | • Establish design rules for large packages at the edges and smaller ones at the center<br>• Use shields for components<br>• Use fixture to absorb heat<br>• Change reflow profile<br>• Use reflective masks for temperature-sensitive components |
| • Components affected by shadow effect and color sensitivity | • Use infrared with convection and longer/medium wavelength IR |

and 12.3 outline technical issues and possible approaches for IR and vapor phase soldering, respectively.

Let us now state some important process issues; in the subsections we will compare performances on such specific issues as cost, flexibility, and solder defects. The two main reflow processes, VPS and IR, differ in their mechanism of heat transfer as discussed earlier. In IR reflow both heating and cooling rates of assemblies can be controlled more precisely than in vapor phase. This has a direct impact on solder defects.

If the predominant heat transfer mechanism is by radiation (near IR), the IR process may be affected by color and by "shadows" of adjacent parts. Also, in the IR process, the packages supplied by some vendors may be susceptible to thermal damage. For example, surface mount connectors may be susceptible in IR because of the longer time in the oven. In addition, some heat-sensitive components require reflective masks to protect them from IR radiation. Finally, in the IR process, the total soldering time, especially in the reflow zone, is longer. This causes formation of a thicker layer of intermetallics, which is not desirable. The intermetallics should be as thin as possible, to prevent formation of brittle solder joints.

Japanese companies use gull wing packages almost exclusively and attach them by means of IR reflow soldering. They are very reluctant to use the IR process for J-lead packages and tend to favor the vapor phase process instead. It is sometimes incorrectly believed that the IR reflow process is more suitable for gull wing type packages than for J lead type packages. Nevertheless, IR can be used successfully for any lead configuration.

**Table 12.3  Technical issues and their resolution in Vapor phase soldering process.**

| ISSUE | APPROACH |
|---|---|
| • Prevention of vapor loss | • Reduce effective throat of in-line system<br>• Use condensing coils<br>• Control exhaust pressure |
| • Prevention of part movement and other defects such as wicking | • Control solder paste characteristics<br>• Control solder profile<br>• Cure paste<br>• Ensure solderability of lead termination and solder pads<br>• Minimize lead coplanarity |

## 12.8.1    Cost and Flexibility

The IR process is more versatile than VPS for surface mounting. For example, the IR oven can be used for curing adhesive, baking paste, and reflow soldering, whereas vapor phase equipment can be used only for soldering. This means that the capital investment requirement for two machines can be halved if IR is selected instead of vapor phase for reflow soldering.

As mentioned earlier, even though the perfluorocarbon fluid for vapor phase costs about $600 a gallon, systems are now available that have reduced operating costs to pennies per board [4].

It is generally suggested that because of its controlled peak temperature, vapor phase is more flexible. For this reason VPS is claimed to be more desirable for subcontract assembly houses in which boards of various types are processed daily. Lambert [11] has suggested that the cost of operating a VPS is lower because there is less machine down time (4 minutes for VPS versus 15 minutes for IR) between profile changes for different products to be processed during a given day. However, even Lambert admits that IR has an edge over VPS if there are minimal board changes per shift [11].

Lower cost in VPS is not necessarily true, however. As we will see later on, the incidence of solder defects is generally higher in the VPS process. If the incidence of repair/rework is at an unacceptable level in vapor phase, the time saved in solder profile development may not matter. Excessive rework not only requires added time but can adversely affect solder joint reliability as well. Hence, taking both the repair and profile set up time into consideration, the IR process may have lower operating cost than VPS.

There are some environmental and safety concerns associated with the use of perfluorocarbon liquids. These substances cause acid buildup within the machine under heavy thermal stress or overheating due to flux buildup. The primary liquid slowly decomposes to form a toxic gas. Stabilized perfluorocarbon liquids that are less prone to decomposition have become available recently [4]. Use of stabilized liquids and regular cleaning of heating elements will reduce the potential for formation of undesirable corrosive by-products. Also, batch VPS uses Freon (secondary vapor) which contains chlorinated fluorocarbons (CFC). Use of CFC will be restricted in the future. (See Chapter 13, Section 13.6.)

## 12.8.2    Solder Profile Development

The definition of a soldering profile for any board product using VPS calls for the establishment of preheat temperature, elevator/conveyor speed, and dwell time in the primary zone. To define a soldering profile for IR, one must set conveyor speed and panel temperatures. These are the major variables in profile development. Experimental boards with representative thermal mass, size, and component density should be used for solder profile development.

Infrared requires an added investment in engineering time to develop custom profiles because of the need to establish conveyor belt speed and settings for various heating panels. Many in the industry develop solder profiles for IR by changing the conveyor speed only, leaving the panel temperature the same for all boards. This is certainly easier, but it may not be the right approach because changing the conveyor speed changes the temperature of the board in every IR zone.

The vapor phase process also requires a custom solder profile, but its development is much more straightforward because the peak fluid temperature is essentially fixed. The only variable that requires adjustment in the in-line vapor phase is conveyor speed, and even this is necessary only if the thermal mass of the board varies significantly.

During the IR process, the surface temperature of the board is not uniform throughout its length and width. Board edges are 10 to 20°C higher than the center. This characteristic requires the designer to locate larger components near the board edges and smaller components near the board center. If larger components are placed in the center, either they will not solder or the smaller components will have to be heated to a higher temperature than otherwise is necessary.

If all the components are of the same size, one will have to either live with nonuniform heating or design some fixtures to shield the edges from heat. The latter option is fairly easy to implement, since fixtures are generally needed anyway, especially if there are any components on the bottom side. However, the effect of fixtures on the temperature profile must be considered.

For VPS, there is no temperature gradient across the surface of the board; therefore no component location design rules for even heating are necessary. Table 12.4 lists the important reflow parameters and their recommended limits.

### 12.8.2.1    Heating Rate

To avoid thermal shock to sensitive components such as ceramic chip resistors, their maximum heating rate, also called the maximum ramp rate of the temperature, should be controlled. For infrared reflow, the maximum heating rate can be maintained below 2°C/second in most in-line IR ovens.

**Table 12.4    Limits for critical reflow profile development.**

| REFLOW PARAMETERS | INFRARED | VAPOR PHASE |
|---|---|---|
| Heating rate | 2–6°C/second | 15–50°C/second |
| Temperature in preheat zone | 100–120°C | 90–100°C |
| Time above melting point of solder | 20–80 seconds | 20–80 seconds |
| Reflow temperature | 210–230°C | 205–215°C |

For batch vapor phase reflow, because of the sharp temperature transition between the primary and the secondary vapor zones, the maximum heating rate cannot be maintained below 50°C/second. With proper elevator or conveyor speed and preheat temperature adjustment, however, the maximum heating rate can be controlled around 15°C/second. For in-line vapor phase, the belt speed should be adjusted to control the heating rate at 15°C/second.

### 12.8.2.2    Peak Temperature in Preheat Zone

During infrared reflow, the boards are preheated in the initial zone of the oven, which is also called the preheat zone. To avoid overbaking the solder paste and exceeding the glass transition temperature of the epoxy in FR-4 boards, be sure that the peak temperature of the boards and the solder paste does not exceed 120°C in the preheat zone. However, this temperature limit is not significant since preheat and reflow zones are integrated within the same machine.

For vapor phase reflow, boards are generally preheated in a separate oven (either a batch convection oven or an in-line infrared oven). In any case, the peak temperature of the boards and the paste during preheat should not exceed 100°C. If the temperature exceeds this limit, solder ball formation, due to oxidation of solder paste, may become a serious problem.

### 12.8.2.3    Time above Solder Melting Point

It is recommended that the solder at the joint be kept above its melting point for a minimum of 20 seconds and a maximum of 80 seconds. The solder should be given sufficient time above its melting point to flow and wet the lands and the leads.

Extended duration above the solder melting point temperature will damage the board and sensitive components. Moreover, excessive intermetallic compound formation between the solder and the base metal of the lands and leads will occur, which may degrade solderability and solder joint fatigue resistance.

### 12.8.2.4    Peak Reflow Temperature

The peak temperature of the solder joint during reflow should be high enough for adequate flux action and solder flow to obtain good wetting. However, it should not be high enough to cause component and board damage or discoloration. In infrared reflow ovens, the peak reflow temperature is primarily controlled by the panel temperatures in the reflow zone and secondarily by the conveyor speed. The peak temperature of the solder joint should remain between 220°C and 230°C. However, peak temperatures as high as 240°C are seen in IR and should be avoided.

In VPS, the peak reflow temperature of the solder joint is determined by the boiling temperature of the primary fluid and the dwell time of the board in the primary zone. The standard primary fluid used for surface mount assemblies has a boiling point of 215°C at sea level. Hence, accounting for elevation changes, the peak temperature of a solder joint during reflow should be limited to the 205-220°C range.

### 12.8.2.5    Cooling Rate and Duration above Glass
###          Transition Temperature

The cooling rate of the solder joint after reflow is important because the faster the cooling rate, the smaller the grain size of the solder, hence the higher the fatigue resistance. If fans are used to cool the boards after they exit the oven, a cooling rate of 10°C/second at the solder freezing point can be easily achieved. However, if there are no cooling fans, the cooling rate at the solder freezing point will be less than 1°C/second.

The time the board temperature is above 150°C should be limited to 2 minutes, maximum. Glass epoxy FR-4 boards can be damaged if kept above this temperature for an extended period because the glass transition temperature of the epoxy is exceeded.

## 12.8.3   Solder Defects

Figure 12.17 compares the solder defects obtained on surface mount boards using IR and VPS. Except in the cases of solder balls and tombstoning, considerably fewer defects were observed after IR than after VPS. Other researchers have also found better results with IR [9]. The causes and remedies of these defects are discussed next.

It must be pointed out that the defect levels of vapor phase soldering can be brought to rough equivalence with IR if all the process variables are fine-tuned. For example, the solder paste used for vapor phase may not necessarily be suitable for the IR process. Other ways to reduce the defect levels in vapor phase involve an adequate preheat profile, tighter lead coplanarity specifications, thicker solder paste printing, and use of

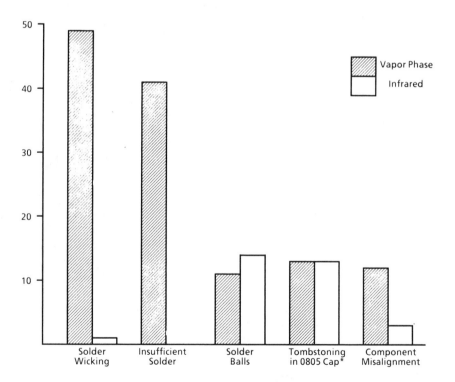

**Figure 12.17   A comparison of the solder defects obtained on a surface mount assembly after each of the two soldering processes: VPS and IR.**

silver in the solder paste. But these steps will help to reduce the defect level in the IR process as well.

### 12.8.4 Solder Opens (Wicking)

As seen in Figure 12.17, solder joint opens and insufficient solder joints are less common in IR than in VPS. Most such defects after VPS occur when the solder wicks up the J leads of PLCC packages. This phenomenon is caused by the lead and the land heating at different rates during VPS, and it is exacerbated by the failure of the lead to touch the land (noncoplanarity of the lead). The lead and land surface temperatures are compared in Figures 12.18 and 12.19 for VPS and IR, respectively.

As shown in Figure 12.18, the surface of the lead reaches the melting point of the solder 16 seconds before the surface of the land during VPS. This causes the solder paste to melt, wet, and wick up the lead before the land becomes hot enough for the molten solder to wet it. By the time the land has reached the melting point of the solder paste, there is no solder left on the land to wet it and form a solder joint.

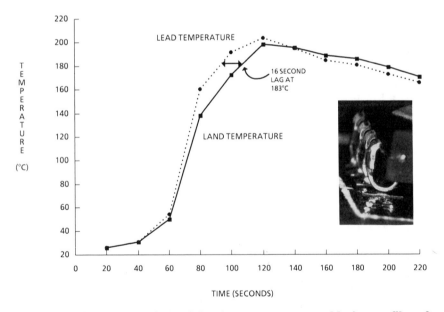

**Figure 12.18 A comparison of the time-temperature soldering profiles of a PLCC lead and land without preheat during VPS in a batch system. The lead and the land reach the solder melting point temperature (183°C) 16 seconds apart. Inset: a solder open caused by the wicking phenomenon.**

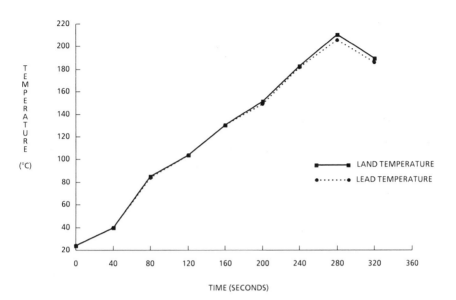

TIME (SECONDS)

**Figure 12.19   A comparison of the time-temperature soldering profiles of a PLCC lead and land during IR. There is no lag between the lead and land temperatures.**

The inset in Figure 12.18 displays the appearance of a solder open caused by wicking, most of the solder is situated on the top part of the J lead. On the other hand, during IR soldering, the lead and the land reach the melting point almost simultaneously, as shown in Figure 12.19. Hence, opens caused by solder wicking are almost nonexistent in a properly pro- filed IR process. In equation (1) $h$ is higher for a vertical surface than for a horizontal surface [5], as discussed earlier (section 12.6.1).

In any soldering process, typically the lead is vertical and the land is horizontal. Therefore, the lead heats faster than the land during the heating phase for VPS. If the land is vertical and the lead is horizontal (as when the board is placed vertically in the primary vapor zone), the heating rate differential between lead and land should diminish. This is confirmed by Figure 12.20, which shows that the period that elapses between the times the lead and the land reach the solder melting point decreases to 3 seconds when the board is vertical. However, keeping the board vertical during VPS is not a practical solution.

It is generally suggested that the practical way to reduce the incidence of solder wicking during the VPS is to preheat the assembly [4,12]. Pre- heating does reduce the problem but will not eliminate it—the lead heats faster than the land even after preheat. This is confirmed by Figure 12.21,

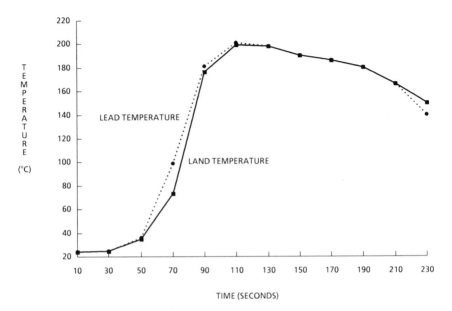

**Figure 12.20** **A comparison of the time-temperature soldering profiles of a PLCC lead and land without preheat during VPS in a batch system. The board was lowered in the primary zone vertically; therefore the lead was horizontal and the land vertical.**

which shows a difference of 10 seconds between the lead and the land in reaching the solder melting point with preheat. As noted earlier, this lag was 16 seconds when no preheat was used (Figure 12.18).

Another approach, suggested by McLellan and Schroen of Texas Instruments, is to change the alloy composition [13]. The basic concept is to use a solder paste with particles made up of pure tin and 10/90 solder (10% tin and 90% lead) instead of particles of eutectic (63% tin and 37% lead) solder particles. However, the proportion of pure tin and 10/90 tin was chosen to make the net composition of the entire paste eutectic. Using this special solder paste, McLellan and Schroen found the following dramatic improvement in solder opens: they were reduced from 250-2000 to 0-130 opens per million joints. In addition, the incidence of voids was reduced when the paste containing 10/90 tin was used.

What is the reason for using two different compositions of solder particles that have higher melting points than the eutectic solder? Since the net composition is still eutectic, the higher melting particles simply delay the melting time, but this is precisely the point. That is, the investigators wanted to delay the melting of paste until the pad had reached

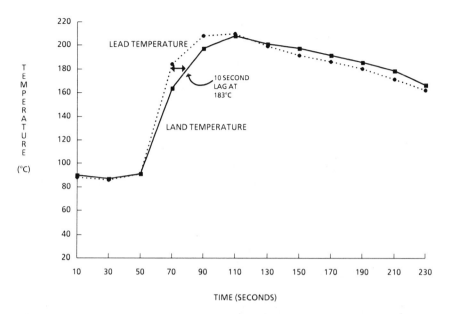

**Figure 12.21    A comparison of the time-temperature soldering profiles of a PLCC lead and land with preheat during VPS in a batch system. The lead and the land reach the solder melting point temperature (183°C) 10 seconds apart.**

the melting point of the solder. Such an approach compensates for the lag time until the lands reach the melting point of solder, as discussed earlier (Figures 12.18 and 12.21). All the particles eventually mix by solid state diffusion. The amount of reflow is delayed by the rate of diffusion, which in turn is controlled by the size of the particles making up the paste.

## 12.8.5    Tombstoning and Part Movement

If the heating rates during reflow soldering are too high, the volatiles in the solder paste will rapidly evolve, possibly causing tombstoning, or standing on end, of components. Design of chip component land patterns also plays an important role in tombstoning. If the land width or the gap between the lands is too large, tombstoning can occur irrespective of the soldering process. The instances of tombstoning reported in Figure 12.17 were due to improper land pattern design.

Since tolerances for component termination widths vary from 10 mils to 30 mils, components with unequal termination widths are fairly common.

These unequal termination widths also can cause unequal upward forces on the component ends, leading to tombstoning. Poor chip component solderability can cause tombstoning, as well, especially if the solderability of one end is inferior to that of the other or if a part is placed out of alignment, even when pad design and termination widths are perfect. Also refer to Chapter 6, section 6.2.

Part movement, a less severe form of tombstoning, is very common in vapor phase soldering. Some of the causes discussed above for tombstoning apply to part movement as well. Equipment design also plays a role in part movement. For example, in vapor phase in-line systems, the incline of the conveyor belt before and after the reflow zone can cause part movement. Other causes include such nonsoldering process variables as fast acceleration and deceleration of the X-Y table of the component placement machine. This applies to both VPS and IR.

## 12.8.6 Thermal Shock on Components

Rapid heating and cooling rates cause thermal shock because there is insufficient time for the center and the surface of the components to reach the same temperature. During application of heat, surface temperature is higher than internal body temperatures, and the discrepancy generates thermomechanical stress. This stress is compounded in materials that are poor conductors of heat, and it can create cracking in brittle materials. Chip resistors and capacitors, which are brittle and poor conductors of heat, are more susceptible to thermal shock than components of other types, as explained in Chapter 14 (Section 14.3.2: Component-Related Defects).

The degree of thermal shock on components is higher in VPS than in IR. The IR soldering profile can be designed to heat the components at a rate of 2 to 6°C per second. Only limited control can be exerted on the heating rate of components and boards in a vapor phase system. The maximum heating rate during VPS is typically between 15 and 50°C per second. Such high rates of temperature increase may damage certain ceramic components, such as chip capacitors and resistors, which are prone to cracking.

## 12.8.7 Solder Mask Discoloration

Some solder masks, especially dry film solder masks, tend to discolor during IR soldering but are unaffected during VPS. This is generally a cosmetic defect only and can occur at a temperature as low as 160°C. In the case of one dry film solder mask, such discoloration was attributed to the oxidation of a polymeric dye in the solder mask. Reduction of the oxygen partial pressure in the reflow zone, by using either a nitrogen envelope or an inert atmosphere, as in VPS, diminishes solder mask discoloration.

New solder masks should be qualified before use to ensure their compatibility with the IR process and to confirm that any discoloration after soldering is cosmetic and will not degrade board functionality.

## 12.9  LASER REFLOW SOLDERING

Laser soldering is a relative newcomer to soldering technology. It complements other soldering processes rather than replacing them. As with vapor phase and IR soldering, it lends itself well to automation. It is faster than hand soldering but not as fast as wave, vapor, or IR soldering. Generally two types of lasers are used for soldering: carbon dioxide gas lasers and YAG-Nd (yttrium-aluminum-garnet/neo-dymium) lasers. The $CO_2$ laser is the more prevalent.

Most of the laser soldering systems have controls for providing different amounts of heat input to different joints on the same board. In selecting a laser soldering machine, features such as wattage, type of laser ($CO_2$, YAG-Nd), and laser beam width control should be considered. Heat-sensitive components that may be damaged in vapor phase or IR can be soldered easily by laser. In addition, moisture-sensitive components that are susceptible to cracking during vapor phase or infrared reflow and may require baking before reflow to prevent package cracking can be soldered by laser without any problems. Refer to Chapter 5 (Section 5.7: Package Reliability Considerations) for an explanation. Laser is also suitable for deleting circuit lines and adding components in a densely populated board without reflowing other joints.

In most commonly used reflow methods such as vapor phase and IR, the glass epoxy substrate is heated above its glass transition temperature (120°C) for 1 to 5 minutes. This can warp or damage the board and form a thick layer of intermetallic. In comparison, laser soldering does not heat the substrate, and the reflow time is only about 300 milliseconds, which is not long enough to form a thicker layer of intermetallic. Thus laser soldering can produce a better quality solder joint.

In every soldering process, there are certain issues of concern that need resolution. Being a relatively new process, laser soldering is no exception. The issues in laser soldering are summarized in Table 12.5

Although laser soldering is capable of producing more reliable solder joints, it is slower than mass soldering methods such as wave and reflow

**Table 12.5 Technical issues and their resolution in laser soldering process.**

| ISSUES | APPROACH |
|---|---|
| • Definition of a method for automating laser soldering | • Investigate method for supplying solder to the connection |
| • Prevention of solder balls | • Pretin mating surfaces before laser soldering |
| | • Fusion of solder paste on pad |
| • Prevention of laser damage to substrates | • Evaluate flux for damage control |
| | • Adjust beam width control mechanisms |
| | • Evaluate laser beam shielding techniques |

soldering. As long as this is the case, cost considerations are likely to prevent its widespread use. However, the laser soldering function has been combined with the laser inspection function, hence according to the equipment manufacturer the technique has lost its cost disadvantage [14]. Despite this improvement, laser soldering is not a widely used process because automated inspection is rarely used. The industry is emphasizing on process control rather than inspection to control quality.

For this lack of emphasis on automated inspection, especially in the commercial market, the laser soldering/inspection system remains a relatively expensive process. In addition to being more expensive, laser soldering entails such process problems as potential damage to adjacent components (if the width of the laser beam is not properly adjusted) and solder ball formation, which can be even more prevalent in laser soldering than in VPS. Solder balls can be prevented by first fusing the solder paste or buying boards with sufficient (0.3 mil) plated or hot air leveled solder coating and then reflowing the solder joints with laser. This solution to a technical problem, however, makes the relatively expensive laser soldering process even more expensive.

Laser soldering has many advantages over other soldering methods, but its use will be confined to specialized applications until the process becomes economically justifiable. For laser soldering to move from the specialized soldering arena into general use, soldering and inspection operations must not only be combined in a single cost-effective system, as suggested above, to make the process relatively faster, hence cheaper, but the automated inspection must be widely used in the industry. And for this to happen the inspection results must be more consistent than is currently the case. As discussed in Chapter 14 (Section 14.5: Solder Joint Inspection), laser inspection results are not always consistent. The new system available from at least one manufacturer shown in Figure 12.22, however, has reportedly made considerable progress [14]. It uses an infrared detector to monitor soldering, and a YAG laser provides the power needed for soldering. As indicated in Figure 12.23, laser energy is targeted at the solder joint to melt it. The infrared detector monitors the infrared energy that is emitted by each joint and forms its unique thermal signature, and compares the thermal signature of each joint to a "known" good joint. The input for the known good joint is programmed in the computer during the developmental phase. The infrared detector directs the computer to shut the laser off when the solder joint melts or when the temperature of the joint is abnormal (high or low). This feature allows the system to mark for repair a joint that has experienced an extreme of temperature, thus the soldering and inspection functions are performed simultaneously.

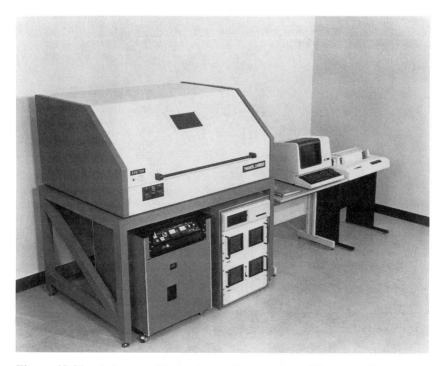

**Figure 12.22   A laser soldering/inspection system. (Photograph courtesy of Vanzetti Systems.)**

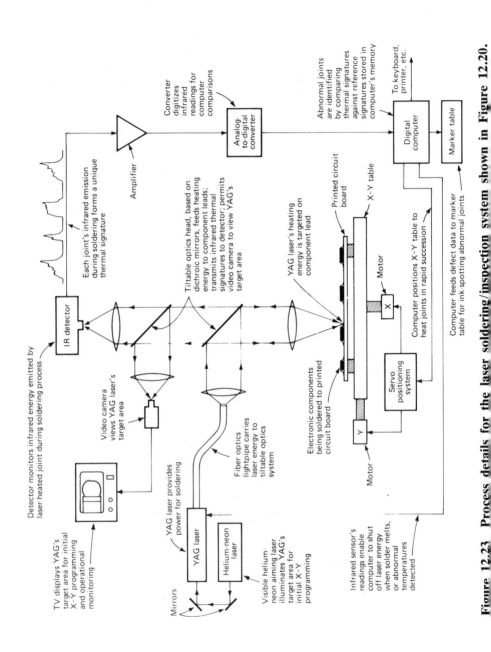

**Figure 12.23 Process details for the laser soldering/inspection system shown in Figure 12.20.**

## 12.10 MISCELLANEOUS REFLOW SOLDERING METHODS

Besides vapor phase, infrared, and laser soldering processes, the other reflow processes are belt reflow soldering, resistance bar reflow soldering, and hot air reflow soldering, which are used for specialized applications. One of the early soldering processes for the reflow soldering of surface mount components was belt reflow. A conveyor belt is heated in an oven tunnel, and the heat transfer is mostly by conduction through the substrate. Only single-sided assemblies can be soldered, since the heat must conduct through the surface of the substrate.

In the hot belt conveyor process, the substrates can get excessively hot before components are able to reflow. Therefore it is generally used for components on ceramic substrates, which do not experience thermal damage. A typical temperature profile for a board in the hot belt conveyor process is shown in Figure 12.24.

The other reflow soldering method, resistance hot bar soldering, has been in use for some time for soldering of flat pack surface mount components on 50 mil centers. Resistance of the heating element and duration of reflow are two of the important variables here. Since no solder paste is applied, sufficient solder plating must exist on the pad.

For soldering of flat packs, solder coating is applied by a hot air leveling process. For good soldering, at least a 0.0005 inch solder coating is required, with the preferred thickness between 0.0008 and 0.0012 inch. With the increasing use of high pin count fine pitch package having pitches of 25 mil and less, reflow soldering of solder paste using either vapor phase or IR is difficult. In some applications, solder paste is applied first and is fused in vapor phase or IR, to prevent bridging, which is commonly seen in the soldering of fine pitch packages.

The fine pitch packages are placed on reflowed pads and soldered using the resistance soldering (hot bar) process. Pai and Thomas have demonstrated through their research that resistance hot bar soldering process is superior to both VPS and IR for soldering fine pitch packages [15]. They have also used this process for rework/repair of fine pitch packages as well. The driving force for using the hot bar process is the susceptibility of fine pitch leads to lead bending/damage during shipping and handling. Since pressure is applied during hot bar soldering process, it can tolerate lead bending/coplanarity better than any other soldering process.

It must be noted that if the screen printing process is controlled properly as discussed in Chapter 9, vapor phase and IR processes can be used to reflow fine pitch packages. This is generally the case if the lead pitch stays at or above 25 mils. For finer lead pitches ($\leq$ 20 mils), the conven-

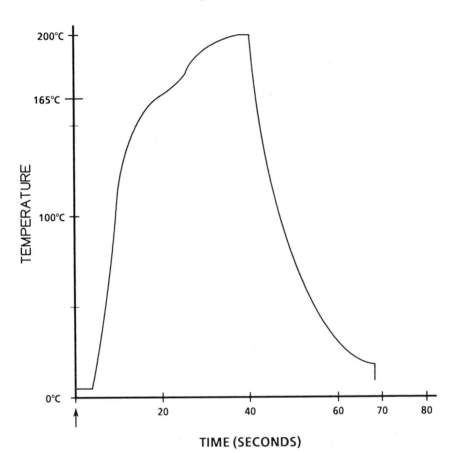

**Figure 12.24 A typical soldering profile for the hot belt reflow soldering process.**

tional process of solder application by screen printing becomes increasingly difficult, but is used by many companies. If a sufficient thickness of solder coating (0.3 to 0.5 mil) can be applied, screen printing paste is not necessary for fine pitch packages, with lead pitches less than 25 mils.

## 12.11 SELECTING THE APPROPRIATE SOLDERING METHOD

No matter which process is selected, the technical and nontechnical issues discussed previously must be resolved. Table 12.6 shows the cost effectiveness of four soldering options in assemblies of various types. For

**Table 12.6    Suitability of different soldering methods for various assembly characteristics.**

| | | COST-EFFECTIVE SOLDERING OPTIONS | | | |
|---|---|---|---|---|---|
| ASSEMBLY CHARACTERISTIC | WAVE | VAPOR PHASE | INFRA-RED | LASER | HOT BAR |
| Throughput | X | X | X | | |
| Surface-mounted components | X | X | X | X | X |
| Through-mounted components | X | | | | |
| Rework and specialized soldering | | | | X | X |
| Temperature-sensitive components | | | | | |
|   Through-mounted | X | | | | |
|   Surface-mounted | | | | X | X |
| Fine pitch (<20 mil lead pitch)/tape automated bonding (TAB) | | | | | X |

most companies, no single soldering process can meet all needs; instead of one particular soldering process, therefore, both wave and reflow soldering are generally used.

Through-hole and surface mount components need both wave and reflow soldering processes. If the components are not very tightly packed, and a strict design rule for chip component layout is followed, conventional wave soldering machines may prove to be adequate. It is much better to use either a dual-wave or vibrating solder wave designed for surface mounting, however.

The selection of wave soldering for surface mounting need not entail much difficulty if a conventional soldering system is already in place, since it may be economical to retrofit the existing machine with a new wave solder pot, as discussed earlier. In such a case the equipment generally is purchased from the present vendor because the wave pot from a different vendor might not be retrofittable.

When it comes to reflow soldering, selection becomes somewhat more difficult. We have discussed in great detail the pros and cons of the vapor phase and IR processes, which will remain the most widely used reflow soldering processes for SMT. However, because of differences in heating mechanisms, the vapor phase process heats the leads and lands nonuniformly, causing solder wicking.

In reflow soldering, nonfocused convection/infrared (C/I) soldering may give better results than either vapor phase or near IR. When the solder profile is properly characterized, other processes can give good results as well. No matter which process is chosen, flexibility should be the key, to allow the handling of a multiplicity of soldering tasks.

Looking at Table 12.6 again, it becomes clear that all soldering processes can be used for generic surface mount components. However, for very fine pitch (< 20 mil lead pitch) and TAB surface mount devices, only resistance hot bar soldering should be used. On the other hand, wave soldering remains the most cost effective soldering process for through-hole components.

Since surface mount technology is essentially in a transition mode, changes in soldering equipment design are continually taking place. There can be great variation from one machine to another, especially in infrared ovens, and this equipment should be evaluated very carefully. Some of the main areas of evaluation in selecting these machines are as follows: defect rates, level of automation features desired, compatibility of a given machine with existing or future equipment, and above all, trace record of the vendor, as measured by customer support and user comments.

Finally, the equipment should be selected on the basis of a specification that establishes performance requirements and sets the acceptance criteria. This is very important because it not only helps fine-tune the equipment requirement but also sets a benchmark against which the equipment can be evaluated at the vendor's facilities before delivery.

## 12.12 SUMMARY

There are many soldering processes available for the soldering of surface mount devices but none is perfect for all applications. The selection of a process depends on the mix of components to be soldered. Because it offers higher yield and lower operating cost, infrared soldering has evolved as the preferred process in the electronics industry.

Vapor phase will not disappear, however. As a matter of fact, there are users who prefer vapor phase over IR for ease of set-up and profile development. In any event, for development activities, the batch vapor phase process remains a popular lab tool because of its flexibility, low equipment cost, and small space requirement. For some specialized jobs, other reflow soldering processes such as laser, hot conveyor belt, and resistance soldering processes are also used. These processes are intended not to replace either vapor phase or IR but to complement them. The process ultimately used should be selected on the basis of the specific requirements of the intended application, the solder defect results, and overall cost.

# REFERENCES

1.  Prasad, Ray, Introduction to Surface Mount Technology, Intel application note AP-403, January 1987, page 14, available from Intel Corporation, Santa Clara, CA.

2.  Biggs, C. "An overview of single wave soldering for SMT applications." *Surface Mount Technology*, June 1986, pp. 8–10.

3.  Avramescu, S., and Down, W.H. "Omega wave: A new approach to SMD soldering." *NEPCON Proceedings*, February 1986, pp. 640–650.

4.  Pignato, J. "New phase in vapor phase." *Circuits Manufacturing*, September 1987, pp. 71–76.

5.  Wenger, G. M. and Mahajan, R. L. "Condensation soldering technology, Part 1: Condensation soldering fluids and heat transfer." *Insulation/Circuits*, September 1979.

6.  Cox, N. R. "Near-IR reflow soldering of surface mount devices." *Surface Mount Technology*, October 1986, pp. 27–30.

7.  Flattery, D. K. "Minimizing solder defect rates through the control of reflow parameters, Part 2." *Surface Mount Technology*, December, 1986, pp. 7–12.

8.  Dow, S. "An examination of convection/infrared SM reflow soldering, Part 1." *Surface Mount Technology*, February 1987, pp. 26–28.

9.  King, S. R. "SMT module improvement with infrared reflow." *Proceedings of NEPCON West*, February 1986, pp. 25–27.

10. Prasad, R., and Aspandiar, R. F. "VPS and IR soldering for SMAs." *Printed Circuit Assembly*, February, 1988, pp. 29–37.

11. Lambert, Leo, Surface mount soldering considerations, *Proceedings of Surface Mount Technology Association/Conference*, August 1988, pp. 133–138, available from SMTA, Edina, MN.

12. Hutchins C., and King, S. "The surface mount solder reflow process." *Printed Circuit Assembly*, April 1987, pp. 20–24.

13. McLellan, R. N. and Schroen, W. H. "TI solves wicking problem." *Circuits Manufacturing*, September 1987, pp. 78–85.

14. Vanzetti, Ricardo and Traub, Alan C., Inspecting while solder-

ing—the laser soldering option, *Proceedings of Surface Mount Technology Association Conference,* August 1988, pp. 149–154, available from SMTA, Edina, MN.

15. Pai, D. K., and Thomas, R. J., Surface mounting fine pitch gull wing VLSI Carriers, *Proceedings of SMART III Conference,* Technical Paper SMT III-25, available from IPC, Lincolnwood, IL.

# Chapter 13
# Flux and Cleaning

## 13.0    INTRODUCTION

The selection of flux and cleaning processes plays a critical role in the manufacturing yield and product reliability of electronic assemblies. The subjects of flux and cleaning are interrelated, one cannot be discussed without the other. The function of flux is to remove oxides and other nonmetallic impurities from the soldering surfaces and prepare a clean surface for joining. After soldering, the flux residues or contaminants must be removed by cleaning. The type of contaminant is determined primarily by the type of flux used, but halides, oxides, and various other contaminants are introduced during storage and handling, as well.

The cleaning process is selected on the basis of type of flux, types of contaminants, and type of assembly—that is, mixed assembly (Type II and III SMT) containing both conventional components and surface mount components or full surface mount (Type I SMT). For example, mixed assemblies may end up using one cleaning process after reflow soldering and another one after wave soldering. The cleaning process selected also depends on the applications. For example, to meet stringent military re-quirements, two- or even three-step cleaning processes may be necessary.

In this chapter we explore the reasons for cleaning and discuss fluxes, cleaning processes, and equipment for both conventional and surface mount assemblies. In addition, we outline the current and future test methods and requirements for cleanliness, to help us determine how clean is clean. Finally, we consider whether the cleanliness test methods and requirements used for through-hole assemblies are valid or even relevant for surface mount assemblies.

## 13.1    CONCERNS IN SURFACE MOUNT CLEANING

The main concern in cleaning of either through-hole or surface mount assemblies is adequate removal of flux to prevent corrosion problems in the field. For surface mount assemblies the job is difficult because flux may be trapped in the tight spaces between surface mount components (SMCs) and the substrate. Also, as a result of higher temperatures and relatively longer times during reflow soldering, the fluxes, especially rosin fluxes, become more tenacious, hence are more difficult to clean.

Whereas typical leaded components, like dual in-line packages (DIPs), are spaced 0.030 to 0.050 inch off the substrates, surface mount components are spaced much closer. Small passive components such as ceramic resistors and capacitors and large active components such as leadless ceramic chip carriers (LCCCs) may be spaced only 0.001 to 0.005 inch off the substrate. Some other active surface mount components, such as small outline integrated circuits (SOICs) and plastic leaded chip carriers (PLCCs), are usually spaced from 0.01 to 0.025 inch off the board, as shown in Figure 13.1. Not all through-hole mount devices have higher standoff heights than surface mount components. For example, as shown in Figure 13.1, axial through-hole components and 14 lead DIPs have lower standoff heights than many surface mount components. Figure 13.1 also shows the area underneath the components along the X axis.

When chip resistors and capacitors are attached with adhesive, the adhsive itself will fill most of the space between the chips and the board, leaving little room for entrapment between the adhesive perimeter and the pad metallization. However, rapid curing of the adhesive may result in

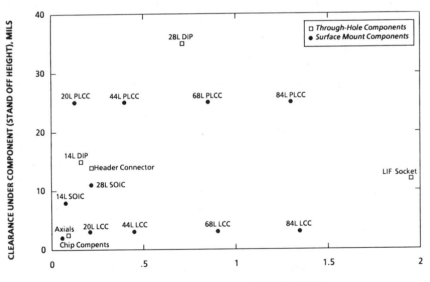

**Figure 13.1** **Plot of relationship between standoff height and the area under components.**

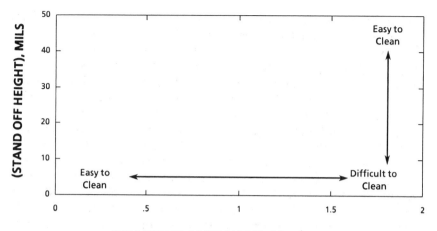

**Figure 13.2   Graphic illustration of difficulty of cleaning versus area and standoff height of components.**

voids in the adhesive and will absorb flux that is almost impossible to remove. Chapter 8 (Section 8.6.1.2: Adhesive Cure Profile and Flux Entrapment) tells how to prevent voids in adhesive during cure. When no adhesive is used, as with LCCCs, SOICs, and PLCCs, flux and other residues are expected to end up under the these components. The problem is especially acute if the gap is tight and the components are large: components the size of LCCCs not only have small standoff heights, they also have a large area underneath, which increases the potential for flux entrapment and makes it difficult for solvents to penetrate beneath them to remove the flux.

The difficulty or ease of cleaning is graphically illustrated in Figure 13.2. Clearly components on the right-hand side (larger area, smaller standoff heights) are more difficult to clean than the components on the left-hand side (smaller area, smaller standoff heights). Thus area and standoff heights should be considered together to ascertain the potential difficulty in cleaning. This is confirmed by Musselman and Yarbrough, who state that cleanability is proportional to the standoff height ($h^4$) and inversely proportional to package area [1].

## 13.2   THE FUNCTION OF FLUX

All metals, except pure gold and platinum, oxidize in air at room temperature. The rate of oxidation can increase with an increase in humidity and temperature. Different metals have different affinities for oxygen, hence they differ in their susceptibility to oxidation, which is a form of corrosion. Some metal oxides, such as those of silver and palladium, revert to pure metals at high temperatures. Oxide surfaces can be either compact or loose. The compact oxides protect the underlying surfaces from further oxidation, as chromium oxide ($Cr_2O_3$) protects stainless steel surfaces. Quite the reverse is true for pure iron, which forms a loose oxide that allows diffusion of oxygen until all the metal has been consumed.

The basic mechanism of joining two metallic surfaces is controlled by the solubility of one metal into the other. As discussed in Chapter 10, the degree of solubility of one metal into other and the temperatures at which they dissolve and form an intermetallic layer can be determined by inspecting the phase diagram of the two metals to be joined. For the formation of an intermetallic bond, two fresh metallic surfaces must come in contact. Since metal oxides are barriers to the formation of such a bond, they must be removed, by the use of flux, regardless of whether the joining process is welding, brazing, or soldering. Soldering differs from the other joining processes only in the type of oxides to be removed and the temperature at which joining takes place.

In soldering, the function of flux is to chemically react with oxides and quickly produce a fresh, tarnish-free surface at soldering temperatures so that intermetallic bonding can take place. When similar surfaces such as solder-coated leads and solder are to be bonded, they simply dissolve into each other. When dissimilar metals such as copper and tin are soldered, a tin-copper alloy is formed as an intermetallic layer.

Although the main function of flux is oxide removals, it must also be able to counter the effects of pollutants collected on the surface during storage and handling and to clean the surface for soldering. Tackiness of flux (in solder paste) is essential for preventing part movement during handling between placement and reflow soldering operations.

## 13.3  CONSIDERATIONS IN FLUX SELECTION

The activity in a soldering flux is generally provided by halides (chlorides, bromides) present in the flux, although there are some halide-free fluxes containing aminoacids or other organic acids. The higher the activity or the halide content in a flux, the more effective it will be in performing its function of removing oxides from the soldering surfaces. Why not use only the active fluxes? The more active a flux, the higher the probability that undesirable corrosive by-products of the flux reaction will be left behind on the board, potentially causing reliability problems in the field. A more aggressive flux may perform its function well by reducing the amount of touch-up, but if the area is not properly cleaned after fluxing, reliability may suffer.

Thus there are two major and contradictory considerations in selecting a flux for electronics product: it should be inactive at room temperatures before and after soldering, but active at soldering temperatures, to promote easy removal of tarnish and oxides. Actually, it should be active slightly below the soldering temperature, to ensure that a fresh surface is ready for soldering at the soldering temperature. If the flux residue were inactive after soldering, it would not jeopardize reliability if left on the board. Unfortunately, the flux that is active at or near soldering temperatures but inactive after soldering does not exist; hence the need to select a flux that offers a good balance between activity and cleanability.

As discussed in Chapter 10, if there is sufficient solder coating on the soldering surface to prevent the exposure of the intermetallic surface, and if the components are properly stored, solderability can be preserved for a long time. If the components are not solderable, they are generally pretinned to restore their solderability. If it is not practical or cost effective to preclean and pretin boards and components before soldering and the solderability of surfaces is questionable, a more aggressive flux is needed. The highly competitive commercial market generally uses a relatively aggressive flux to accommodate these conditions. It is tempting to use the most active flux to achieve the highest possible yield in manufacturing, but one must choose carefully according to the application. For example, in many military applications pretining is performed at increased cost to restore solderability. An active flux is generally not used because of reliability concerns.

## 13.4 FLUX CLASSIFICATION

Fluxes are classified based on their activity and their constituents (e.g., inorganic, organic; rosin, synthetic materials), which determine activity. Flux activity, in turn, is an indication of its effectiveness in removing surface contaminants. Keeping flux activity as a measure, the fluxes are generally classified as inorganic acid, organic acid (OA), and rosin. There is a special fourth category called synthetic activated (SA) flux.

If we use corrosiveness as a measure, these fluxes can be classified as highly corrosive (inorganic acid fluxes), mildly corrosive (OA, SA, and super activated rosin [SRA] fluxes) and noncorrosive (rosin fluxes). Table 13.1 summarizes different types of flux and their corrosiveness as adapted from Reference 2. The classification of fluxes based on corrosiveness can be misleading because even the so-called noncorrosive rosin fluxes can be corrosive if not properly cleaned. Besides corrosiveness is hard to define. Degree of corrosiveness does indicate, however, the activity of various categories in relation to each other.

The highly corrosive inorganic acid fluxes are seldom used in the

**Table 13.1 Properties of Fluxes** (as adapted from Reference 2)

| PER-FORM-ANCE | ROSIN | SYN-THETIC RESIN | SA[a] | OA[b] |
|---|---|---|---|---|
| Activity | Fair/good | Fair/good | Very good | Very good |
| Corrosion of residues | Noncorrosive | Noncorrosive | Mildly corrosive | Corrosive |
| Residue removal | Not critical but required. | Not critical but required. | Solvent wash required | Water wash required |

[a]Synthetic activated, solvent soluble.
[b]Organic acid, water soluble.

electronics industry, and the mildly corrosive fluxes are generally used only in commercial electronics. The SA and SRA fluxes in the mildly corrosive category have activity comparable to that of the OA fluxes and are designed for solvent cleaning, whereas the OA fluxes are meant for aqueous cleaning.

The rosin fluxes are divided into four types: rosin (R), rosin mildly activated (RMA), rosin activated for military application (RA mil), and simply rosin activated (RA) for nonmilitary applications. The rosin fluxes can be cleaned by either aqueous or solvent methods. For reflow soldering, the most commonly used fluxes (for either military or commercial application) in solder paste are the RMA types. RA and OA fluxes are rarely used in solder paste. However, for the wave soldering portion of mixed surface mount assemblies, RMA, RA, OA, SA, and SRA fluxes are used for commercial applications, and rosin only (R, RMA, and RA) for military work. In the pages that follow we discuss the inorganic, organic, rosin, and synthetic fluxes in more detail.

## 13.4.1 Inorganic Fluxes

The inorganic fluxes are highly corrosive, being comprised of inorganic acids and salts such as hydrochloric acid, hydrofluoric acid, stannous chloride, sodium or potassium fluoride, and zinc chloride. These fluxes are capable of removing oxide films of ferrous and nonferrous metals such as stainless steel, Kovar, and nickel irons, which cannot be soldered with weaker fluxes.

The inorganic fluxes are generally used for nonelectronics applications such as the brazing of copper pipes. They are, however, used sometimes for lead tinning applications in the electronics industry. Inorganic fluxes should not even be considered for electronics assemblies (conventional or surface mount) because of potential reliability problems. Their major disadvantage is that they leave behind chemically active residues that can cause serious field failures.

## 13.4.2  Organic Acid Fluxes

The organic acid fluxes are stronger than rosin fluxes but weaker than inorganic fluxes. They provide a good balance between flux activity and cleanability, especially if their solids content is lower. These fluxes contain polar ions, which are easily removed by a polar solvent such as water. Because of their solubility in water, the OA fluxes may be environmentally more desirable (see Section 13.6, item 8: Environmental Considerations). As the name implies, the OA fluxes contain organic acids commonly found in food and dairy products, such as citric acid, lactic acid, and oleic acid. Since fluxes of this type are not covered by government specifications, their chemical content is controlled by the suppliers. OA fluxes are available with or without the use of halides as activators.

The OA fluxes generally have been shunned even for the conventional assemblies because of the term "acid" flux. As indicated previously, however, the fact is that even the so-called noncorrosive rosin fluxes contain halides that will cause corrosion if not properly removed.

The use of OA fluxes may be justified for mixed assemblies (Type II and III) for both military and commercial applications. It is incorrectly believed that a change from OA to rosin-based fluxes (RA and RMA) is mandatory when wave soldering Type II and III SMA boards. Since the surface mount components reside closer to the printed board, it is difficult for aqueous solutions to adequately flush the OA flux and other contaminants from the tight spaces. However, using rosin instead of OA flux is not without its problems. Soldering defects can be increased, and strict control of the board and component solderability becomes critical.

Contrary to popular belief, OA fluxes have been successfully used on military programs. Examples include electronic assemblies for the Air-Launched Cruise Missiles and Airborne Early Warning Control Systems manufactured by the Boeing Aerospace Company. Many other leading companies in the commercial, industrial, and telecommunications sectors are using OA fluxes to wave solder surface mount chip components glued to the bottom of the board. OA fluxes have been found to meet both military and commercial requirements for cleanliness [3]. (The results of the study cited as reference 3 are discussed in Section 13.10.3.)

There is no technical reason for not using OA fluxes in solder paste. These materials have served successfully as flux coatings for the solder doughnut used in the reflow soldering of leaded through-hole components. Even after having gone through the reflow soldering operation, they are easy to clean with water. However, solder pastes with water-soluble flux are not widely used. They are not as tacky as rosin-based fluxes, and tackiness in solder paste is necessary to prevent part movement during placement. Because of environmental concerns associated with the cleaning of rosin-based pastes with solvents, however, it may be worthwhile for the industry to evaluate the use of OA fluxes in solder paste.

### 13.4.3    Superactivated Fluxes (SRA and SA)

Super-rosin-activated (SRA) and synthetic-activated (SA) fluxes fall within the mildly corrosive category. The SRA fluxes are the more active formulations of the rosin fluxes discussed in the next section. The SA fluxes, consisting of alkyl acid phosphates and activators, have flux activity comparable to that of OA fluxes but are designed for solvent cleaning.

The SRA and OA fluxes have much in common. For example, neither is generally used in solder paste, and both are used for wave soldering of either through-hole or mixed surface mount assemblies. The SRA fluxes are also very active and can be used to join difficult-to-solder materials such as Kovar and nickel. They do not meet the Mil-F-14256 requirements, hence are used only in commercial applications.

### 13.4.4    Rosin Fluxes

Among all the fluxes discussed above, only the rosin fluxes are used for both wave soldering and in solder paste for reflow soldering. Rosin or colophony is a natural product that is extracted from the stumps or bark of pine trees [4]. The composition of rosin varies from batch to batch, but a general formula is $C_{19}H_{29}COOH$; it consists mainly of abietic acid (70-85% depending on the source) with 10 to 15% primaric acids. Rosins contain several percent of unsaponifiable hydrocarbons; thus for removal of rosin flux, saponifiers (a form of alkaline chemical) must be added.

Rosin is inactive at toom temperatures but becomes active when heated to soldering temperatures. The melting point of rosin is 172 to 175°C (342-347°F), or just below the melting point of solder (183°C). A desirable flux

**Table 13.2 Temperatures at Which Flux Decomposition Becomes Significant [2]**

| SUBSTANCE | TEMPERATURE (°C) |
|---|---|
| Rosin | 273 |
| RMA flux | 275 |
| FSW32 flux[a] | 258 |
| Synthetic resin | 375 |
| RMA-type synthetic resin flux | 373 |

[a]German specification.

should melt and become active slightly below the soldering temperature. A flux is not effective if it decomposes at soldering temperatures, however. This means, as shown in Table 13.2, that synthetic fluxes can be used at higher temperatures than rosin fluxes, since the former decompose at higher temperatures [2].

In general rosin fluxes are weak, and to control their fluxing action, Mil-F-14256 allows the use of halide activators. Rosin flux with a halide content of about 0.2% meets the requirements of the military specification, which does not give a specific halide content, but states only the water extract resistivity of the flux.

Based on the water extract resistivity of fluxes, there are three types of rosin flux allowed for military applications: rosin (R), rosin mildly activated (RMA), rosin activated (RA mil). There is also an RA for nonmilitary applications with higher halide content. These four categories differ basically in the concentration of the activators (halide, organic acids, amino acids, etc.). Except for R and RMA types, it would be certainly misleading to call rosin fluxes noncorrosive, hence safe, RMA fluxes are not even cleaned in some applications. However, without cleaning, the reliability of assemblies may be compromised because the sticky rosin can attract dust and harmful contaminants from the environment.

## 13.5 CONTAMINANTS AND THEIR EFFECTS

After soldering, various residues are left behind on the board surface. The overriding reason for removing them is to prevent potential electrical failures due to electromigration. The impact of different types of contaminant on electronic assemblies is summarized in Table 13.3. The general classification of the residues to be removed can be broken down into three main categories: particulate, nonpolar, and polar.

**Table 13.3  Impact of Contaminants on Printed Circuit Boards**

| CONTAMINANT | EFFECT |
|---|---|
| • Polar or ionic (chloride activators) | —Dielectric breakdown<br>—Electrical leakage<br>—Component/circuit corrosion |
| • Nonpolar or nonionic (rosin) | —Cosmetic<br>—May attract ionic contaminants through dirt<br>—Poor conformal coat adhesion<br>—Mealing<br>—Poor electrical contact during ATE/bed-of-nails testing<br>—Poor electrical contact with surface mount or edge connectors |

## 13.5.1  Particulate Contaminants

Dust, lint, and solder balls are examples of particulate materials that must be dealt with in surface mounting. These residues are best removed by mechanical action such as use of spray or nozzle pressure and ultrasonic cleaning.

Solder balls are a type of solder defect, not a flux residue. They are included as "contaminants" because they can be removed by cleaning. Solder balls are fairly common in surface mounting and could cause electrical shorts if equipment vibration results in the collection of many small balls at one location. The causes of solder balls were discussed in Chapter 9.

## 13.5.2  Nonpolar Contaminants

As summarized in Table 13.4, nonpolar and polar contaminants require nonpolar and polar solvents, respectively. Some fluxes produce only polar contaminants, and others produce both types. For example, rosin fluxes contain both polar and nonpolar contaminants; hence their complete removal calls for a solvent that is both polar and nonpolar. Examples of nonpolar contaminants are rosin residues, oil used in wave soldering, grease or wax from placement or insertion machines, and makeup or hand lotions picked up from operators during handling. Contaminants of this type are

**Table 13.4  General Classification of Contaminants and Solvents**

| CONTAMINANT | SOLVENT |
|---|---|
| • Polar or ionic (activators) | • Polar solvents (alcohols, water) |
| • Nonpolar or nonionic (rosin) | • Nonpolar solvents [1,1,1-trichloroethane, fluorocarbon 113 (e.g., 1,1,2-trichloro-1,2,2-trifluoroethane)] |

removed by nonpolar solvents such as trichloromethane or trichlorotrifluoroethane (Freon), which dissolve them (see Table 13.4).

If not removed, nonpolar contaminants will attract dust and other foreign particles in the field. The dust itself may bear deleterious polar contaminants that could cause corrosion or electromigration under the influence of temperature and humidity. Nonpolar flux residues left on the surface are an indication of poor process control in cleaning. Moreover, they may hide such defects as board damage or solder joint opens.

Nonpolar flux residues are insulating by nature. They can make testing of assemblies difficult by preventing good electrical contact between the test probe and the test nodes during bed-of-nails testing. Flux on gold edge fingers or selectively gold-plated pads will also interfere with good electrical connections with surface mount connectors.

Any nonpolar flux residue left on the board will also cause poor adhesion of conformal coating, visible when the board is subjected to thermal cycling and under high humidity as a mildewlike appearance called mealing or vesication. To avoid mealing, the board surface must be very clean before the conformal coating is applied.

## 13.5.3  Polar Contaminants

Examples of polar contaminants are activators, such as halides, acids, and salts. They are the main causes of concerns in electromigration, which can degrade the insulation resistance between conductors or, at worst, cause complete corrosion of circuits. In electromigration, under certain conditions of contamination and voltage, metallic positive ions are transported across conductors.

Electromigration produces dendritic growth between conductors and causes shorting. Here is an easy experiment to determine whether this phenomenon has occurred: apply a low voltage between two circuit lines having some flux residues; if there is any contaminant present, the dendritic growth causing an electrical short can be seen. As discussed later, electromigration is the basis for one of the key cleanliness measurement methods, namely surface insulin resistance (SIR). The higher the SIR value (i.e., no electromigration), the cleaner the assembly.

In extreme cases, failure to remove corrosive flux residues may lead to corroded circuit lines. There are various mechanisms of electrical failures. For example, the lead (Pb) in solder oxidizes and combines with the hydrochloric acid (HCl) in the flux to form lead chloride ($PbCl_2$), as explained by equations (1) and (2) below. This is consistent with the function of the flux noted earlier—that is, to remove oxides, in preparation of a fresh surface as needed for a good intermetallic bond. However, if the lead chloride is not removed during cleaning, it reacts with the moisture ($H_2O$) and carbon dioxide ($CO_2$) in the air to form lead carbonate ($PbCO_3$), and the hydrochloric acid is regenerated, in accordance with equation (3) below. The regenerated hydrochloric acid keeps the chemical reaction going until all the lead in the solder has been consumed. This weakens the solder joint and will cause eventual electrical failure. The chemical reaction proceeds as follows [5]:

$$Pb + \tfrac{1}{2} O_2 = PbO \tag{1}$$
$$PbO + 2HCL = PbCl_2 + H_2O \tag{2}$$
$$PbCl_2 + H_2O + CO_2 = PbCO_3 + 2HCl \tag{3}$$

The hydrochloric acid also attacks copper, where present, to form copper chloride [5]:

$$CuO + 2HCl = CuCl_2 + H_2O$$

Most important of all, the board surface needs to be nonconductive to ensure insulation between conductors. Conductive material on the surface of the board may interfere with the performance of the circuit.

## 13.6 MAJOR CONSIDERATIONS IN SOLVENT SELECTION

The selection of solvents involves consideration of the economics of usage, compatibility with the substrates and components, performance, soil capacity, surface tension, streaking characteristics, toxicity, and environmental factors. These important issues are outlined below.

1. *Economics of usage.* Cost should be the deciding factor only if more than one solvent meets performance and other technical require-

ments. For example, aqueous cleaning may appear to be more economical than solvent cleaning, but Manko [6] found that aqueous cleaning may cost 20% more than solvent cleaning if all the costs are taken into account. Costs of solvent cleaning also vary. Ellenburger [7] found one chlorocarbon solvent to be twice as expensive as a fluorocarbon solvent when annual operating costs were compared.

2. *Compatibility with equipment and components.* The compatibility of the solvents with any equipment one already has could be the starting point. For example, if the tubing in the existing equipment is plastic, a solvent that requires stainless steel tubing would not be considered for reasons of incompatibility.

3. *Performance:* If a solvent does not perform its intended function, all other qualities, including cost, are meaningless. For surface mounting, performance is even more critical because of the tighter gap between components and the substrate. Hence the user may want to emphasize cleanliness tests as a measure of performance to determine the efficacy of the cleaning solution before committing it to prodution. For example, Sanger and Johnson [8] found that cleanliness results depend more on flux than on solvent. The results of their extensive solvent evaluation project are summarized in Table 13.5, which shows the residual contaminant as measured in micrograms of NaCl equivalent per square centimeter after cleaning. The assemblies were soldered with RMA and RA fluxes, which met the requirements of Mil-F-14256D. The acceptable limit was set at 1.55 µg of NaCl equivalent per square centimeter. Table 13.5 shows that for RA fluxes, ethanol azeotrope did not produce the desired results. However, methanol azeotrope or tricholorinated solvents with 7% alcohol produced acceptable results. Methanol is a more effective polar solvent than ethanol. If the results of Table 13.5 are any guide, the selection of a particular solvent should be made after careful evaluation of the cleanability of the flux.

4. *Soil capacity or kauri butanol value:* Soil capacity is a measure of solvent performance that refers to a standard solution of kauri gum in n-butyl alcohol. It is the number of milliliters of solvent that must be added to 20 grams of a standard kauri gum solution at 25°C to produce the same turbidity as 100 milliliters of benzene [9]. The idea is that higher the number, more effective the solvent will be in dissolving contaminants such as rosin. A reverse and more meaningful test may be simply to measure the amount of sodium chloride (ionic contaminant) and rosin (nonionic contaminant) left on the board after cleaning, to compare the effectiveness of different solvents.

5. *Surface tension:* The lower the surface tension of the cleaning solution, the more easily it can penetrate the tight spaces between the surface mount components and the substrates. For reference, the surface tension of water is 72 dynes/cm. When saponifier is added to the water, the surface

**Table 13.5 Cleanliness study results[a] on rosin fluxes adapted from Reference 8**

| SOLVENT BASIS | OVERALL AVERAGE | RMA FLUXES | | | RA FLUXES | | |
|---|---|---|---|---|---|---|---|
| | | AVERAGE | MAXIMUM | MINIMUM | AVERAGE | MAXIMUM | MINIMUM |
| F-113/ethanol azeotrope | 3.06 | 1.48 | 6.59 | 0.148 | 4.65 | 7.91 | 0.464 |
| F-113/ethanol azeotrope | 2.35 | 1.38 | 4.77 | 0.371 | 3.32 | 5.55 | 0.626 |
| F-113/ethanol azeotrope | 1.73 | 1.01 | 4.20 | 0.086 | 2.44 | 4.17 | 0.192 |
| F-113/methanol azeotrope | 0.954 | 0.877 | 3.83 | 0.034 | 1.03 | 1.71 | 0.096 |
| 1,1,1-Trichloroethane | 2.64 | 1.63 | 6.87 | 0.066 | 3.65 | 7.33 | 0.199 |
| Unknown chlorinated | 3.03 | 1.99 | 7.63 | 0.197 | 4.06 | 6.85 | 0.290 |
| 1,1,1-Trichloroethane/7% alcohol | 1.34 | 1.45 | 6.53 | 0.037 | 1.22 | 2.12 | 0.055 |
| Average of averages | 2.15 | 1.40 | 5.77 | 0.134 | 2.91 | 5.09 | 0.275 |

[a]Residual contamination in micrograms per square centimeter of NaCl after cleaning 5 Mil-F-14256D RMA fluxes and 5 Mil-F-14256D RA fluxes in seven commercially available solvents. The acceptable limit value of the test was considered to be 1.55 mg/cm² [8].

tension is reduced to 30 to 35 dynes/cm. Most solvents have a surface tension of about 20 dynes/cm. Surface tension alone is not important, however. The solvents for cleaning should not only penetrate the tight spaces between the component and the substrate, but also should dissolve the residue at these locations. These dissolved residues, of course, must be removed, but physical factors such as low surface tension, which draws the solvent into tight spaces, also opposes its removal from these areas. Hence, high pressure sprays are necessary to remove solvents and the dissolved residues from under the components.

6. *Streaking properties:* Streaking is any visible dirt that remains on the board after cleaning; the property is related to the boiling points of solvents. Lower boiling solvents cause more streaking than higher boiling solvents because they tend to evaporate faster, leaving behind some of the dirt.

7. *Threshold limit value:* The threshold limit value (TLV), a guideline for human exposure to solvents, is expressed as time-weighted average (TWA), which represents parts per million of vapor in air. According to the regulations of the federal Occupational Safety and Health Administration (OSHA), an employee's exposure to solvents cannot exceed the TWA exposure limit in an 8 hour period or 40 hour work week. Unnecessary exposure should be avoided even if one is well under the limit. Generally, the TLV-TWA values are used as an indication of solvent toxicity. They are not sufficient for this purpose, however, because toxicity is also a function of volatility. For example, a solvent with higher TLV value but faster evaporation rate may be more hazardous than a solvent with lower TLV value and slower evaporation rate. Maximum allowable concentration (MAC) is used instead of TLV in some countries to determine levels of solvents to which workers may be exposed in a 40 hour work period.

8. *Environmental considerations:* In addition to OSHA regulations, one must consider the forthcoming requirements of the Environmental Protection Agency (EPA) with respect to chlorinated flurocarbons (CFCs) usage. CFCs are non-flammable, inert chemicals of low toxicity. Their typical uses are as follows:

- Airconditioning and refrigeration systems     50%
- Flexible foams for furniture, bedding, foam insulation and packaging     34%
- Sterilization of medical supplies and instruments     4%
- Cleaning of electrical and electronics components     12%

The specific CFCs that harm the environment are Freon 11 (Trichlorofluromethane) and Freon 12 (Dichlorodifluoromethane) used in refrigerants. Freon 114 (Dichlorotetrafluoroethane) and Freon 115 (Chloropentafluroethane) are also harmful but they are rarely used. Freon 113 (Trichlorotrifluroethane) is of main concern to the electronics industry.

In formation of ozone layer first the oxygen molecules ($O_2$) are split into two oxygen atoms (O) when they absorb sun's ultraviolet radiation. These oxygen atoms then combine with oxygen molecules to form the ozone ($O_3$) layer. Ozone is a pungent, slightly blue gas—a close cousin to molecular oxygen ($O_2$). About 90 percent of the earth's ozone is located above the earth's surface in a frigid region of the atmosphere known as the stratosphere. This ozone layer acts as a shield against ultraviolet radiation.

Chloroflurocarbons (CFCs) have been proven to deplete the ozone layer. When they reach the stratosphere, the ultraviolet radiation breaks the chlorine atom from the CFCs. The freed chlorine atom reacts with ozone molecule ($O_3$) to form chlorine monoxide (CIO) and oxygen molecule ($O_2$). This is the basic mechanism of ozone layer depletion. However, the chlorine atom is regenerated from chlorine monoxide to repeat the process. Oxygen atoms (O) formed by splitting of $O_3$ by the ultraviolet radiation mentioned earlier, combines with the chlorine monoxide to form $O_2$ and a free chlorine atom. Now the same chlorine atom is free again to repeat the formation of chlorine monoxide to repeat the cycle. So the freed chlorine atom repeatedly reacts with the ozone layer causing its depletion. It is believed that a single chlorine atom causes depletion of 100 ozone molecules. Since some of the CFCs have a life time of 120 years, one can only imagine the extent of environmental damage caused by CFCs.

The ozone layer protects living things by absorbing the solar ultraviolet light emissions. Depletion of the ozone layer increases skin cancer, increases cataracts, suppresses human immune system, and harms aquatic systems and biological organisms. Depletion of ozone also compounds the green house effect which heats up the climate.

CFC 113, is a widely used component of solvents for cleaning of electronics assemblies using rosin fluxes. CFC 113 is valued becuse of its stability and compatibility with materials used in the electronics industry. However, the very stability of this compound causes depletion of the ozone layer. Other chlorofluorocarbon compounds, also deplete the ozone layer, but they are used in refrigeration systems, not in the solvent cleaning of electronics assemblies. Halons, used in fire extinguishers, also deplete the ozone layer.

On September 16, 1987, twenty-four countries, including the United States and the members of European Economic Community, signed the Montreal Protocol to control the use of CFC compounds that deplete the ozone layer [10]. On March 14, 1988, the U.S. Senate ratified this treaty, which covers CFCs 11, 12, 113, 114, and 115 and Halons 1211, 1301, and 2402. Now the EPA will be issuing guidelines for the control of these five CFCs and three halons compounds that deplete the ozone layer, and state and local governments may issue additional (even stricter) guidelines to control these compounds. It must be noted that the electronics industry contributes only about 12% of the CFC emission as mentioned earlier.

Both Halons and CFCs are being controlled. The halon consumption will be frozen at 1986 levels by 1993. The controls on CFCs are more stringent, however. Figure 13.3 summarizes the provision of the Montreal Protocol to reduce consumption of CFCs in three stages to minimize the impact of the agreement on the industry. The first step involves freezing the consumption of CFCs to the level used in 1986. The second step is to reduce the consumption by 20% by mid-1993, and the third step is to reduce consumption by 50% by mid-1998. For the second and third stages, the CFC consumption in 1986 will be used as the reference point. This means that the supply of solvents containing CFCs will be reduced, causing price increases in the future. Consumption will be minimized by improving equipment design to cut solvent vapor loss, using alternate solvents with lower or no CFCs, aqueous cleaning and so on. Many companies may switch completely to aqueous cleaning. Clearly, then, both the flux and the solvent should be selected with this environmental constraint in mind.

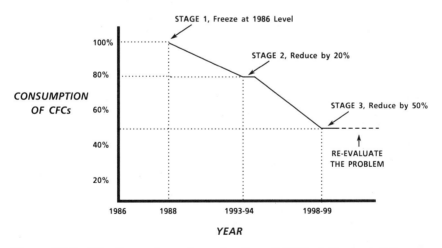

**Figure 13.3   The three stages for reduction of CFCs (chlorinated fluoro-carbons) to control depletion of the ozone layer as specified in the Montreal Protocol.**

## 13.7 COMMONLY USED SOLVENT TYPES

As discussed earlier, the type of solvent to be used depends on the type of contaminant to be removed. For example, as shown previously in Table 13.4, the nonpolar contaminants such as oils, grease, and rosins require nonpolar solvents. Nonpolar solvents can be either chlorinated or fluorinated. An example of a nonpolar chlorinated solvent is 1,1,1-trichloroethane ($CH_3CCl_3$), also known as methyl chloroform. It is one of the least aggressive chlorinated solvents.

Fluorocarbon F-113 ($C_2F_3Cl_3$) is a nonpolar fluorinated solvent used to dissolve nonpolar contaminants. Unlike chlorinated solvents, which can be purchased under their chemical names, fluorinated solvents are generally sold under their trade names (e.g., Freon TE).

There are pros and cons for both chlorinated and fluorinated solvents. The former have better solvency; moreover, with their higher boiling points (166°F versus 118°F for fluorinated solvents), they are more effective in the removal of contaminants from the tight spaces. Generally, however, they are not compatible with plastic piping in the cleaning equipment. For example, chlorinated solvent blends were found by Ellenberger to be incompatible with the PVC defluxer still, piping, and component markings [7].

Since most contaminants are a mixture of both polar and nonpolar substances, mixtures of polar and nonpolar solvents known as azeotropes are used to ensure complete removal. An azeotrope, from the Greek word meaning "to boil without change," is a blend of two or more solvents that behaves as a single solvent. It has one boiling point like any single-component solvent, but it boils at a lower temperature than either of its constituents. The constituents of the azeotrope cannot be separated by distillation. Only azeotropes should be used in systems such as vapor degreasers that use vapor as the cleaning medium.

F-113 forms an azeotrope with many solvents with the addition of 2.8 to 6.1% alcohol. The azeotrope 1,1,2-trichloro-1,2,2-trifluoroethane ($Cl_2FCCClF_2$), commercially available from Allied Corporation as Genesolv DFX, is shown in Table 13.6. Another commercially available azeotrope, Dupont's Freon TMS, has a similar composition except that it has neither acetone nor isohexane. Azeotropoes also contain a small percentage [0.2%] of stabilizers, such as nitromethane, to prevent corrosion of metallic surfaces; the alcohols in azeotropes (ethanol, methanol, isopropanol, etc.) are intended to remove ionic contaminants. Methanol, being more polar than ethanol, is more effective in removing ionic contaminants.

Azeotropic solvents can be either fluorinated or chlorinated. Examples

**Table 13.6  Properties of Some Commercially Available Azeotropic Solvents**

| SOLVENT | BOILING POINT (°F) | SURFACE TENSION (dynes/cm) | MAC or TLV | SPECIFIC GRAVITY |
|---|---|---|---|---|
| FC-113 (1,1,2-trichloro-1,2,2,-trifluoroethane) | 117.6 | 17.2 | 1000 | 1.57 |
| Genesolv DFX | 103.8 | 17.7 | 480[a] | 1.43 |
| Freon TMS | 103.6 | 17.4 | 475[a] | 1.51 |
| Prelete | 164.0 | 25.2 | 197[a] | 1.28 |

[a]Calculated.

of fluorinated azeotropes are Freon TMS and Genesolve DFX. Prelete (1,1,1-trichloroethane) is a chlorinated azeotrope. Because of ozone depletion problems discussed earlier, solvents with lower CFCs are becoming commercially available. An example is Freon SMT from Dupont. Because of its lower boiling point (106°F) this solvent is susceptible to cause explosion. Naturally this solvent is not currently being used. But other solvents with lower CFCs are expected to be introduced in the future.

# 13.8    SOLVENT CLEANING EQUIPMENT

There are basically two types of cleaner, batch and in-line. Batch cleaners are used in the laboratory or for small volume production, as well as to clean reworked assemblies. For high volume production, in-line cleaners are used; because in-line systems are not operator dependent like the batch systems, they make it possible to achieve consistent and repeatable results. For both batch and in-line systems, the liquid solvent is generally heated by immersed heating elements. For significant savings in energy cost, however, one can use a refrigerant in a closed-loop heating and cooling system instead of relying on immersion heating elements. An ultrasonic cleaning unit can be added to cleaners of either type.

## 13.8.1    Batch Solvent Cleaning Equipment

Batch solvent cleaners, also known as vapor degreasers, are available in quite a few variations. In the schematic shown in Figure 13.4, cooling

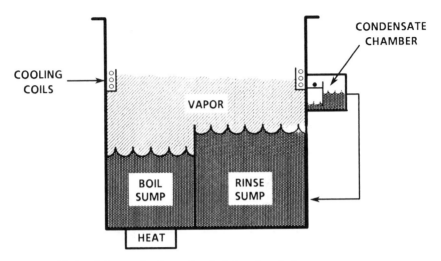

**Figure 13.4 Schematic view of a batch solvent cleaner.**

coils condense the vapor and recycle the liquid to the sump. In other systems, the vapor is collected in a recovery sump, and still other systems have a cascading flow of solvents, such that the assembly starts in the dirtiest part of the cascade and moves toward the cleaner solvents. Batch systems should have top covers to minimize loss of vapor.

No matter which system is used, the principle of cleaning is the same. A relatively cooler assembly is lowered in the vapor of the solvent. The vapor condenses on the assembly, dissolves the contaminants, and then evaporates, bearing with it the dissolved contaminants. The cleaning cycle is repeated by moving the assembly back and forth between a relatively cooler sump and hot vapor.

Because there are so many cleaning steps, batch systems are highly operator dependent. This factor may be decreased somewhat by adding a hoist to the system to move the basket containing assemblies to different zones. Most vapor degreasers are equipped with a foot-controlled spray wand to mechanically dislodge hard-to-remove nonionic contaminants. When using batch solvent cleaning, the following guidelines should be followed.

1.  Start vapor degreasing immediately after reflow and while the surface is warm. If this is not possible, start flux residue removal within one hour of soldering. With a longer time lapse, the rosin becomes harder to remove.
2.  Place the assembly in vapor for a minimum of 2 minutes or until vapor condenses on the surface. Generally 5 minutes in vapor is the maximum.

3. Spray the assembly with the spray wand. The spray nozzle should be at least 6 inches below the vapor level.
4. Cool the assembly by immersing into the condensate sump for a minimum of half a minute and a maximum of 2 minutes.
5. Hold the assembly in the vapor for a minimum of 1 minute and maximum of 2 minutes. (This is a repetition of step 2.)
6. Repeat steps 3 and 4.
7. Slowly remove the assembly from the vapor, allowing the liquid to drain off and the asembly to cool.
8. If visible residue remains, repeat step 2 to 4.
9. Dry with compressed air or wait until all the solvent has evaporated.

## 13.8.2  In-Line Solvent Cleaning Equipment

In-line systems use the same principle of cleaning as batch systems—that is, hot vapor condenses on the assembly for cleaning, immersion is used to cool the assembly, and finally the assembly passes through a hot vapor zone. The in-line process, however, is automated for high volume production, and there are no operator-dependent variables. In-line cleaning equipment is much preferred over batch vapor degreasers for surface mount assemblies.

Because in-line cleaners are closed systems, vapor losses also tend to be much lower. External or internal solvent stills can be used, but external stills allow better control of solvent levels, which constantly fluctuate because of vapor and dragout losses.

In in-line cleaning systems the assembly generally passes through three different zones on a horizontal or inclined conveyor: preclean spray, immersion, and final spray. Some systems are available with as many as eight different variations of these three basic cleaning zones [11]. Inclined conveyor systems are preferred because they allow better drainage.

A self contained cleaning system shown in Figure 13.5 shows the following feature for cleaning surface mount assemblies:

- Preliminary vapor soak zone in the entrance tunnel.
- Recirculating preclean hot spray (25 to 30 psi).
- Boiling immersion with 25-30 psi circulating overhead and underneath sprays.
- Recirculating distillate hot spray flush (25-30 psi).
- Drain and vapor dry through the exit tunnel. Final pure distillate spray flush (5-10 psi) can also be added to the schematic shown in Figure 13.5.

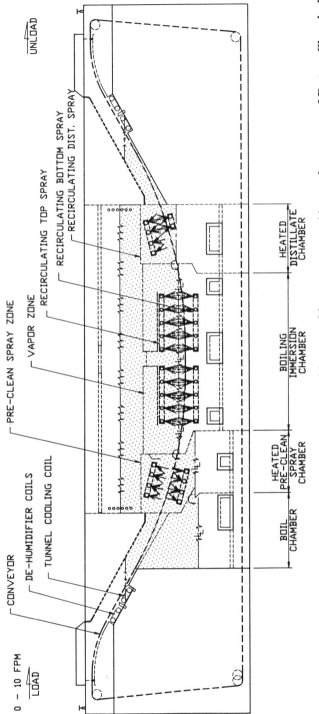
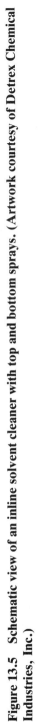

**Figure 13.5   Schematic view of an inline solvent cleaner with top and bottom sprays. (Artwork courtesy of Detrex Chemical Industries, Inc.)**

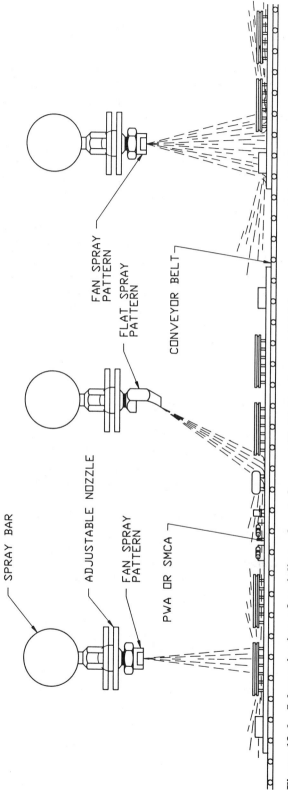

**Figure 13.6  Schematic view of an inline solvent cleaner with high volume and high pressure angled (middle) and fanned (right) sprays. (Artwork courtesy of Detrex Chemical Industries, Inc.)**

For specially hard to clean assemblies, another hot spray zone after the immersion zone, ultrasonics and super high pressure sprays (50-200 psi) can also be added [11]. Also, as shown in Figure 13.6, by adding and lowering nozzles that are angled or fanned and by increasing pressures up to 200 psi, most surface mount assemblies can easily be cleaned.

Whereas it is difficult to apply high pressures through the spray wand nozzles in a batch system because of the vapor containment problems inherent in an open system, in-line systems, being closed, allow the use of higher pressures, which are very important in removing not only flux residues but also solder balls if present.

Lermond demonstrated that with 200 psi nozzle pressure, all visible flux residues between the glass slides spaced at just 1 mil clearance between 3 square inch surfaces can be removed in just 9 seconds [12]. This is a realistic test because it simulates the worst condition with respect to standoff height—namely, that of a large LCCC package. In-line solvent cleaning machines with 200 psi pressure are commercially available.

Lermond also found that if the sprays are directed at immersed assemblies, considerably more time (150 seconds) and higher pressure (400 psi) are required to achieve similar results [12].

## 13.8.3    Ultrasonic Cleaning Equipment

Ultrasonic units are available during the immersion cycle, as an option for thorough cleaning. Although ultrasonic cleaning is controversial at this time because of potential damage to the internal wire bonds in components, is being used very successfully by many companies in the United States and in Japan.

The glass slide experiment mentioned above was extended to include ultrasonic agitation. When applied at 55 kHz and 2.4 W/inch in the boiling solvent, 15 seconds was required to achieve cleanliness levels comparable to those obtained after 9 seconds in a 200 psi pressure spay [12]. Considering the potential for damage to components in ultrasonic systems, it is clearly preferable to use a high pressure system.

## 13.9    AQUEOUS CLEANING

Aqueous cleaning can be used for cleaning both OA and rosin fluxes for removing particulate, polar, and nonpolar contaminants. Table 13.7 summarizes the effectiveness of aqueous cleaning in the removal of particulate, nonpolar, and polar contaminants. The effectiveness of aqueous cleaning on mixed surface mount assemblies is discussed in Section 13.10.3.

**Table 13.7   Relative Effectiveness of Aqueous Detergents [13]**

| CONTAMINANTS/ RESIDUES | CONTAMINANT REMOVAL | | |
| --- | --- | --- | --- |
| | EFFEC-TIVE | MODERATELY EFFECTIVE | INEFFEC-TIVE |
| **Polar** | | | |
| Fingerprint salts | X | | |
| Rosin activators | X | | |
| Activator residues | X | | |
| Cutting oils | | X | |
| Temporary solder masks/solder stops | X | | |
| Soldering salts | X | | |
| Residual plating | X | | |
| Residual etching salts | X | | |
| Neutralizers | X | | |
| Ethanolamine | X | | |
| **Nonpolar** | | | |
| Resin fixative waves | X | | |
| Waxes | | X | |
| Soldering oils | X | | |
| Cutting oils | X | | |
| Fingerprint oils | X | | |
| Flux resin, rosin | X | | |
| Markings | | X | |
| Hand cream | | X | |
| Silicones | | | X |
| Tape residues | | | X |
| Temporary solder masks/solder stops | X | | |
| Organic solvent films | X | | |
| Surfactants | X | | |
| **Particulates** | | | |
| Resin and fiberglass debris | | | X |
| Metal and plastic machining debris | | | X |
| Dust | X | | |
| Handling soils | | X | |
| Lint | | X | |
| Insulation dust | | X | |
| Hair/skin | | X | |

When water-soluble fluxes are used, contaminants that are not soluble in water are produced as a result of thermal decomposition of the flux. Neutralizers containing amines or ammonia and surfactants (i.e., surface-active agents) are added to provide better cleanliness levels. Surfactants are used to lower the surface tension of water. A detergent contains surfactants and other agents needed for a specific application. Soap is a familiar surfactant.

Aqueous cleaning for rosin fluxes is also used especially when there are environmental constraints. For example, if solvent cleaning is forbidden in a locality, aqueous cleaning may be the only option. As discussed in Section 13.6, the use of solvents may be restricted in the future. Hence it is prudent to evaluate aqueous cleaning now.

Because rosin is not soluble in water, when aqueous cleaning is used for rosin fluxes, alkaline chemicals called saponifiers are added to the water. These amine-based agents change the acid rosin into a form of soap and lower the surface tension of water. The content of rosin that cannot be saponified must be either removed with water-soluble glycol ether or physically scrubbed.

Water contains various compounds of magnesium and calcium that make it "hard"; hardness is measured in parts per million of calcium carbonate. Softeners are added to allow soap to be effective and to prevent the formation inside cleaning systems of scales that could block the nozzles. In addition, halides, sulfides, and other contaminants are present in water. Hence, to be an effective solvent, the water itself must be pure. In addition, since water is not a natural solvent for contaminants such as rosin and the by-products of soldering, additives are used to dissolve contaminants that come from sources other than water itself.

For purification of water, two processes—deionization and reverse osmosis (RO)—are generally used. In deionization, water is passed over resin beds to remove ionic contaminants through an ion exchange system. Deionized water, which is very expensive, is corrosive to most metal surfaces. However, it should be used when the cleanliness level it provides is justified. In RO systems, the water is passed through tubes of special materials at high pressures to separate contaminants by osmotic activity (diffusion of fluids through porous partitions).

## 13.9.1 Aqueous Cleaning Equipment

Batch and in-line systems are the most common types of aqueous cleaning system in use. A batch aqueous system is very much like a household dishwasher: the assemblies are loaded vertically like dishes, detergents are added, and cleaning and drying are accomplished in different timer-controlled cycles. Batch systems are good for low volume production. Because of higher pressures available during cleaning, moreover, batch systems may provide even better results than in-line cleaning. Aqueous batch systems are not operator dependent like solvent batch systems; instead, as indicated above, the cleaning cycles are automatically controlled as in household dishwashers.

In-line systems are very much like batch systems except that the board passes through different modules rather than different cleaning cycles as in batch systems. A typical in-line aqueous cleaner is shown in Figure 13.7 [13]. In-line systems are used for high volume production. The different modules of an in-line system are as follows: incline or horizontal loader, prerinse, recirculated wash, final rinse, air knives, drying stage, and assembly unloader. A schematic of the different modules appears in Figure 13.8.

The prerinse module, which is used for the removal of gross contaminants, is kept separate from the recirculating stages to prevent contamination. The recirculating wash needs detergents for rosin fluxes. Water heated to 140 to 160°F (60-72°C) is used. Generally additives are not required for OA fluxes. There may be additional recirculating stages if detergents are used, if faster conveyor speeds are required, or if higher levels of cleanliness are necessary. It is very critical to have high pressures at different angles in the recirculating stage for effective removal of contaminants. Pressures from 10 psi to 30 psi are used. Pressures in the high end of the range (30 psi) are preferred if available.

The assembly goes through the final rinse to remove not the flux contaminant but the contaminated water, which should not be allowed to dry on the assembly. The clean water is blown away with air knives (more than one air knife is generally used). Finally, the clean assembly is dried and unloaded. The final air knife and the drying stages can be combined to reduce the energy cost. The board must be completely dry before it emerges from the cleaner so that it cannot attract any contaminants.

## 13.10 CLEANLINESS TEST METHODS AND REQUIREMENTS

There are three different cleanliness test methods commonly used in the electronics industry. As we proceed to cover visual examination, solvent extraction, and SIR, we will also discuss the validity for surface mount assemblies of each approach.

### 13.10.1 Visual Examination

Boards are inspected under an optical microscope at 2 to 10X magnification for flux residue and other contamination. The main disadvantage

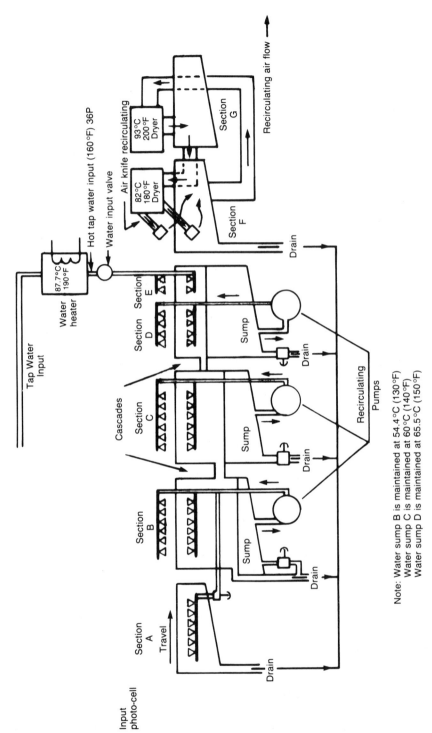

**Figure 13.7 Schematic view of an in-line aqueous cleaner [13].**

# Control Panel #

| | Description of Controls |
|---|---|
| A | Emergency stop, reset, low level indicator (wash), gross rinse, auto-off, wash |
| B | Wash (con't), air knife, pre-rinse/rinse |
| C | Pre-rinse/rinse (con't), dual air knifes, first dryer |
| D | Second dryer, conveyor, resistivity meter |
| E | Emergency stop, reset |

# Module #

| | Description |
|---|---|
| 1 | Incline/horizontal loader |
| 2 | Pre-rinse (gross rinse) |
| 3 | Wash (recirculated) |
| 4 | Air knife top |
| 5 | Pre-rinse/rinse |
| 6 | Dual air knives |
| 7 | Dual Dryer |
| 8 | Drive unload |

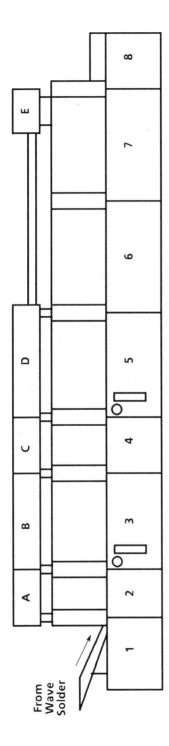

**Figure 13.8   Schematic view of an aqueous cleaner.**

of this method is that the flux residues trapped under large components cannot be inspected microscopically. For process characterization, therefore, the components must be removed to allow visual examination. The visual method is qualitative and is valid for gross contamination levels only. It is difficult to visually detect and quantify minute amounts of residue. The criterion for the visual method is that no flux or paste be visible under the components or on the board.

## 13.10.2   Solvent Extraction

The solvent extraction method involves immersing the board in a test solution and then measuring its ionic conductivity in terms of micrograms of NaCl equivalent per square unit of board area. For this method to be effective, the test solution (isopropyl alcohol and deionized water) must remove the contamination from under every component.

Without sufficient agitation of the solution, it is questionable whether all the flux residues, or indeed any, are being removed from under the components that reside close to the board surface. Recently, equipment has become available that allows agitation of solvents during the test. Commercial names for such equipment include Omegameter and Ionograph. This method is commonly used to monitor the cleanliness of conventional assemblies.

For process characterization, the component should be removed after cleaning and subjected to the solvent extraction test. The test should be conducted on a sampling of actual products, as well. If the Omegameter is used, the NaCl equivalent per square inch of board area should be less than 18 mg. Some companies have a lower requirement (10-14 µg).

## 13.10.3   Surface Insulation Resistance (SIR)

The primary advantage of the SIR measurement method over other cleanliness measurement tests for surface mount assemblies is that it is direct and quantitative. Unlike the Omegameter method, which averages the contamination present over the entire board surface area, SIR can detect the presence of flux even in localized regions of an assembly. The major disadvantage of SIR measurements is the need to design additional circuitry on the surface layers of the printed board to conduct the measurements effectively. (See Section 13.11: Designing for Cleaning.)

The SIR measurements give very useful results when used on boards with an aggressive flux. If the boards are not properly cleaned, they fail the 100 megohm SIR requirement. SIR tests can also flag adhesive cure problems. For example, if adhesive is cured too rapidly, the voids generated in adhesive may entrap flux, which will cause SIR failures.

The SIR test method is used widely for determining the insulation or moisture resistance of laminates and other printed board materials [14-16] and the compatibility of fluxes with printed board materials [17], as well as for cleanliness testing of printed board assemblies [18]. This method has also been used to measure the presence of flux under components [3,19].

The trace pattern on which the SIR is measured for the above-mentioned applications is standardized to either a typical Y pattern or a typical "comb" pattern [20]. Since the different component sizes make it difficult to standardize on one particular SIR pattern on the board, to obtain a fair indication of the contamination under the components, as much of this area as possible needs to be filled with the SIR pattern.

A Y pattern (Figure 13.9) can be employed under components of small area, such as chip capacitors and resistors. Comb patterns (Figure 13.10) are used for larger components, such as PLCCs and SOICs. The comb pattern shown in Figure 13.10 does not surround the pads, where flux may be present in excessive quantity. For this reason, another comb pattern (Figure 13.11) may be used. The pattern shown in Figure 13.11 surrounds all the pads where the paste and the flux are present.

### 13.10.3.1    SIR Measurement Test Conditions

There is no standard for SIR measurement test conditions for the evaluation of ionic contamination levels under components on surface

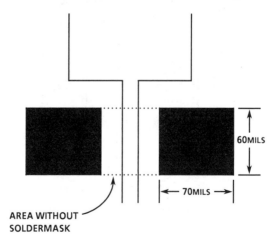

**Figure 13.9    Y pattern for SIR measurements under chip components.**

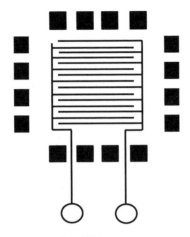

**Figure 13.10    Comb pattern for SIR measurements under PLCC/SOIC components.**

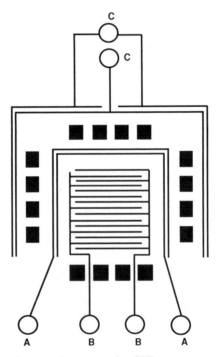

**Figure 13.11    Another comb pattern for SIR measurements under PLCC/ SOIC components.**

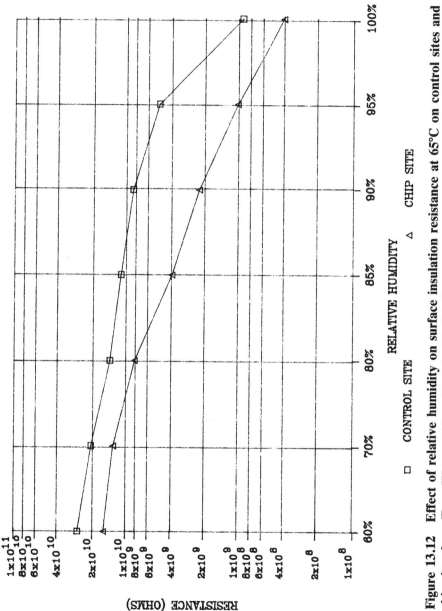

**Figure 13.12** Effect of relative humidity on surface insulation resistance at 65°C on control sites and chip sites for a Type III assembly soldered with OA flux.

mount boards. However, the measurement test conditions specified in IPC standards for moisture and insulation resistance determinations on laminates and insulators [14-16] can be used.

The SIR values are dependent on test conditions. For example, they go down as the relative humidity (RH) is increased, as shown in Figure 13.12. Also, as shown in Figure 13.12, the controls have higher SIR values than the chip sites. Therefore, tolerances should be established for SIR test conditions. The test conditions that will have an important impact on SIR values are relative humidity and temperature in the test chamber, time in the chamber, purity of the deionized water used in the chamber, and the test voltage.

The IPC standards measurement conditions are listed in Table 13.8. These test conditions are applicable for process qualification only. Some form of process control for cleanliness is necessary, especially when aggressive fluxes are used. The test conditions shown in Table 13.8 could not be used in a manufacturing environment on a production board. For example, a product could not be left in a humidity chamber for 7 days, if the SIR test were to be used for daily process monitoring. The following test conditions are recommended for monitoring SIR values in a manufacturing environment on real product or dummy test samples:

1. RH = 90 ± 5%.
2. Temperature = 65 ± 2°C for Class III (high reliability products)
3. Temperature = 50 ± 2°C for Class I and II (commercial and industrial products)
4. Time in chamber = 1 hour minimum.
5. Test voltage = 100 V dc.
6. Only pure water (deionized) with resistance greater than 4 megohms should be used in the humidity chamber. Otherwise, there may be a SIR failure due to the dirty water used for the test, not due to a bad board.

**Table 13.8   SIR measurement conditions stated in IPC specifications [14-17]**

| PARAMETER | CLASS I | CLASS II | CLASS III |
|---|---|---|---|
| Temperature, °C | 35 | 50 | 65 |
| Relative humidity, %RH | 90+ | 90+ | 90+ |
| Test voltage, volts | 500 | 500 | 500 |
| Polarization voltage, volts | 100 | 100 | 100 |
| Temperature conditions | Static | Static | Cyclic |
| Duration, days | 4 | 7 | 7 |

7.  Polarization voltage (application of voltage while the board is in the chamber) is not necessary. However, if polarization voltage is used, the polarity must be reversed when taking the readings.

8.  SIR values should be measured at both control and chip sites. A failure at the control site indicates either a serious cleaning problem or malfunctioning of the test equipment.

The equipment for measuring SIR values consists of high resistance meters, generally referred to as megohm meters, and humidity chambers. To minimize operator-dependent variables, one can select equipment that combines the humidity chamber and the resistance meters in one system. If such a system is not used, an operator must monitor test chambers to ensure that relative humidity and temperature requirements are consistently met (see next section).

How clean is clean? Table 13.9 shows the cleanliness requirements for Class I, II, and III assemblies as established in various IPC specifications. The SIR values measured at different SIR patterns with different trace geometries can be compared by normalizing the measured SIR values. Normalization usually is achieved by reporting the SIR as ohms per square [18,21].

SIR in ohms per square is equivalent to surface resistivity or sheet resistance. The measured resistance in ohms is converted to resistance per square by multiplying it by the total length of one trace pattern set and dividing it by the average spacing between the two sets of traces. For example, if the trace is 0.060 inch long and the gap between the traces is 0.012 inch, the ohms per square is given by ohms (measured) times (0.060/0.012). Thus for this case, ohms per square is 5 times the measured resistance.

**Table 13.9   IPC requirements for minimum SIR values for Class I (commercial), Class II (industrial), and Class III (military/high reliability) applications [14-17]**

| CLASS | SIR (OHMS/SQUARE) |
|-------|-------------------|
| I     | $1 \times 10^8$   |
| II    | $1 \times 10^8$   |
| III   | $5 \times 10^8$   |

*13.10.3.2 Application of the SIR Test*

It is wrongly believed by some in the electronics industry that OA fluxes, hence water cleaning, should not be used for the wave soldering of mixed surface mount assemblies because these assemblies cannot be adequately cleaned. It is incorrectly argued that because of its high surface tension, water cannot penetrate the tight spaces between the surface mount components and the board. On the contrary, however, thousands of mixed surface mount assemblies have been shipped to customers for commercial applications with no reported problems to date [22]. For conventional assemblies, OA flux with aqueous cleaning with deionized water has been successfully used for more than a decade by leading U.S. companies such as AT&T, Boeing (yes, for military applications), Intel, IBM, Motorola, and Xerox, to name just a few. Because of the harmful effects of CFCs on the ozone mentioned earlier, the current trend especially in the commercial sector, is to move towards aqueous cleaning [23].

As shown in Figures 13.13 and 13.14, OA fluxes can give cleanliness results similar to those of RMA rosin fluxes for mixed surface mount assemblies [3]. Table 13.10 gives the test conditions for the results shown in Figure 13.13 and 13.14. These test conditions are identical to that of the IPC test conditions (Table 13.8) for Class III (military) products. The results also meet the cleanliness requirements of Class III products.

The results shown in Figure 13.13 and 13.14 are derived from a Type II SMT assembly that had surface mount active components and through-hole components on the top side and surface mount passive components on the bottom side. The top side components were reflow (vapor phase and infrared) soldered using RMA flux in the paste. After reflow soldering of the top side surface mount components, wave soldering was used to solder the through-hole components on the top side and the chip components glued to the bottom side. Both RMA and OA fluxes during wave soldering were used to compare the cleanliness results.

Figure 13.13 and 13.14, respectively, show the SIR test results under chip components using the Y SIR test pattern and the results under PLCCs and SOICs using a comb pattern similar to the one shown in Figure 13.9. The results in Figure 13.13 and 13.14 are the average of 28 SIR patterns (12 SIR comb patterns under the PLCCs, 12 comb patterns under SOICs, and 4 Y patterns under 1206 chip resistors). The SIR patterns were not covered with solder mask. When taking SIR measurements, no polarization voltage was applied.

The SIR results discussed so far are generally valid for water-soluble fluxes only. For example, Bredfeldt [24] found that rosin fluxes do not show SIR failures. It is believed that rosin tends to repel water and that

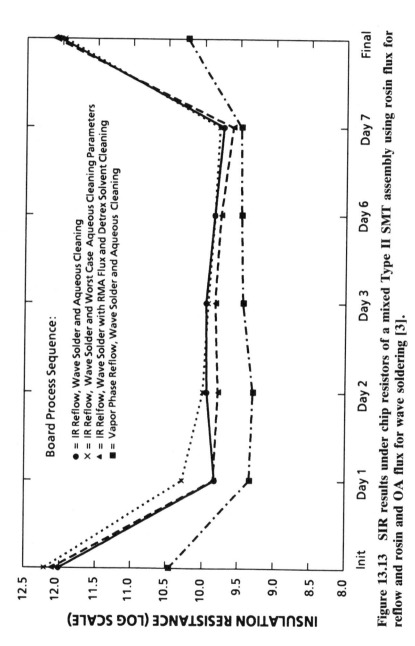

**Figure 13.13  SIR results under chip resistors of a mixed Type II SMT assembly using rosin flux for reflow and rosin and OA flux for wave soldering [3].**

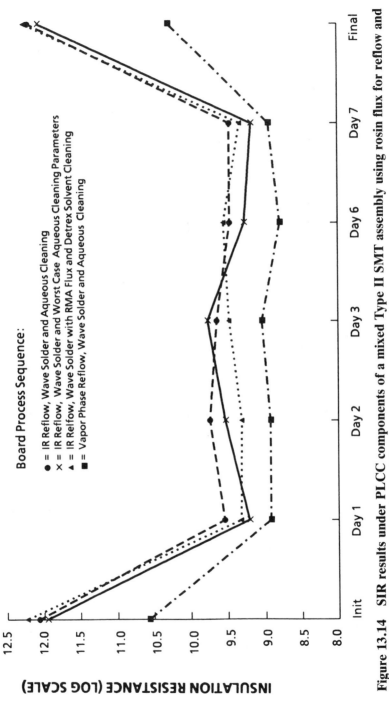

**Board Process Sequence:**

● = IR Reflow, Wave Solder and Aqueous Cleaning
✕ = IR Reflow, Wave Solder and Worst Case Aqueous Cleaning Parameters
▲ = IR Relfow, Wave Solder with RMA Flux and Detrex Solvent Cleaning
■ = Vapor Phase Reflow, Wave Solder and Aqueous Cleaning

INSULATION RESISTANCE (LOG SCALE)

**Figure 13.14** SIR results under PLCC components of a mixed Type II SMT assembly using rosin flux for reflow and rosin and OA flux for wave soldering [3].

**Table 13.10   SIR measurement conditions for test results shown in Figures 13.13 and 13.14 for SMT Type II assemblies Compared with conditions for Type III [22]**

|  | ASSEMBLY TYPE | |
| --- | --- | --- |
| PARAMETERS | TYPE III | TYPE II |
| Temperature, °C | 70 | 65 |
| Relative humidity, % RH | 85 | 90 |
| Test voltage, volts | 10/100 | 100 |
| Polarization voltage, volts | 0/40 | 0 |
| Temperature conditions | Static | Cyclic |
| Duration, days | 10 | 7 |

electromigration does not take place when the assembly is subjected to high humidity conditions. Thus the SIR test may not be suitable for rosin fluxes. Nevertheless various industry and government task forces are currently using the SIR test method for evaluating the effectiveness of various solvents with varying CFC content. Using this test, they are searching for effective solvents with minimum or no CFC content. Note that it is possible to fail SIR test on a rosin flux if the halide content is high. It should also be noted that, the test boards that may pass the SIR requirements for rosin flux may fail the visual and the solvent extraction tests. For this reason, all three (visual, solvent extraction, and SIR) cleanliness tests should be used. After SIR measurements, components should be removed and visually inspected for entrapped flux. Naturally such a test can be conducted on process control or dummy test boards only.

## 13.11   DESIGNING FOR CLEANING

The first element in designing for cleaning entails proper component orientation on the board when the assembly enters the cleaning equipment. Since the conveyor belts tend to be inclined at the entrance and exits, the components should be oriented such that drainage of solvents is possible. Designing for cleanliness also should take into account shadowing of adjacent components. The guidelines for component orientation and inter-package spacing recommended in Chapter 7 (Sections 7.5 and 7.6) apply for cleaning as well.

The second element of designing for cleanability entails the use of solder mask between surface mount lands and over traces (conductors). Using solder mask between lands of packages with 50 mil centers is not a problem. However, it is not common to provide solder mask between the lands of fine pitch packages. Hence, with only bare substrates and spacing of less than 0.012 inch, electromigration is a real possibility when fine pitch packages are used. Extreme care is necessary for cleaning sites that do not have the protection of solder mask. Also refer to Chapter 7 (Section 7.8. Solder Mask Considerations).

This brings us to the last and most important element of designing for cleanliness, namely, daily process monitoring. Some form of process control for cleanliness is necessary, especially when using aggressive fluxes. For example, based on some sampling plan, a given number of boards (either dummy test boards or actual products) should be checked for cleanliness.

It is easy to check for cleanliness by visual and solvent extraction methods. However, use of SIR patterns must also be considered to ensure reliable products. Two SIR patterns, as shown in Figures 13.9, 13.10, or 13.11, should be used. The traces used for SIR testing must not be covered with solder mask, because the mask will prevent the flux from reaching the traces. A test pattern covered with solder mask will always pass.

At least two SIR test patterns must be used on each board. One pattern serves as the control and should have no component mounted on it. Only the second test pattern should be mounted with a passive device, such as a 1206 resistor. If the SIR readings are unacceptably low (<100 megohm) on the control test pattern, either the test is invalid or the cleanliness process is really out of control.

The SIR test patterns should not be connected to any circuit. Incorporation of these test patterns will require inclusion of one additional nonfunctional passive component on the schematic and other necessary documentation used in the company. A passive 1206 component is generally selected because it is cheap and represents the worst-case flux entrapment problems because of its low standoff height.

SIR test patterns are not really necessary for reflow soldering because milder RMA flux is used in the solder paste. If activated rosin flux (RA) or organic acid fluxes are used in wave soldering, cleanliness test patterns should be provided on the secondary side of the board.

Insulation resistance as measured with two parallel traces at least 0.005 inch apart under the device should meet the requirements of Table 13.14 (>100 megohms or 500 megohm square after cleaning and after a 10 day temperature-humidity test) as measured under test conditions shown in Table 13.8.

## 13.12    SUMMARY

We discussed the function of flux, the principal types of flux, and the motivation for cleaning, as well as cleaning processes and equipment, and cleanliness test methods and requirements. These are basic areas for a theoretical and practical understanding of the subject.

It is generally thought that cleaning of surface mount assemblies is very difficult because for example, standoff heights between the surface mount components and the board are small, offering a tight gap that may entrap flux, which may be hard to remove during cleaning. This concern may be more imagined than real, however, for actual cleanliness data do not support this concern. Indeed, if proper care is taken in selecting the cleaning processes and equipment, and if the soldering and cleaning processes are properly controlled, cleaning of surface mount assemblies should not be an issue even when aggressive fluxes are used. It does need to be emphasized however, that a good process control is essential for using agressive water soluble fluxes.

The section of a particular cleaning process depends on the types of contaminant present, and these are determined by the type of flux and by the soldering and handling methods. Generally, for reflow soldered surface mount assemblies, rosin fluxes are used. For wave soldering of either mixed or through-hole mount assemblies, rosin and water-soluble fluxes can be used. The water-soluble fluxes require water for cleaning, whereas the rosin fluxes are generally cleaned with azeotropic solvents. The rosin fluxes can be cleaned with water if saponifiers are added, however. Selection of a cleaning solvent should follow a careful evaluation of technical, economic, and environmental considerations, but plans should be made to use aqueous cleaning in the relatively near future, because of environmental restrictions scheduled to begin in 1993 (see Figure 13.3).

Among the three methods for determining the cleanliness of surface mount assemblies, surface insulation resistance may be most relevant, especially when aggressive fluxes are used in mixed assemblies. The conventional methods of cleanliness tests such as solvent extraction must also be used because we are still in mixed mode for surface mounting and there are some indications that the SIR test may not be valid for rosin fluxes. One must take steps to ensure shipment of clean products and to prevent any potential problems in the field.

It must be noted, however, that there is no such thing as the best flux, the best cleaning method, or the best method for determining cleanliness. We cannot even state how clean is clean. These variables depend on the application. Thus, using the guidelines discussed in this chapter, the user must establish requirements for flux, cleaning, and cleanliness testing based on empirical data for a particular application. This means that the cleanliness tests (SIR, solvent extraction, and visual) should be performed on randomly selected assemblies cleaned during one day as a check on the process. There is no substitute for good process control because if a bad board passes the cleanliness test, the failed assembly lot cannot be recalled, recleaned, or retested.

## REFERENCES

1. Musselman, R.P., and Yarbrough, T.W., "The fluid dynamics of cleaning under surface mounted P. W. As and hybrids." *Proceedings of NEPCON West*, February 1986, pp. 207–220.

2. Rubin, W. "An inside look at flux formulations." *EP&P*, May 1987, pp. 70–72.

3. Aspandiar, R., Prasad, P. D., and Kreider, C. "The impact of an organic acid wave solder flux on the cleanliness of surface mount assemblies." *NEPCON Proceedings*, February 1987, pp. 277–289.

4. Wherry, R. W. "Injections of reagent stimulate yield of rosin in pine stumps." *Adhesive Age*, September 1981, pp. 34–36.

5. Romenesko, B. M. Cleaning, Surface Mount Technology. International Society for Hybird Microelectronics Technical Monograph Series 6984-002, 1984, Reston, Va, pp. 239–243.

6. Manko, H. M. *Soldering Handbook for Printed Circuits and Surface Mounting*. New York: Van Nostrand Reinhold, 1986, pp. 232 ff.

7. Ellenberger, C. K. "TI tests find best cleaner, Part II," *Circuits Manufacturing*, December 1986, pp. 34–44.

8. Sanger, D., and Johnson, K., A Study of Solvent and Aqueous Cleaning of Fluxes: Naval Weapon Center, China Lake, (1983).

9. Ellis, B. N. *Cleaning and Contamination of Electronics Components and Assemblies*. Ayr, Scotland: Electrochemical Publications, 1986.

10. Elliot, D. A. "Cleaning electronic assemblies." *NEPCON Proceedings*, February 1988, pp 510–518.

11. Gerard, D. R. "Eight different cycles for cleaning surface mount assemblies." *Circuits Manufacturing*, March 1987, pp. 48–50.

12. Lermond, D. S. "SMT flux cleaning." *Circuits Manufacturing,* September 1987, pp. 86–92.

13. ANSI/IPC-AC-62. *Post Solder Aqueous Cleaning Handbook.* IPC, Lincolnwood, IL. Approved Feb. 5, 1987, as an American National Standard.

14. IPC-CC-830. Qualification and Performance of Electrical Insulating Compounds for Printed Board Assemblies." IPC Standard, Lincolnwood, IL, January 1984.

15. ANSI/IPC-SM-840A. Qualification and Performance of Permanent Polymer Coating (Solder Mask) for Printed Boards. IPC Standard, Lincolnwood, IL, July 1983.

16. IPC-ML-950B. Performance Specification for Multilayer Printed Wiring Boards. IPC Standard, Lincolnwood, IL, December 1977.

17. IPC-SF-818. General Requirements for Electronic Soldering Fluxes IPC Standard, Lincolnwood, IL, August 1983.

18. Tautscher, C. J. "Measuring cleanliness of printed wiring: what really counts." *Proceedings of NEPCON West,* March 1983, pp. 556–564.

19. Wargotz, W. B. "Physical design for irrigability in aggressive flux assembly of printed wiring." *Proceedings of NEPCON East,* June 1985, pp. 23–36.

20. IPC-TM-2.6.3. Moisture and Insulation Resistance, Rigid, Rigid/ Flex and Flex Printed Wiring Boards. IPC Test Method, Lincolnwood, IL, February 1986.

21. Klein Wassink R. J., *Soldering in Electronics.* Ayr, Scotland: Electrochemical Publications, 1984, p. 140.

22. Aspandiar, R. F., Piyarali, A., and Prasad, P. "Is OA OK?" *Circuits Manufacturing,* April 1986, pp. 29–36.

23. Tuck, J. Round Table on cleaning-renewed interest in water cleaning, Circuits Manufacturing, August 1988, pp. 48–52.

24. Bredfeldt, K. "How well can we qualify cleanliness for surface mount assemblies?" *NEPCON Proceedings,* February 1987, p. 165.

# Chapter 14
# Quality Control, Repair, and Testing

## 14.0 INTRODUCTION

The dominance of the consumer electronics, steel, automotive, semiconductor, and computer industries by Japan is largely attributed to the manufacture by Japanese firms of products of consistent quality. Thus, quality control is the key to effective competition in the international market. An actual difference in cost effectiveness will result when manufacturable designs are produced in tightly controlled processes.

How should one proceed to accomplish this goal? First, statistical quality control (SQC) should be used to minimize the defects in assemblies. We shall introduce the concept of SQC, and then discuss the establishment of quality requirements, which generally are laid down in the areas of workmanship and material and process specifications.

Either visual or automated inspection is used to ascertain the quality of products in accordance with established requirements. When those requirements are not met, repair is needed to ensure compliance. Defects that slip by during incoming and assembly inspections are generally found by electrical tests.

The issues of quality control, inspection, testing, and repair are interrelated, and we must treat them as a system. If we are to manufacture reliable products at the lowest possible cost, we cannot view these areas in isolation, the resources spent in pursuit of goals in one dimension directly affect the others.

## 14.1 STATISTICAL QUALITY CONTROL

It is almost universally agreed that statistical quality control (SQC), also known as statistical process control (SPC), is the best way to collect data, analyze data, and take corrective action. SQC began as a conscious discipline in the 1920s, with the in-house applications at Bell Labs invented by Walter Shewhart. During World War II the U.S. government adopted SQC wholeheartedly, and private applications proliferated. Thereafter, general disillusionment with government regulations led many American companies to discard the SQC system as a generator of unnecessary paperwork.

In Japan however, strict regulations and the associated paperwork were seen as the path to overcoming a reputation for poor quality; SQC was not only adopted, but became a source of national pride. Now more than one-third of major Japanese companies practice SQC. The remarkable postwar recovery of Japanese industry and its subsequent domination of worldwide markets are largely attributed to SQC.

Among other things, the Japanese followed the advice of an American named W. Edwards Deming. Dr. Deming taught Japanese managers that SQC can be a powerful tool that allows the user to measure process/product variation and then to correct a given process. The major benefit derived form SQC comes from identifying, measuring, and reducing process complexity.

Dr. Deming has proven that with the use of SQC, a slack manufacturing operation can be transformed into a smooth and efficient one, while reducing defects and increasing productivity. Japanese and American companies that have used SQC consistently have found that they can do the following [1,2]:

1.  Reduce defects by one or two orders of magnitude.
2.  Shorten manufacturing times by a factor of 10.
3.  Cut inventories by a factor of 2.
4.  Cut floor space by a factor of 2.
5.  Cut labor required on a specific job by a factor of 2.

What is SQC? According to Peter Gluckman, a student of Dr. Deming [3], SQC consists of three major elements:

Process analysis to understand the system
Inductive reasoning to measure the system
Leadership to change the system

Let us look at the definition of SQC in another way. The three elements of SQC are statistics, quality, and control. "Statistics" can be defined as a body of concepts, methods, and tools used for making decisions. The second element of SQC is quality. What is quality? It is conformity to requirements or fitness for use. To measure the system (the second component of Gluckman's definition), the requirements must be established first, as discussed later in this chapter. The third element of SQC is control, which Gluckman calls leadership to change the system. It involves the administration and management of systems, processes, operations, people, and machines.

To practice all three elements of SQC, Gluckman advises thinking of manufacturing operations as a series of processes rather than as a collection of unique events. He believes that reducing process complexity is faster and less costly than increasing process efficiency.

Dr. Deming believes that only 15% of an organization's problems are related to employees; the rest are built into the process and only management can address them. A simple process may involve a series of separate steps. By using SQC to gain an understanding of the various process steps, we can take action to eliminate unnecessary or unproductive ones. SQC can expose the roles played by process steps in several ways. Here we discuss the use of control charts, omitting such other SQC tools as Pareto analysis and cause-and-effect diagrams.

The control chart, a graphic record of data is used to monitor the natural precision of a process by measuring its process average and the amount of variation from that average. Upper and lower control limits are defined based on the average. Examples of common SQC analysis tools, including control charts, are shown in Figure 14.1.

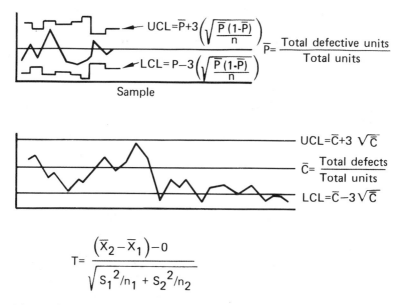

**Figure 14.1   Examples of common SQC analysis tools: P-Charts (top) showing fraction defective plot, ($n$ = samples size) and $C$ − Charts (bottom) showing defects per unit plot.**

The two commonly used SQC tools are the P chart (fraction defective plot) and the C chart (defects per unit). The formulas for determining upper control limits (UCL) and lower control limits (LCL) for P and C charts are shown in Figure 14.1. The details on constructing a control chart can be found in the American Society for Testing and Material (ASTM) standard STP-15D [4]. The effectiveness of SQC is best described by providing some actual examples.

## 14.2 APPLICATION OF SQC: A CASE HISTORY

SQC can be used in various applications, such as administration and management of systems, processes, operations, people, and machines. One practical application in electronics was used by the author to troubleshoot soldering defects. We will take this case history to illustrate the use of SQC to bring a soldering process under control. The example happened to involve a through-hole board, but it can be instructive with respect to SMT applications as well. In this study, in addition to determining the various causes of wave solder defects, the human variables in inspection were identified.

First we regrouped major solder defects in various categories and identified all possible machine-controllable and non-machine-controllable process variables. We used SQC to determine the relationships, to the extent possible, between wave solder defects and wave solder variables. We collected the data most likely to be related to wave solder defects and sifted it to see which factors were correlated with fluctuations in quality.

One program, designed program A here, had defects outside the control limits (Figure 14.2); in other programs, defects were within the control limits (Figure 14.3). Hence, we focused on program A, and three differences appeared: (a) the boards were designed with a metal core that was not present with other programs, (b) the component leads were not pre-tinned before soldering, as was the case in the other programs, and (c) the bare boards were purchased from outside suppliers, whereas the other boards were made in-house.

Further analysis of defects showed a predominance of voids in program A boards. Plated through-hole quality was found not to be the cause of voids. When the preheaters were adjusted to raise the top side board temperature from 200°F to 230°F, however, voids dropped by 50%. Program A boards had much longer leads protruding from the bottom side. Experiments were run with both long and short leads. Whereas the defects for larger leads were outside the control limits (Figure 14.4), the defect level dropped within control limits for boards with shorter leads (Figure 14.5).

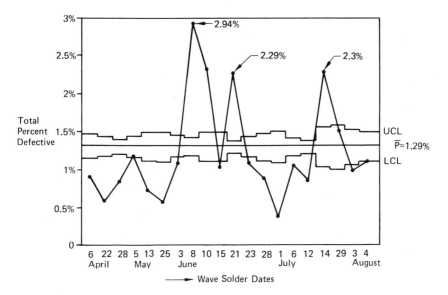

**Figure 14.2 Control charts for program A: April to August, 1983. [5]**

We also found that the lot that was out of control limits had poor board solderability and the lot within control had good board solderability. The details of this study, conducted at the Boeing Company, have been published elsewhere [5].

## 14.2.1 Implementing Statistical Process Control

How can SQC be implemented on the shop floor? The help of a statistician may be necessary in setting up control charts, although the people who use them daily do not have to be statisticians. (One does not have to be a mechanic to drive a car.) Because many people are intimidated by statistics, Dr. Warren Evans, at Intel Corporation, preaches statistics-free SQC. He has found that developing SQC classes for a vertically integrated team consisting of managers, engineers, and operators responsible for the quality of product is the most effective method, based on five principles [6]:

1.  Perfect training materials cannot compensate for improper management support.
2.  The simplest SQC tools (control charts, Pareto analysis, cause-

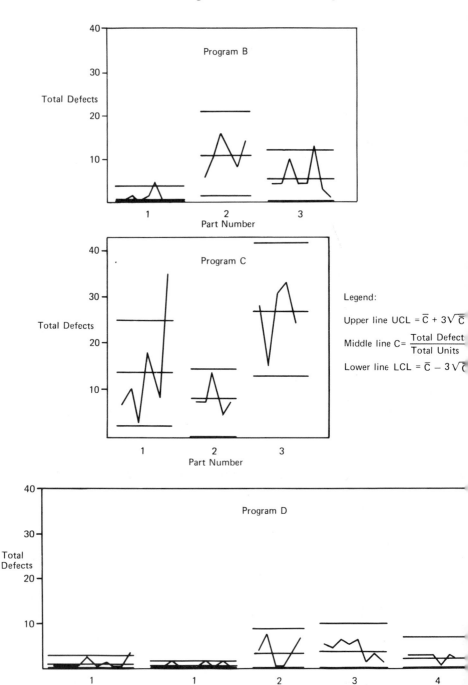

**Figure 14.3 Control charts for programs with process under control. [5]**

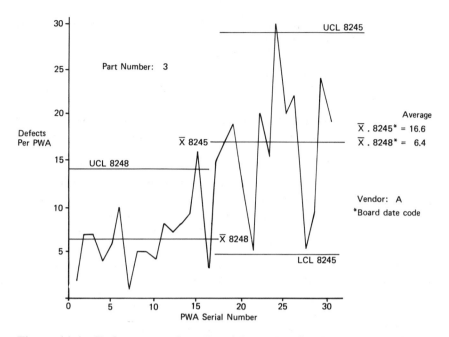

**Figure 14.4 Defect versus board serial number for PWAs with 0.3 inch lead extension. [5]**

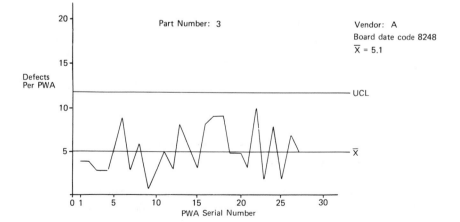

**Figure 14.5 Defect versus serial number for PWAs with 0.1 inch lead extension. [5]**

and-effect diagrams) identify 80% of the process problems, but problem-solving methods fix the problems.

3. You don't have to be a statistician to use statistics.
4. SQC requires teamwork, so people should be trained in teams. Teams need to be empowered by management to accomplish change.
5. Training should be practical and relevant and should reinforce the fact that calculating data is a minor part of process improvement.

With these guidelines, Dr. Evans found that one team was able to reduce defects so substantially that its work station was reduced from 12 people to 2 people in 15 weeks, saving the company hundreds of thousands of dollars in rework costs [6]. Similar improvements in process are reported elsewhere [7]. This is possible when the team is motivated and is fully supported and most important, assured by top management that no jobs will be lost. The real improvement starts with people, because automating an out-of-control process only produces defects faster. This is especially true in SMT because automation is natural for the technology.

The road to total SQC implementation in American companies is not an easy one. The cost of data collection and reporting is generally considered to be cost effective. This misconception can be eliminated by the appearance of tangible benefits like reducing defect rates and such intangible benefits as improved quality and better customer relations. It is better to change the process so that there is less spilled milk than to spill the milk and then try to save it. Prevention is better than cure.

In the rest of this chapter we discuss inspection, testing, and repair, three process steps that add no value to the product. If we can get the process under control, inspection, repair, and testing will be largely unnecessary. It is not realistic to asume that these functions can be entirely eliminated. However it is realistic to assume that fancy and expensive inspection, repair, and test systems will not be necessary if the potential benefits of SQC can be tapped.

## 14.3 DEFECTS RELATED TO MATERIALS AND PROCESS

A good quality product calls for the establishment of design standards, materials and process standards, and solder joint quality standards. The design issues were discussed in Chapters 5, 6, and 7, the material issues (adhesive and solder paste) were detailed in Chapters 8 and 9, and the process issues were discussed in Chapters 11 (placement), 12 (soldering), and 13 (flux and cleaning).

Solder joint quality standards set the goal for quality, defining what is acceptable and what is not. The design, materials, and process requirements provide the road map for achieving that goal. Some of the materials and process requirements apply to incoming materials supplied by the vendors, and others apply to in-house manufacturing.

No matter which category defects fall into, their cause and their impact on reliability and cost should be investigated. Even minor defects may be an indication of a serious problem and should be corrected no matter what the application. According to Ralph Woodgate, it is no more expensive to produce a perfect joint than it is to produce a bad joint [8].

In through-hole mount components, the plated through hole provides electrical connection and the added mechanical strength. Surface mount components, on the other hand, rely on solder joints on the surface of the board for their mechanical integrity. Thus it is of paramount importance that every solder joint be of good quality.

# 14.3.1    Substrate-Related Defects

Generally speaking, the substrate is more a custom product than the components. Since, moreover, unlike the supplier of components, the substrate supplier does not enjoy the benefit of fine-tuning his process over very large volumes, users should expect more quality problems with the substrates than with components.

For substrates, the major areas of concern are trace width and spaces, finished via hole sizes, selective gold plating (if used), and the type of quality of solder mask to be used. Surface mounting generally forces one to use finer trace widths and smaller vias, especially for densely packed assemblies that contain primarily logic devices.

In substrates, warp and twist are of special concern for surface mounting. Warpage of 1%, which is generally acceptable for conventional assemblies, compounds the coplanarity problem for surface mount assemblies. Excessive warpage makes Z axis adjustment difficult for the pick-and-place machine and can cause solder skips in wave soldering.

If the board is used with dry film solder mask, solder skips can be expected because of the cratering effect of the thicker solder mask film around the passive devices glued to the bottom side. Dry film solder mask is also susceptible to absorption of halides or activators in wave soldering and may adversely affect the cleanliness of the board. This is especially critical if water-soluble organic acid (OA) flux is used.

Since it is difficult to apply solder mask between the lands of fine pitch packages, these areas may be free of solder mask. If OA flux is used for the wave soldering of mixed assemblies containing fine pitch devices, absorption of OA flux in the laminate material and reduced spacing between adjacent lands may result in corrosion between the lands.

When hot air leveling is used for the application of solder to the board, solder thickness can vary between 50 and 1400 microinches. The thinner coatings can cause solderability problems, and the thicker coatings can compound problems of lead coplanarity.

Thus to minimize substrate-related defects, warpage in boards should be less than 0.5%, dry film solder mask should not be used if the assembly is going to be wave soldered, and whenever possible, plating instead of hot air leveling should be used for substrate finish. The requirements for warpage, twist, solder mask, and hot air leveling should be specified in close coordination with the supplier to ensure that they can be met.

## 14.3.2 Component-Related Defects

Both active and passive devices pose unique problems in surface mounting. For passive components the critical requirements are a nickel barrier underplating, solderability, and component markings. For active devices, one of the critical issues in surface mount packages is moisture induced package cracking. The plastic packages may need to be baked before reflow soldering to prevent this problem. (See Chapter 5, Section 5.7 for details.) Complanarity of leads is also an issue for active devices. Nonplanar leads beyond 4 mils can cause solder opens or solder wicking. The lead coplanarity problem, hence the solder wicking problem, is compounded by improper component handling. Tapes and reels are generally preferred over other forms of shipping containers to preserve factory-shipped lead coplanarity.

Poor board or lead solderability can also cause open solder joints. Solder-dipped terminations or leads provide good solderability but may exacerbate the coplanarity problem. Either dipped or plated solder coatings are acceptable as long as they meet coplanarity and solderability requirements. Refer to Chapter 3 (Section 3.7) and Chapter 10, (Section 10.9) for additional details on component procurement and component lead finish, respectively.

One commonly experienced component-related defect for passive components is the cracking of ceramic multilayer capacitors. The cracks are generally so small that they are difficult to see, but they can grow in service and cause failures. (See Chapter 3 [Section 3.2.3 and Figure 3.4]). There are various causes of cracking, but component quality plays an important role. Some dielectric materials (e.g., Z5U) are more susceptible to cracking than others (e.g., X7R). Some pick-and-place equipment exerts excessive pressure on components and may crack them. Susceptibility to cracking is also determined by the thickness of the dielectric (thicker capacitors are more susceptible).

Another cause of cracking in ceramic capacitors is thermal shock during soldering. Different materials differ in their resistance to thermal shock. Bobrowsky [9] and others have used thermal shock parameters for engineering materials as defined by the following formula.

$$\text{thermal shock parameter} = \frac{K\sigma}{\alpha E}$$

where $K$ = thermal conductivity of material
$\alpha$ = coefficient of thermal expansion (CTE)
$\sigma$ = tensile strength
$E$ = effective modulus of elasticity

Materials with a high thermal shock parameter can withstand thermal shock without cracking.

A multilayer cermic chip capacitor consists of metallic electrodes, nickel-silver termination, and ceramic dielectric (see Figure 3.3). The higher thermal conductivity of the electrodes and terminations causes these components to heat up more rapidly than the ceramic body. In addition, the stresses generated by CTE mismatch can crack the component because the tensile strength of ceramic is low and its modulus is high. Ceramic, that is, has a very low thermal shock parameter value. However, good quality ceramic materials with higher thermal shock parameters (higher thermal conductivity, lower elastic modulus, and higher tensile strength) can reduce susceptibility to cracking.

Elliott [10] states that "most manufacturers of chip capacitors use the highest quality materials available and these components easily survive the three seconds in the solder wave since they are designed to withstand soldering temperatures of 260°C (500°F) for up to 10 seconds. However, Elliott continues, some manufacturers of chip capacitors either do not have the same process control in the fabrication of their chips, or they are using lower cost and inferior materials which cannot withstand the thermal shock in the wave. As a result, some chip manufacturers specify 2°C per second ramp rate during preheat and a 100°C (180°F, not 212°F) maximum difference between the end of preheat and wave temperature." Various IPC specifications also require that chip components must withstand 260°C (ambient to 260°C solder dip) for 10 seconds.

There is no problem in meeting vendors' recommendations for ramp rate and 100°C differential for infrared reflow. However it may not be possible to follow these recommendations for every soldering process. For example, for vapor phase the reflow temperature cannot be controlled because the fluorocarbon fluid boils at a fixed temperature (215°C), and if the preheat temperature is much higher than 100°C, solder balling may reach unacceptable levels. This means that the difference between the preheat and the reflow temperatures in vapor phase stays about 115°C, and the user can do very little to reduce this differential without risking excessive solder ball problems.

This temperature difference may be even higher (120-125°C) in vapor phase, since the board cools down somewhat after preheat, which is not an integral part of the vapor phase system as is the case in wave soldering. If a ceramic cracking problem is found in vapor phase work, therefore, poor quality of the component is the likely cause. In other words, the vendor needs to improve resistance to thermal shock by selecting materials with a high thermal shock parameter.

In wave soldering, it is possible but difficult to maintain the 100°C differential between preheat and wave temperatures. For example, in wave soldering the solder pot temperature is generally held around 500°F (260°C). Lowering of the solder pot temperature below 475°F (246°C) for eutectic solder or raising the *top side* preheat temperature above 250°F (bottom side temperature around 260°F, or 126°C) will reduce thermal shock but may cause other problems—for example, a lower solder pot temperature may increase solder defects. Also, a preheat temperature above 126°C may exceed the glass transition temperature of the laminate material, causing board warpage.

In any event, since the component body and substrate do not reach the wave pot temperature due to short exposure (generally about 3 seconds), 100°C differential requirement can be met on most assemblies for 250°F maximum top side preheat and 475°F wave pot settings.

A temperature differential of about 120°C between preheat and wave pot settings should be acceptable for wave soldering if component quality is good. This temperature difference is similar to what is experienced in vapor phase soldering. Thermal shock generally is not an issue for IR. Refer to Chapter 12 for the soldering profiles of various soldering processes.

The cracking problem can be prevented by a concerted effort by component vendors, placement equipment suppliers, and users. When the land pattern design is correct (excessive solder fillet may also cause cracking), component quality is properly controlled by the vendor, pick-and-place equipment does not exert undue pressure on the components, and a correct soldering profile is used, cracking should not be a problem. One should follow the vendor's recommendation by keeping the differential between preheat and reflow as low as possible but maintaining a 100°C differential is not necessary. Many major companies do not maintain 100°C differential (it ranges from 120-140°C) but do not have cracking problems. For this to happen, most important of all, only components that can meet the requirements of the manufacturing process should be used. For example, X7R instead of Z5U dielectric should be used when feasible because the former, which contains more electrodes, heats more evenly. Refer to Chapter 3, (Section 3.2.3: Ceramic Capacitors) for details on dielectric materials.

## 14.3.3    Adhesive-Related Defects

In surface mount assemblies that use chip components glued to the bottom side for wave soldering, adhesive-related defects are predominant. The defects stem from such adhesive properties as stringing characteristics, viscosity, and dispensing characteristics. Related to dispensing characteristics is the dispensing of improper amounts of adhesive: too much or too little.

Too little adhesive causes loss of chips in the wave. The components may even fall off if the solder mask is changed to a formulation that is not compatible with the adhesive being used. Whenever a new solder mask is introduced on the board, its compatibility with the adhesive should be checked.

The most common adhesive-related defect is dispensing of too much adhesive. The excess adhesive can get on the lands and prevent formation of solder fillets. The circuit and part terminations should be free of adhesive. Ease of application without stringing reduces adhesive-related solder defects. Thus it is important to be aware that any variation in dispensing characteristics or use of an adhesive with an expired shelf life can cause the adhesive to string, increasing solder defects significantly.

One of the most critical but less commonly known adhesive-related defects is board corrosion. If adhesive is cured too rapidly, voids may be formed during the curing cycle. Open voids in adhesive can trap flux that is almost impossible to remove during cleaning. If flux is not completely removed, however, exposed conductors on the assembly may corrode and cause failure in service. An appropriate adhesive cure profile is needed to prevent such a failure. See Chapter 8 (Section 8.6.1.2: Adhesive Cure Profile and Flux Entrapment). Also refer to Chapter 8 for the adhesive properties required to prevent other defects caused by this material.

## 14.3.4    Defects Related to Solder Paste

The solder paste requirement is critical and will vary depending on the manufacturing process [7]. Solder balls are the most common of the defects due to solder paste. They are more prevalent in vapor phase soldering than in infrared soldering. Solder balls are generally caused by excessive oxide content in the paste, but rapid heating of paste before reflow is sometimes responsible for this phenomenon. Because the quality of solder paste changes very quickly, it must be tested daily to guard against balling.

Other important properties of the paste are its viscosity and slump characteristics, which affect screening and dispensing. For example, if the viscosity is too high, inconsistent or incomplete printing may result. If solder paste deposits appear to be more rounded than square, the paste may have poor slumping properties caused by low viscosity or low metal content. These properties are even more critical when printing paste for fine pitch devices. If the screen printing properties are not properly controlled, there will be excess solder bridging during soldering. Refer to Chapter 9 for the tests to be performed either by the vendor or the user to qualify a solder paste.

## 14.3.5 Process-Related Defects

The defects that are attributable to soldering processes include solder opens, solder wicking (solder open), tombstoning, bridges, misalignment, part movement, and solder balls. There are many causes for these defects— for example, poor solderability, lead coplanarity, poor paste printing, component misplacement, and poor soldering profile. There may be one cause, or there may be numerous causes. Solder wicking, as discussed in Section 12.8.4, is a good example: it can be caused by lead coplanarity, by inadequate paste thickness, or by improper soldering profile.

Wicking is expected if lead coplanarity exceeds 0.004 inchs, but vapor phase soldering may compound the problem, and the same component may provide an acceptable solder joint in an infrared process. Is this a component-related problem or a soldering problem? Or did insufficient paste thickness cause the problem? The correct answer in many cases involves a combination of all these.

Part movement, tombstoning, and misalignment are related with respect to their causes. All these defects are seen after soldering, but the root cause could be at the upstream processes such as solder paste printing or component placement. For example, if the paste was not tacky, a part might have moved during placement. Or the pick-and-place machine might not have been placing components correctly. Or the heating profile might have been too rapid.

Without belaboring the point, it is obvious that conditions generally thought to indicate a soldering problem may indeed represent either a design problem, a component problem, or a materials problem. Refer to all the relevant chapters, cited at the beginning of Section 14.3 for details before deciding on the cause of a given defect and implementing corrective action.

Some of the commonly seen defects, their acceptance criteria, and their potential causes are discussed next.

## 14.4  SOLDER JOINT QUALITY REQUIREMENTS

Workmanship standards set the goal for quality or — what is acceptable and what is not. Woodgate suggests that the only acceptable joint is the perfect joint. Anything else requires corrective action [8]. What is a perfect joint? A perfect joint is a joint showing complete wetting of the lead and the pad. This requirement applies to surface mount and through hole, and to commercial as well as military applications: reliability is important in any application. If a joint is not perfect, do not worry about repair but find the cause of the problem and fix it.

A simple solder joint standard that allows only perfect solder joints does not exist and will remain far from reality, at least in the near future. Thus we have to deal with more specific solder joint criteria. Solder defects are generally classified as minor, or cosmetic. Major defects, such as bridges and dewetted or nonwetted joints, affect circuit performance and must be corrected to permit the circuit to operate as intended. (Dewetted or non-wetted joints may pass the electrical tests, but they are sure to fail in the field.)

Minor defects, such as discoloration of solder mask, white spots or measles, solder balls, component shift on pads, and insufficient or excessive solder when joints are fully wetted, may not cause failure in circuit performance. However, minor defects for one application may be major defects for another. This is where philosophical differences generally arise. For some, the so-called minor defects are an indication of process gone haywire and are considered to be in fact major. For others, the same conditions are not serious defects, and accept/reject criteria may depend on customer requirements.

Defects in the cosmetic category generally entail some overlap with defects in the minor class. Cosmetic defects include rough and grainy joints, exposed weave textures on the substrate, and exposed (i.e., not covered with solder) lead ends. Some examples of solder defect acceptance criteria are discussed below. These criteria can be easily met if the design and process guidelines discussed in the preceding chapters are followed and the processes are in control.

The acceptance criteria for LCCCs are summarized in Figure 14.6; the description of defects discussed below (Figures 14.7-14.11) applies to

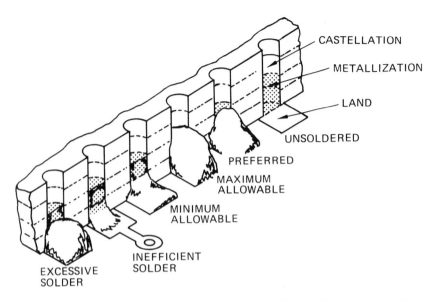

**Figure 14.6 Summary of accept/reject criteria for leadless ceramic chip carrier solder joints. [adapted from IPC specifications]**

these chip components (top figures), gull wing and J lead devices (middle figures) and LCCCs (bottom figures).

The commonly followed acceptance criteria for solder joints of these devices are as follows. The term "termination" discussed below applies to terminations of passive (resistors, capacitors etc.) and active (gull wing, J lead, and LCCC castellation) devices.

1. Solder joints should show evidence of solder wetting all along the component-to-land interface. The acceptance criteria for wetting were discussed in Chapter 10 (Section 10.4.1).

2. In a preferred solder joint (Figure 14.7), the termination is completely wetted with solder and exhibits a concave fillet indicative of good wetting. Preferred solder joints are acceptable.

3. In a maximum allowable solder joint (Figure 14.8), the termination is completely wetted with solder and exhibits a convex curvature. Wetting of the solder can be seen extending to the very edge of the circuit and component terminations. Evidence of solder reflow under the component is not grounds for rejection. Solder does not reduce the spacing between the termination and adjacent conductors. Such a joint is acceptable.

4. In a minimum allowable solder joint (Figure 14.9), metal sur-

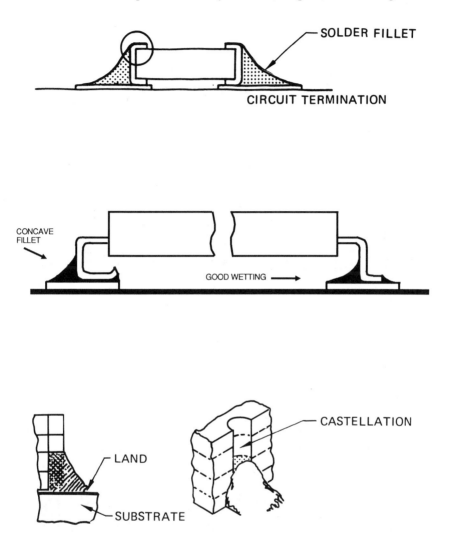

**Figure 14.7** **Preferred solder joint in chip resistors/capacitors (top), leaded (middle) and leadless (bottom) packages. (Acceptable)**

faces of the component terminations are wetted. On chip resistors and capacitors, and LCCCs the solder wets 25% of the part's height and 25% of castellation respectively. On SOICs and PLCCs, the solder fillet is formed from the land to the top of the foot's edge on at least two sides. Minimum allowable solder joints are acceptable.

    5.    In an insufficient solder joint (Figure 14.10), the solder fillet is not evident, or a gap is observed at the component termination and land

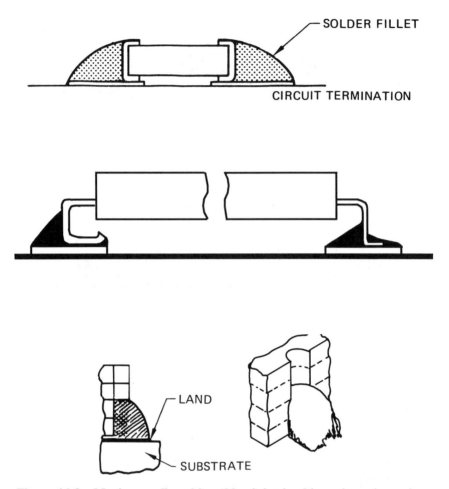

**Figure 14.8 Maximum allowable solder joint in chip resistors/capacitors (top), leaded (middle) and leadless (bottom) packages. (Acceptable)**

interface. On chip resistors, capacitors, and LCCCs solder wets less than 25% of the solderable termination height. On SOICs and PLCCs, solder does not wet to the top edge of the leads.

Insufficient solder is the opposite of bridging, it is also a less severe case of wicking, and the causes for wicking as discussed above will apply. Again, the problem may be related to poor printing or inadequate paste thickness. In wave soldering, insufficient fillet is generally caused by wrong component orientation over the wave.

Insufficient solder joints are not acceptable, especially if they are

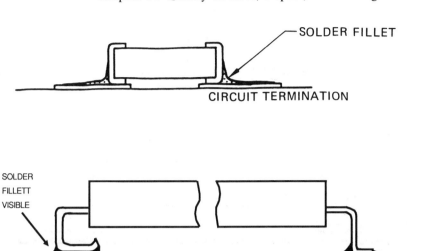

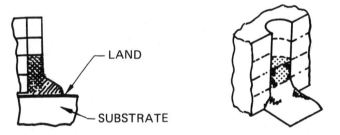

**Figure 14.9** **Minimum allowable solder joint in chip resistors/capacitors (top), leaded (middle) and leadless (bottom) packages. (Acceptable)**

caused by insufficient wetting. They should be carefully distinguished from minimum solder joints. Figure 14.11 shows photographs of preferred (top) and insufficient (bottom) solder joints in a fine pitch gull wing lead device.

6. In an excess solder joint (Figure 14.12), on chip resistors, capacitors, and LCCCs the solder bulges past the edge of the land and exhibits poor wetting at the land edge of the top of the part. On SOICs and PLCCs, solder bridges between the part body and the land, or eliminates the stress relief bend.

In the wave soldering of chip components, excess solder on trailing terminations is common if the components are not properly oriented during wave soldering. See Chapter 7 for component orientation requirements.

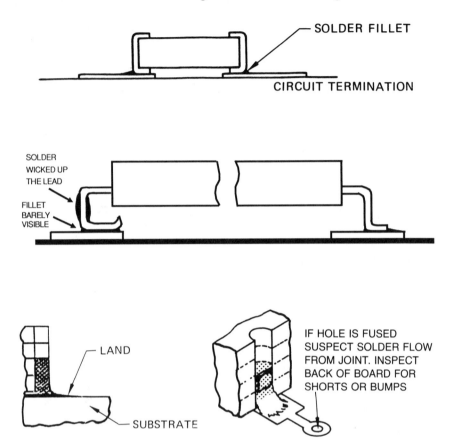

**Figure 14.10 Insufficient solder joint in chip resistors/capacitors (top), leaded (middle) and leadless (bottom) packages. (Not acceptable)**

7.    Component misalignment on the pad should not exceed 25% of the component termination width. This requirement should be taken into account when specifying test pads because misaligned components reduce the gap required for test probes.

The best way to prevent component misalignment is to use an accurate placement machine. Rapid heating of assemblies during reflow soldering can also cause misalignment.

8.    Solder bridging, another form of excess solder, is a major defect that must be corrected. A common cause for the bridging of adjacent conductors is insufficient spacing between terminations, or lands or conductors.

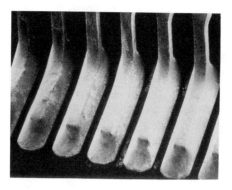

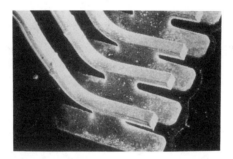

**Figure 14.11 Photographs of preferred (top) and insufficient (bottom) solder joints in a fine pitch gull wing leaded device (Photographs courtesy of Control Data Corporation).**

In reflow soldering, bridging may be caused by excessive paste thickness or excessive metal content. Alternatively, the paste may be slumping or the paste viscosity may be low. Misalignment of component or misregistration of paste can also cause bridging.

In wave soldering bridging may be design related (e.g., the conductors are too close together) process related (e.g., the conveyor speed is too slow, the wave geometry is improper, or there is an inadequate amount of oil in the wave or insufficient flux). Flux specific gravity and preheat temperature also influence bridging. Solder mask is generally used to achieve smaller gaps between conductors without causing bridges.

For reflow soldering, solder mask is less critical for preventing bridging. For fine pitch devices (25 mil center pitch), screen printing is the most critical process variable for preventing bridging.

    9.    Icicles are generally seen on the lead ends of through-hole mount

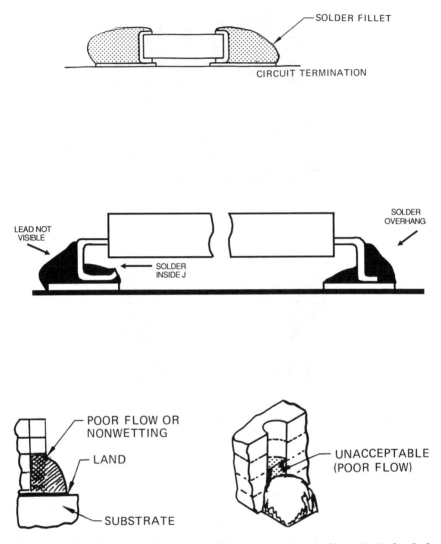

**Figure 14.12 Excess solder joint in chip resistors/capacitors (top), leaded (middle) and leadless (bottom) packages. (Not acceptable)**

devices. Their cause is generally the same as for bridges. The most effective way to eliminate icicles is to use the appropriate wave geometry (see Chapter 12, Section 12.2).

10. Tombstoning or drawbridging is a severe form of component misalignment in which the components stand on their ends. Uneven lead termination width, bad land pattern design, or improper solder paste are

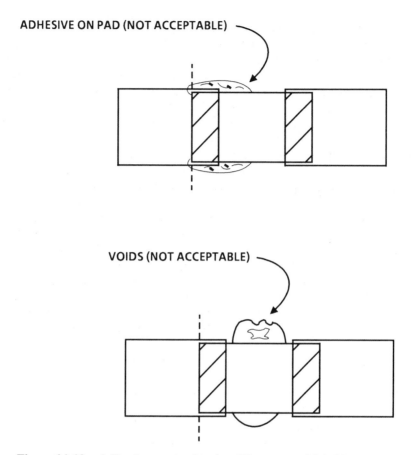

ADHESIVE ON PAD (NOT ACCEPTABLE)

VOIDS (NOT ACCEPTABLE)

**Figure 14.13    Adhesive contamination (Not acceptable). Note adhesive on pad (top) and adhesive with voids (bottom).**

some of the causes. Variations of this defect include components that turn upside down or on their sides. (Refer to Chapter 6, Section 6.2.)

11.    In solder wicking, (see inset in Figure 12.18) the paste, during reflow, travels up the lead, and very little if any is left on the pad to form a good fillet. In extreme cases, solder wicking can cause a solder open, which becomes a major defect. Solder wicking is commonly seen in vapor phase soldering. We discussed the cause of wicking in Section 14.3.5. (Also refer to Chapter 12, Section 12.8.4.)

12.    Adhesive contamination (Figure 14.13) is one of the major defects in wave soldering devices glued to the bottom of the board. Adhesive that has poor dispensing characteristics (stringing/viscosity) may end up on solder pads, causing a lack of solder fillets. The same defect

may result from poor dispensing accuracy in the placement/dispensing system, movement of components on the adhesive after placement, or simply an excess quantity of dispensed adhesive that ends up on the lands (Figure 14.13, top). Other adhesive related defects caused during cure cycle are voids (Figure 14.13, bottom). Refer to Chapter 8, Section 8.6.1.2.

13.   There is lot of concern about voids in solder joints. If solder joints are cross-sectioned, it is very common to find voids both in surface and through-hole mount joints. In general, voids in solder joints are unacceptable if the bottom of the void is not visible or if x-ray inspection during process qualification shows that more than 10% of the area directly under a part is composed of voids. (X-ray inspection is not required in manufacturing.)

Only solder pastes that are not prone to void formation should be selected. The paste formulation discussed in Chapter 12, Section 12.8.4, (Reference 13) not only minimizes solder wicking, but also minimizes formation of solder voids. In wave soldering, voids may be an indication of serious problems in the plated through holes, such as cracked barrels or moisture in the substrate. Commonly, boards are baked before soldering to prevent moisture-related voids. Voids may also indicate a dewetting problem. (See Chapter 10, Section 10.4.3.) Voids in a through-hole joint are shown in the next section.

Accept/reject criteria for other areas of assemblies are as follows:

1.   Splicing wire or lead conductors with solder is not acceptable.
2.   Parts, wire insulation, or printed circuit board resins that have been charred, melted, or burned should not be accepted.
3.   Heater bar, probe, or hold-down tool marks should not be causes for rejection.
4.   Lifted lands should not be accepted.

When in doubt, the normal practice is to touch up the joint. This is not recommended. Instead, if the joint shows evidence of good wetting, it should be accepted. The conditions leading to rejection should be solder dewetting, bridges and icicles, solder balls, solder wicking, component misalignment, and tombstoning.

Defects should be continually monitored and appropriate corrective action taken in three major areas; design, in-house processes, and vendor material quality (as exemplified by good design for manufacturability practice [See Chapters 5, 6, and 7], board and component solderability, lead coplanarity, and adhesive and solder paste quality). As discussed earlier, materials and process requirements must be established to prevent these defects.

## 14.5    SOLDER JOINT INSPECTION

Solder joint inspection is an after-the-fact step. A more effective practice is to take preventive action: that is, to implement process control to ensure that the problem does not occur. Does this mean that inspection is not necessary? Far from it. Inspection will have to continue to complete the loop on defect collection, to monitor the process, and to implement corrective action.

The most common and widely used method of inspection is visual inspection at 2 to 10X magnification. This approach, however, is a subjective one. For example, in the SQC case history study mentioned earlier [5], the same operator performed all soldering operations but at least 10 inspectors inspected the assemblies. When it appeared that these inspectors would report different quality levels even if looking at the same assembly, we analyzed the data for 3 months (Figure 14.14) and found that inspector D reported three times more defects than the average of all inspectors. When this analysis was expanded to cover defect data for 7 months, a wide variation in reported defects still existed (Figure 14.15).

Further analysis of defect data was conducted by charting defects reported by each inspector for different part numbers of the same program. The results of only four inspectors are plotted in Figure 14.16 for clarity. Again inspector D reported consistently more defects than the others. Inspector E consistently reported fewer defects than others, and the reports of inspectors C and G showed wide fluctuations.

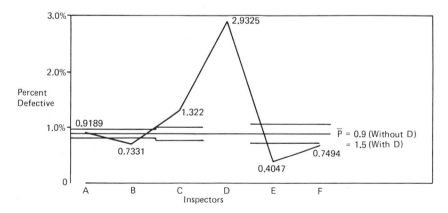

**Figure 14.14    Defect rate versus inspection for program A, June to August, 1983. [5]**

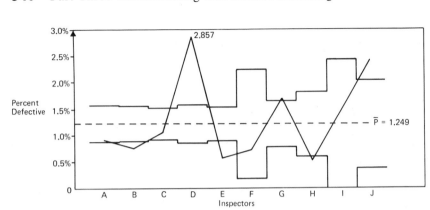

**Figure 14.15  Defect rate versus inspectors for program A, April to October, 1983. [5]**

A normal response to this situation would be to reduce the human factor and switch to one of the many automated inspection systems, such as IRT, Nicolet, and Four Pi, which use x-ray fluoroscopy and a vision system. These systems have their share of problems, too, however, as we find in discussing the popular automated laser inspection system marketed by Vanzetti.

The inspection system patented by Vanzetti uses a laser to heat the solder joint. An IR detector monitors the peak solder joint temperature. Connections that become hotter than a known good joint are considered

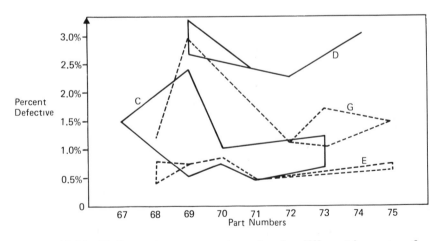

**Figure 14.16  Defect rate versus part number for different inspectors for program A, April to October, 1983. [5]**

to be unacceptable. The Vanzetti system is very effective for soldering (see Chapter 12), but one evaluation [11] of the Vanzetti laser inspection system showed inconsistent results compared with visual and x-ray methods. For example, the solder joint in Figure 14.17 shows no solder fillet, and x-ray inspection revealed internal voids. This solder joint is unacceptable, by the foregoing criteria, but the laser system passed it as a good solder joint.

The joint in Figure 14.18, on the other hand, was found to be visually acceptable, but x-ray and laster system rejected it because of excessive voids. Voids are a fairly common phenomenon, especially if oil is inter-mixed in the solder wave. Voids are also commonly seen in surface mount assemblies.

Also, since the results of a laser system are tied to volume of solder, a perfectly wetted joint with minimum solder may show up to be hotter, hence unacceptable. Thus the automated systems may be "subjective" as well. Considerable development work is necessary to make the laser system, or any other system, effective.

Since the completion of the evaluation above, considerable progress has been made in the Vanzetti system. For example, it now combines inspection and soldering functions in a single unit, to provide instantaneous

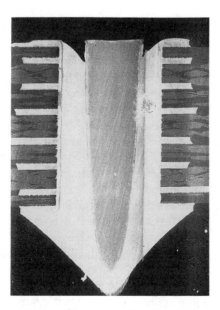

**Figure 14.17    An example of unacceptable solder joint by X-ray and visual inspection but considered good by laser inspection. [11]**

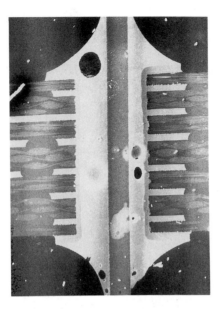

**Figure 14.18  An example of visually acceptable solder joint considered poor by laser and X-ray inspection. [11]**

feedback if a solder joint is poor. (See Chapter 12, Section 12.9). This is a relatively new system and industry data are sparse. However, since the military is emphasizing process control more than rework now, this or a similar system looks promising.

Despite the availability various types of inspection systems, many companies in the industry use automated testing equipment not only for testing but also as an alternate inspection system. The limitation of such a system is that it is effective only in finding opens and shorts. However, it can be used as an alternate automated inspection system if used in conjunction with a good process control system discussed earlier.

In any event, the problem in inspection is not the inherent "subjectivity" of visual or automated methods, but a lack of definition and understanding of what constitutes an acceptable solder connection [12]. This problem is compounded because the accept/reject criteria are generally not based on product performance.

Data collection can be done with either visual or automated inspection. The changeover to an automated system is primarily a financial decision, but the technical issues remain to be resolved, as pointed out above.

Whether inspection is visual or by an automated system, should the goal be to point out the need for corrective action or to achieve process control? It is often suggested that inspection is for process control (i.e., preventive action), especially when an automated inspection system is used.

Yet inspection is performed in a manner that leads to corrective action, so that the defect does not compromise product integrity.

It is generally believed that automated inspection systems can be used for process control by changing the appropriate variables to correct defects on a real-time basis. This may be wishful thinking at present, since many of the changes necessary to prevent the problems that are revealed require human intervention.

There is no system that can really pinpoint a given defect, hence identify its cause. Analysis of a defect requires human intervention and engineering judgment. For example, the Vanzetti system largely depends on the mass of solder (insufficient, excess, or none). Defects having more than one cause, however, could never be placed into only one of these three categories to meet the needs of an automated inspection system. For example, insufficient solder can be caused by voids or by insufficient fillet. The corrective actions for these two defects are different, calling for human judgment and intervention.

## 14.6   REPAIR EQUIPMENT AND PROCESSES

Despite all the effort put into design, process, and material control, there will always be some assemblies that fail to meet the requirements and must be repaired. This is true in every company in every country. Even with the conventional through-hole process, which has been around for more than three decades, some level of repair is performed by every company. Surface mounting is no exception.

No matter how well the material quality at the vendor and the in-house process are controlled, there will always be some defects at the end of the line in surface mounting, as is the case in through-hole assemblies. Fortunately, contrary to popular belief, the repair and rework of surface-mounted components is easier and less damaging than it is for through-hole mount. There are many reasons for this SMT advantage. For example, repair of through-hole assemblies requires considerable operator skill, especially for removing multileaded through-hole devices, and the process itself is prone to cause board damage [13]. The major causes of thermal damage in repair/rework of through-hole assemblies are pressure on the pad, and uncontrolled temperature and time of desoldering and soldering. For example, as shown in Figure 14.19, depending on the size of the hand soldering iron tip, the tip temperature can vary widely from the set temperature. And a soldering iron tip temperature above 700°F has much higher probability of thermal damage than a tip temperature below 700°F [13].

Hand soldering, especially soldering temperatures above 700°F can

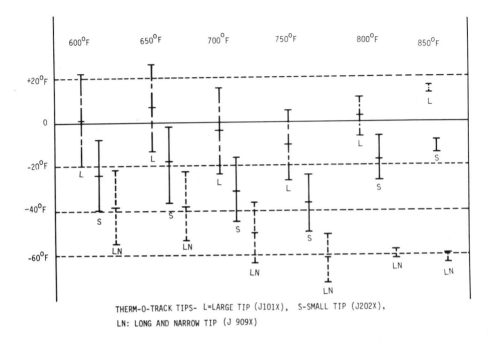

THERM-O-TRACK TIPS- L=LARGE TIP (J101X),   S-SMALL TIP (J202X),
LN: LONG AND NARROW TIP (J 909X)

**Figure 14.19   Variation in soldering iron tip temperatures for different sizes of tips at dial settings from 600–850°F. [13]**

cause charring of board surface and delamination around the pad as shown in Figure 14.20 (top figure). The bottom figure shows the cross-section of the board and clearly shows internal delamination. Pressure on the pad is the most harmful. In surface mounting, however, most of the repair systems use hot air, which totally eliminates this pressure.

Also in through-hole assemblies, most of the damage is caused during removal of components. When the leads are removed from the plated through holes, they exert mechanical and thermal stress on the internal connections and can cause internal failure [13], as shown in Figure 14.21. The top of Figure 14.21 taken at 40X, shows the cross-section of a through-hole joint. The bottom figure taken at 100X, clearly shows break between the pad and the circuit line. In surface mount assemblies, however, there are no plated through holes, hence no damage due to removal of leads from the board.

In addition to being less damaging, repair of surface mount devices is easier than repair of through-hole assemblies because operator-dependent variables are reduced in the hot air systems used in SMT. Once the repair variables have been defined, it is easier for an operator to make repairs with acceptable quality consistently.

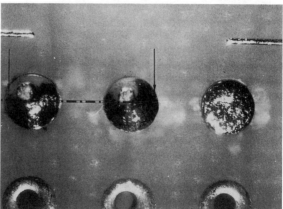

**Figure 14.20    Thermal damage on board due to temperature and pressure of soldering iron. The bottom photograph shows the cross-sectional view of the top photograph. Note delamination in the bottom photograph. Due to absence of pressure, such damages are rare during rework of surface mount assemblies with hot air systems. [13]**

There are some problems associated with the repair of surface mount assemblies, nevertheless. For example, the interpackage spacings on surface mount boards are much smaller than those encountered on conventional boards. This means that when removing or replacing surface mount components with hot air, the correct size of hot air nozzle must be used to ensure that adjacent devices are not adversely affected.

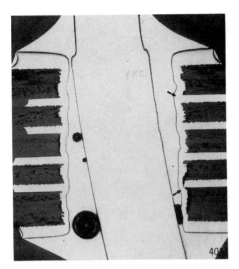

**A
Internal Pad-Tubelet
Separation**

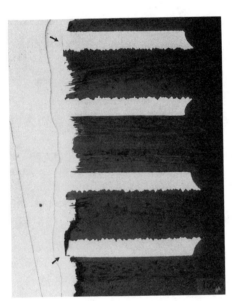

**B
Magnified (100X) view of A**

**Figure 14.21** **Internal pad-tubelet separation: at 40X magnification (top)
and 100X magnification (bottom). [13]**

## 14.6.1    Repair Requirements

The repair equipment and processes for surface mounting fall into two main categories: hot air devices and conductive tips. The latter are not as common, but they are inexpensive, whereas the cost of hot air systems can easily exceed $10,000. We will discuss some of the specific rework equipment in Section 14.6.2. No matter which equipment is used, the following requirements should be observed to minimize thermal damage on the board.

When using hot air devices, care should be taken to prevent thermal damage to adjacent components or to the substrate. When using tip attachments to soldering irons or when using other resistance-heated conductive tips, pressure on the surface mount lands should be minimized. The basic idea is to prevent thermal damage to the substrate (Figures 14.20 and 14.21) or adjacent parts.

The solder fillets resulting from the repair/rework procedures should meet the requirements of the solder joint criteria discussed in the preceding section. The following specific guidelines should be used to prevent thermal damage and to produce an acceptable solder joint after repair/rework.

1. It is recommended that the number of times a part is removed and replaced be kept to a minimum (maximum of two times) to prevent internal thermal damage in the printed circuit board [13].

2. Preheating of the substrate for 30 $\pm$ 10 minutes at 94°C (200°F) is recommended during component removal. An oven or a built-in preheating system in the repair machine may be used for this purpose. Preheating is generally recommended for removal of devices, especially if they are connected to large heat sinks such as power or ground planes, or metal cores, or if they have thicker leads.

3. The tip temperature of the tip attachments or conductive tips should not exceed 370°C (700°F). Potential for thermal damage increase significantly above 700°F [13]

4. Desoldering time (heat source on land) should not exceed 3 seconds when using conductive tips. Figure 14.22 shows the temperatures of soldering iron tip, and the temperature inside and outside [50 mils away] of the plated through hole (PTH). The locations of thermocouples are illustrated in Figure 14.23. A careful observation of Figure 14.22 shows that for any tip temperature between 550 and 850°F, the solder inside the plated through hole melts within 1.5 seconds. This is illustrated by the inflection points shown in all the curves of Figure 14.22 at 361°F (melting point of entectic solder). So staying any longer than 3 seconds is not necessary for accomplishing the reflow of the solder joint. Hence unnecessarily

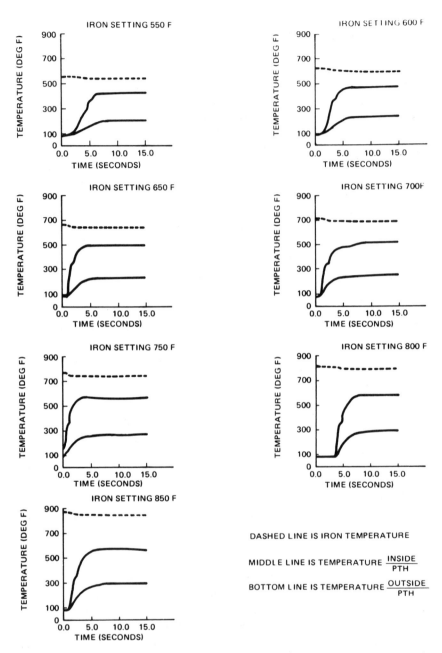

**Figure 14.22**　**Relationship between soldering iron tip and plated through-hole temperature. Note that the solder joint melted within 1.5 seconds (inflection points in the middle curves) for all tip temperatures. See Figure 14.23 for locations of thermocouples. [13]**

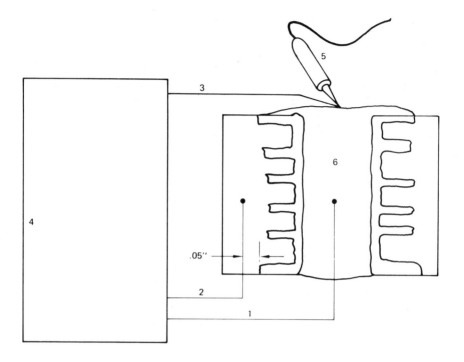

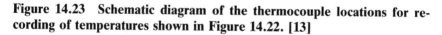

1. THERMOCOUPLE INSIDE PTH.
2. THERMOCOUPLE OUTSIDE PTH.
3. THERMOCOUPLE WELDED TO IRON TIP.
4. TEMPERATURE RECORDER CONNECTED TO H.P. 9825A SYSTEM.
5. SOLDERING IRON.
6. SOLDER PLUG.

**Figure 14.23 Schematic diagram of the thermocouple locations for re-cording of temperatures shown in Figure 14.22. [13]**

longer soldering or desoldering time with iron tip simply increases the potential for thermal damage.

5. When using hot air removal/replacement equipment, an appropriate attachment should be used to direct the flow of hot air to the component to be removed or replaced. The time and temperature for desoldering/ soldering should permit the rapid removal of parts (<20 seconds) without charring or burning flux or damaging the board. Some larger PLCCs require longer than 20 seconds for removal. This may result in charred flux, which is difficult to remove. Some surface mount packages may need as much as 60 seconds for reflow by hot air. Longer times tend to discolor board surface. (Refer to Section 14.6.4: Rework Profiles.)

6. Components should be cleaned immediately after rework.

7. Only RMA flux should be used for repair/rework.

8. Wires of 30 gauge or thinner should be used to accommodate changes in design. Whenever possible, via holes should be used to solder in jumper wires. If this is not possible, the J or gull wing wire leads of the surface mount packages should be used to solder the jumper wires. The wires should be hooked onto the leads, but the solder fillet in the hook of the wire should be as small as possible, to avoid stiffening the leads.

Butt joints or lap joints on the solder pads can also be used instead of hooks on J or gull wing wire leads. However, in this case, the jumper wire must be glued to the board with adhesive to prevent detachment or breaking of the butt or lap joints.

9. Parts that are bonded to the substrate with adhesive should be removed with a tool that will cut through the adhesive bond or by holding the part rigid and twisting the part body, so that the bond between the part and the adhesive is sheared, keeping the part body intact. A gentle rocking or prying action should be used to pry the part loose from the board.

Residual adhesive left after part removal may interfere with the new part replacement. Therefore, the residue should be carefully scraped from the surface of the substrate. A uniform, thin layer of residue may be acceptable if it does not interfere with the placement of the component.

All adhesive should be removed from the lands to achieve an acceptable solder fillet.

## 14.6.2   Soldering Irons for Surface Mount Repair

Soldering iron tip attachments are used in conjunction with the soldering iron itself for the desoldering of components from the substrate. An example was shown in Chapter 7, Figure 7.15 (left). The attachments should be made of copper and tin-plated after forming; they can be purchased or built in-house to fit the specific part to be desoldered. The tip attachments should be shaped such that all the solder joints on a particular device are heated simultaneously to the melting point of the solder, and the part should be lifted either by the tool itself or by other aids, such as tweezers.

The size and shape of the attachments will vary with the size and shape of the part to be removed. Thus, chips and SOICs have a heating attachment on two sides of the package, as shown in Figure 7.15 (left), but PLCCs need an attachment on all four sides. Using the appropriate soldering iron tip, heat is applied to the surface mount component part to be desoldered, until all the solder melts. The part is then removed with a twisting motion, to minimize damage to the leads for leaded parts.

## 14.6.3    Hot Air Systems for Surface Mount Repair

Hot air is blown on the leads/terminations/castellations and the solder joints of the part to be reworked, and the part is pulled away from the board when the solder on all solder joints is molten. The hot air is usually directed on the leads by the design of the nozzle on the hot air reflow equipment. Hence, the temperature reached by the package body during rework is much lower than during infrared and vapor phase reflow. The package body is heated indirectly via conduction along the leads, convection via the hot air impinging on the ends of the package near the leads, and to a minor extent even by radiation from the nozzle.

Initially, the leads are preheated with the nozzle some distance away (typically an inch or more) from the package body. Then the nozzle is lowered to a point just above the package body and the temperature of the leads increases sharply, till it reaches a peak. At this point the package is removed from the printed circuit board.

Hot air removal systems are the most common. The basic idea is to heat the components with hot air as quickly as possible. There are few hot air systems for the removal of passive devices, but there are many systems for active devices on the market today, and new ones are continually being introduced.

The Weller Pick-a-Chip apparatus shown in Figure 14.24 is an example of a rework machine used for the removal of capacitors and resistors only. The higher temperature and reflow time for hot air equipment can be tolerated because, unlike conductive tips, no pressure is applied to the land. Generally, hot air up to 400°C can be used. Figure 7.15 (right) in Chapter 7 shows an schematic view of this equipment.

The Pick-a-Chip operates by blowing hot air onto the chip, which melts the solder fillets at its ends; the chip can then be removed with the attached tweezers without any extra motion by the operator. When using this equipment, the stream of hot air is directed at the part to be desoldered, while the tweezers are placed carefully around the part. When the solder melts, the part is removed with a twisting motion.

Many commercial systems for active devices are the Hot Air Repair Terminal (HART) from Nu Concept and the KRAFT 100 from Pace. Some allow easy alignment of components for both removal and replacement, and others require tweezers and are more cumbersome. Some come with a microscope for easy component placement and others do not. Two commercial hot air repair systems from AirVac and Unger are shown in Figures 14.25 and 14.26 respectively. This equipment is expensive. Relatively inexpensive equipment from Unger is shown in Figure 14.27.

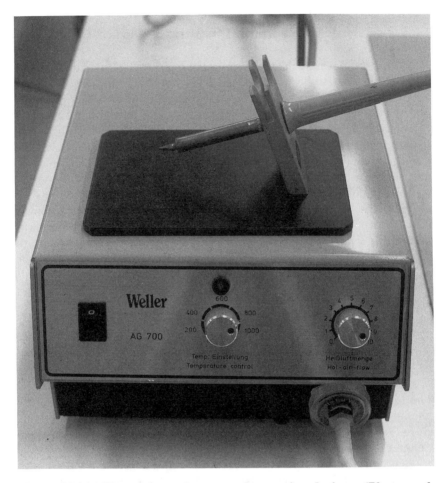

**Figure 14.24 Hot air rework system for passive devices. (Photograph courtesy of Weller.)**

A system should be selected for easy component alignment for both removal and replacement, and there should be some degree of automation to pick up devices as the solder joints melt. Total automation really is not necessary, however. First, such systems are extremely expensive, and also, one should not gear up for an assembly line of rework; rather, the goal is to control the process to minimize the defects. Rework tools are for occasional use, not for mass production.

When using a hot air system, appropriate nozzles should be used to prevent melting of adjacent component joints. In addition, silicone caps can be placed over neighboring components to shield them from the hot

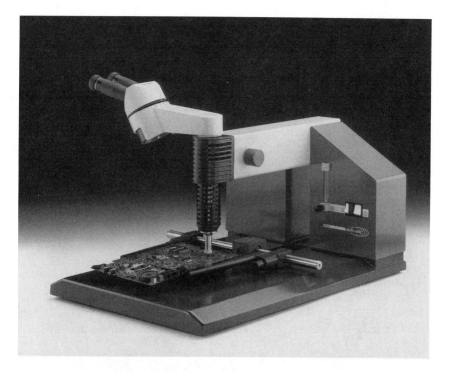

**Figure 14.25    Hot air rework system for active devices. (Photograph courtesy of Air-Vac.)**

air. These molded silicone caps are inexpensive, thin walled to accommodate spacing constraints, and infinitely reusable.

A system recently introduced by the Austin American Technology, the Smart System (SRT), uses vacuum to remove hot air (so that solder joints of adjacent components do not reflow) and bottom preheat to expedite component removal. This system uses glass nozzles, which can break in handling or if dropped, but they do allow a clear view of the solder joints as they melt.

## 14.6.4    Rework Profiles

Some systems heat all four sides of a PLCC more uniformly than others. It is important to control the heating rate in rework. The quickest ramp rate in the hot air rework profile occurs when the nozzle is lowered onto the package after preheat. This temperature ramp rate should be restricted to less than 8°C/second, to avoid excessive thermal shock on the package body.

**Figure 14.26   Another hot air rework system for active devices. (Photograph courtesy of Unger.)**

The peak temperature of the leads during rework obviously should be greater than the melting point of the solder in the solder joint. If the peak temperature is too high, however, the package or the printed board may be damaged by excessive heat application. For eutectic tin-lead solder, this peak temperature, should be in the range 200-230°C. Different equipment has different rework profiles which vary with the type of component to be removed. Figure 14.28 shows rework profiles for the Air-Vac hot air system (Figure 14.25) for removing active devices.The required removal time decreases with an increase in air flow rate. Also note that the larger devices require a longer time for removal. Similar rework profiles should be developed for all rework equipment to determine the optimum equipment settings. Refer to Section 14.6.1, Item 5 for removal time require-

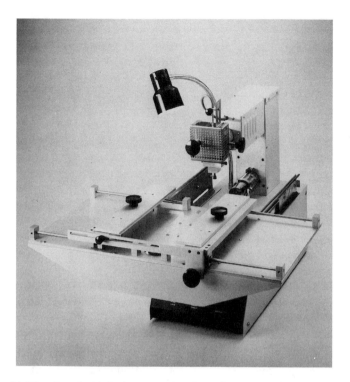

**Figure 14.27    A relatively inexpensive hot air system for active devices. (Photograph courtesy of Unger.)**

No matter which system is used, the nozzle design is critical to obtaining an effective rework profile using hot air rework equipment. Generally, the leads on the package corners heat up more quickly than the leads toward the center. Consequently, the corner leads reach the melting point of the solder before the center leads on each side of the package.

The temperature lag between the corner and center leads is greater for larger packages—as much as 20°C for poorly designed nozzles. However, with a nozzle that blows more air on the center leads, this temperature difference can be reduced to less than 5°C. An excessive differential in temperature between the center and the corner leads actually means that more heat than necessary is being supplied to the component body. It is better to have a uniform temperature throughout to prevent overheating the component body.

# 14.7    ASSEMBLY TESTING

Surface mount is a relatively new technology, and process yields tend to be lower than in conventional assemblies. With higher defect rates, the

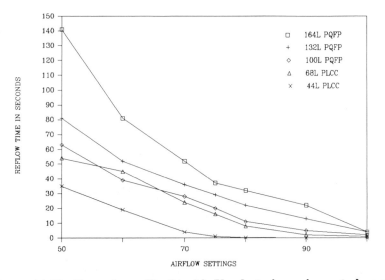

**Figure 14.28  Rework profile for Air-Vac hot air equipment shown in Figure 14.25 for removing PLCCs and fine pitch (25 mil lead pitch) devices.**

cost of test and repair can be a very significant part of the total cost of assembly. Design for testability directly affects real estate by consuming space that could have been used for additional functions on the board. However, a board that cannot be tested in a cost-effective manner will incur very significant costs of repair and test.

Design for testability is more important than ever for surface mounting. For this reason, the test engineer along with the manufacturing engineer must be integral parts of the product team. Requirements for manufacturability and testing should be considered at the design stage.

The specific guidelines and rules for design for manufacturability and testing and repair were discussed in Chapter 7. In this section, we consider only issues that concern the test engineer in production. It is assumed that the product meets the guidelines presented in Chapter 7.

Functional tests in which assemblies are run through the connector at speed and temperature provide moderate fault coverage, which can be improved if the board is designed for the system level test. Generally, the diagnostic accuracy of functional or system level tests is limited to functional blocks on the board instead of component or solder joint level (i.e., manufacturing) defects. Since the newness of the process increases the potential for this type of defect, functional tests are not considered to be very effective for SMT boards.

The defects encountered in SMT—solder opens, solder wicking (another form of open), tombstoning, bridges, misalignment, part movement, solder balls, and so on—are caused by solderability and lead coplanarity problems, poor paste printing, bad placement, and improper soldering process profiles; some of these defects may be even worse in some reflow processes than in others. These phenomena translate into two major defect types: opens and shorts.

Automated equipment test (ATE), also known as in-circuit testing, bed-of-nails testing, and pogo pin testing, is the best way to detect the manufacturing defects discussed above. For ensuring that boards can be tested by ATE, access to test nodes must be provided in the design.

In ATE, the test probes are sharp and can easily damage the surface if improperly applied. They must not touch the brittle top passivation layer of the resistors; not only might the surface be damaged, but if some of the resistive material is chipped away, even the resistance value might change. Also, to guard against false-positive results, the test probes must not touch the leads or terminations of the components. If the test probe is pressed against an improperly aligned part, the probe itself may be bent or be damaged. Figure 14.29 shows good (top two figures) and poor (bottom two figures) practices for probing with test probes.

## 14.7.1   Fixtures for ATE Testing

The fixtures for ATE can be either single sided or double sided (clam shell). The latter are expensive, but since nodal access is provided by both sides, they use less real estate. Single-sided ATE fixtures are simpler and cheaper but require nodal access from one side and consume more real estate. They are a minor variation of the existing technology and are most commonly used.

Test fixtures use 50 mil probes for surface mount devices, and the less fragile 100 mil probes for through-hole devices. As shown in Figure 14.30, some probes consist of three parts: the probe itself, a receptacle, and a wire plug. These three elements are generally found only in the current generation of 50 mil probes. The tips of the test probes vary considerably as shown in Figure 14.31. The figure also summaries the functions of various types of probe tips.

Although 50 mil probes can be five times as expensive and half as reliable [14] as the probes used for through hole, fairly good reliability can be achieved if the 50 mil probes are handled properly. Their use should be minimized, never the less: that is, they should not be used if 100 mil probes would work on that location without compromising board real estate.

There are test fixtures for all three categories of surface mount device. Type III SMT test fixtures, which are essentially the same as conventional

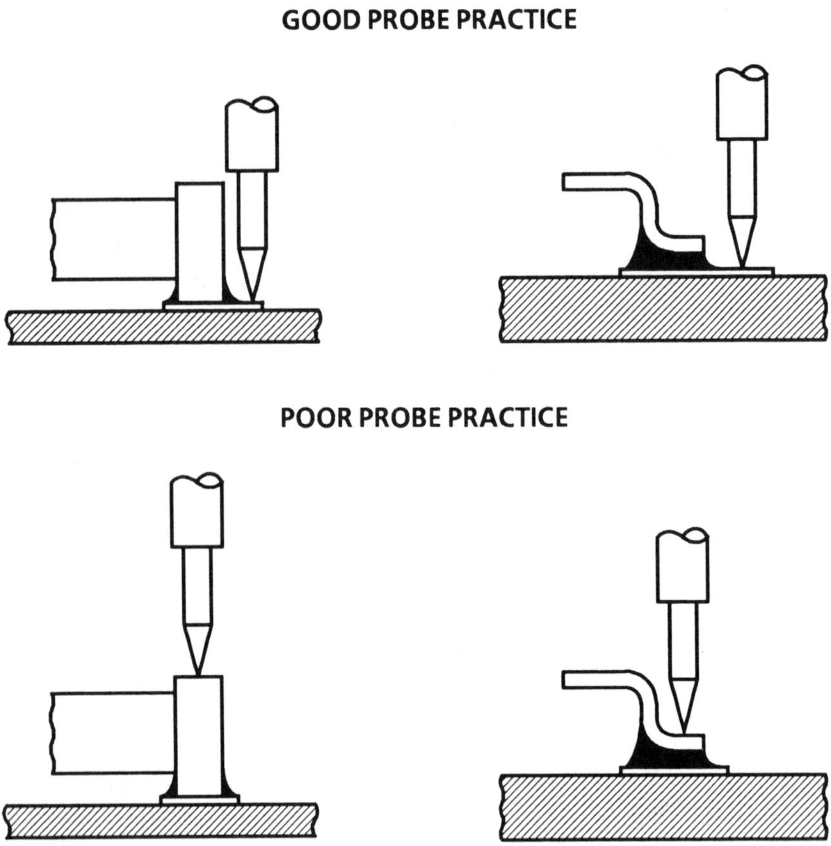

**GOOD PROBE PRACTICE**

**POOR PROBE PRACTICE**

**Figure 14.29** **Good (top) and poor (bottom) practices for probing during test.**

test fixtures, use 100 mil test probes. Type I SMT test fixtures, which use only 50 mil test probes, serve for memory boards only because of restraints imposed by component availability. Type II SMT test fixtures, which use 100 mil probes for through-hole devices and 50 mil probes for surface mount devices, tend to present more problems in debugging. If the densities of the 100 mil and 50 mil probes are not appropriately controlled, the needed contact may not be achieved. This happens especially when the density of probes is so high that the vacuum pressure is insufficient to pull the board onto the fixture to make contact.

Type III and II assemblies do not have any vacuum leakage problems because all the test vias are sealed during wave soldering. There may be vacuum leakage problems with Type I assemblies, since the test vias are

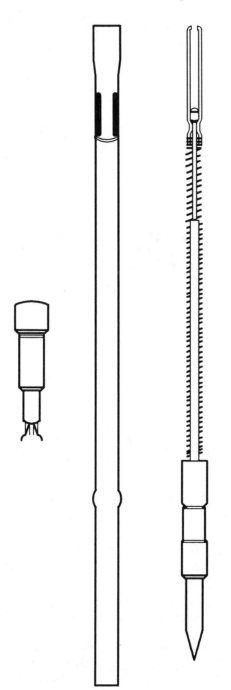

**Figure 14.30    Typical elements of a test probe: plunger (right), receptacle (center), and wire plug (left).**

**RADIUS TIP** -- The smooth radius tip will not scratch circuitry but requires clean surface.

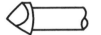 **PYRAMID TIP** -- Used for contacting plated through holes.

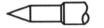 **POINT TIP** -- Sharp single point penetrates through surface contaminants for good electrical contact.

 **4 POINT TIP** -- Sharp points penetrate surface contaminants for good electrical contact.

 **TRI-POINT TIP** -- Sharp points penetrate surface contaminants for good electrical contact.

**Figure 14.31    Various types of test probe tips for automated test equipment.**

open. In a single-sided SMT board the holes can be plugged with a dry film solder mask from one side, but the holes may entrap contamination. Wet film solder mask is not very effective in sealing via holes, but they may be plugged by screening solder paste from one side. The other side of the via hole should remain open to allow the cleaning solvent to flush out any entrapped flux.

Care should be taken to ensure that solder mask does not overlap onto test pads. This may be very critical for the small test pads (25–35 mils) used for 50 mil probes. When the test pads are bare and the other vias are covered with solder mask, it is easy to identify the test pads.

Keeping the test pads square in shape and other pads round also helps in distinguishing them. The square pads also provide more area than round pads. Square pads are especially desirable if the test pad size is under 32 mils diameter. A test map, created at the design stage, is also necessary, however, to ensure the identification of individual test pads.

If additional test probes become necessary, the user cannot easily drill the test fixture and add new probes. Thus any fixture modification is generally done by the vendor. One critical issue in test fixtures is the proper alignment of the test probes on the small test pads. Having a test fixture specification helps to some extent, as we see in the next section.

## 14.7.2   Issues in ATE Testing

With increased functional density, hence reduced interpackage spacing (see Chapter 7 for actual recommended numbers), the use of 50 mil probes is essential. These probes are not only more expensive, however, they are less reliable. Overall, the small probes tend to increase the cost of the test fixture. In addition to cost and reliability, there is the question of accuracy in the test fixtures. The smaller probes allow smaller test pads (30–32 mils with 18 mil holes). Even 25 mil pads with 12 mil holes have been used by some companies. However, it is very difficult to accurately align the test probes on the smallest test pads.

The alignment problem on smaller test pads is especially compounded because, unlike board fabricators and assembly vendors, the test fixture vendors are not used to following a test fixture specification. Moreover, very few companies have an in-house test fixture specification, and there is no industry standard. Even if a fixture specification exists, it is a Herculean task to determine whether a given fixture meets it.

The accuracy problem is further compounded by the method of inserting receptacles—the manual process of hitting fragile small test receptacles with a mallet can easily bend them. Hardly anyone auto inserts even 100 mil receptacles. The accurate insertion of 50 mil receptacles is a matter of the right "touch," representing a combination of experience and care. So much for the high tech world of SMT testing.

Since almost all the vendors use the same method of receptacle insertion, the alternative of finding another "right" vendor does not exist. One must work very closely with the selected vendor, who will cooperate in the development of a workable fixture specification to meet the requirements of both the user and the supplier.

When the users and the vendors have agreed on a fixture specification, they must establish the method for determining both accuracy and compliance to the specification. Some methods for measuring accuracy include video-equipped coordinate measurement, milling machines, and transparent alignment plates.

## 14.8 SUMMARY

Having dealt extensively in this book with the technical issues related to design and processes, it is important to consider the quality, testability, and yes, the repair of surface mount assemblies at a competitive cost.

Statistical process control is used to ensure compliance of the materials and process specifications. This preventive measure has been very successfully used in many companies to manufacture products with consistent quality over time. One must develop in-house material, process, and solder joint quality specifications. The solder joint quality specifications establish the final requirements, and the material and process specifications provide the road map to the achievement of the intended quality.

The function of statistical process control is to show all concerned including operators, engineers, and management, how the company is doing with respect to quality objectives. A problem may be caused by the vendors or by the in-house manufacturing processes (machine or non-machine dependent) or by the design. By pinpointing the problem, quality objectives make it possible to implement corrective action. When the root of the problem is known, however, the basic cause can be eliminated permanently, not just the symptoms.

Even though many automated inspection tools are available, the visual method remains popular. The subjectivity of visual inspection cannot be totally overcome by the automated systems as long as the inspection criteria remain subjective.

There are many automated tools available for rework and repair, making it easier and less damaging to repair a surface mount assembly. Repair of conventional through-hole assemblies is very operator dependent, hence is prone to cause additional damage.

The completed assemblies need to be tested before shipment to customers to ensure that they meet all requirements. Given the newness and complexity of the technology, automated test equipment or bed-of-nails testing is considered more effective than functional or systems testing to detect manufacturing defects.

No matter how effective the materials and process specifications and the SQC tools, some level of defects is expected in an imperfect world. Therefore, one must plan for some inspection, repair, and testing, but continue to aim for zero defect through continuous process improvement in order to compete successfully in the international market place.

# REFERENCES

1.  Roome, D. R. "Cutting through complexity, How SQC can help." *Printed Circuit Assembly*, July 1987, pp. 16–18.
2.  Keane, T. P. "JIT and SPC, a checking." *Manufacturing Systems*, October 1987, pp. 22.
3.  Gluckman, P., "Introduction to statistical quality control." *PC Fab*, March 1983.
4.  ASTM Manual on Presentation of Data and Control Chart Analysis, ASTM Standard STP-15D. Philadelphia: American Society for Testing and Materials, 1976.
5.  Prasad, P. and Fitzsimmons, D. "Troubleshooting wave soldering problems with statistical quality control. *Brazing and Soldering*, Vol. 10, No. 4, Summer 1984, pp. 13–18.
6.  Evans, W. D. "Statistics-free SPC." *Manufacturing Systems*, March 1987, pp. 34–37.
7.  Roos-Kozel, B. "Process driven solder paste selection." *Printed Circuit Assembly*, July 1987, pp. 19–24.
8.  Woodgate, Ralph. "Solder joint inspection criteria." *Printed Circuit Assembly*, February 1987, pp. 6–10.
9.  Manson, S. S., Thermal stress and low cycle fatigue, Bobrowsky thermal shock parameter, Section 7.1.4, Page 283, Robert E. Krieger Publishing Company, Malabar, Florida, 1981.
10.  Elliott, D. A., Concerns in wave and infrared soldering of surface mount assemblies, Proceedings of Surface Mount Technology Association Conference, August 1988, pp. 177–188, available from SMTA, Edina, MN.
11.  Prasad, R. P. and Vaa, R. K., Internal Evaluation report, Boeing Electronics Systems Division, July, 1983
12.  Davy, J. G. "The prospects for automated PWA solder defect inspection." *Brazing and Soldering*, No. 12, Spring 1987, pp. 9–15.
13.  Prasad, R. P. "Contributing factors for thermal damage in PWB assemblies during hand soldering." *IPC Technical Review*, January 1982, pp. 9–17.
14.  Small, C. H. "Surface mount technology forces engineers to follow testability guidelines. *EDN*, May 14, 1987, pp. 93–97.

# Appendix A*
# Surface Mount Standards

This appendix lists various standards, specifications, and guidelines that have an impact on surface mounting from U.S. and international bodies, including the U.S. Department of Defense. The documents can be obtained from the following sources.

- Central Offices of the IEC (International Electrotechnical Commission)
  3 Rue de Varembe
  1211 Geneva 20, Switzerland
- EIA (Electronics Industries Association)
  2001 Eye Street, N.W.
  Washington, DC 20006
  202-457-4930
- IEC documents available from ANSI:
  American National Standards Institute
  1430 Broadway
  New York, NY 10018
- IPC (Institute for Interconnecting and Packaging Electronic Circuits)
  7380 North Lincoln Avenue
  Lincolnwood, Illinois 60646
  312-677-2850
- Naval Publications and Forms Center
  5801 Tabor Road
  Philadelphia, PA 19120

Note: The Surface Mount Council may also be approached either at the IPC or EIA address. Even though the council does not write standards, it

*Adapted from Surface Mount Council SMT Status and action plan January 1989, available from IPC and EIA

can help develop new standards or change existing ones through industry participation. The council is a management body chartered by the IPC, EIA, military and other organizations to promote surface mount technology through standards and other means. See Figures A-1 and A-2 for IPC and EIA standardization flow charts respectively.

## 1.0 SURFACE MOUNT COMPONENT COMMITTEES AND STANDARDS

EIA-PDP-100    Mechanical Outlines for Registered and Standard Electronic Parts (Passive Devices)

EIA-JEP-95    JEDEC Registered and Standard Outlines for Semiconductor Devices

EIA-RS-481    Taping of Surface Mount Components for Automatic Placement

## 1.1 CAPACITORS (EIA-Specifications)

IS-28    Fixed Tantalum Chip Capacitor, Style 1, Protected, Standard Capacitance Range

IS-29    Fixed Tantalum Chip Capacitor, Style 1, Protected, Extended Capacitance Range

IS-36    Multilayer Ceramic Chip Capacitor

Bulletin CB-11    Guidelines for the Surface Mounting of Multilayer Ceramic Chip Capacitors

EIA-469-A    Standard Test methods for Destructive Physical Analysis of High Reliablity Ceramic Monolithic Capacitors (embraces chips)

EIA-510    Standard Test Methods for Destructive Physical Analysis of Industrial Grade Ceramic Monolithic Capacitors (embraces chips)

# IPC STANDARDIZATION FLOW CHART*

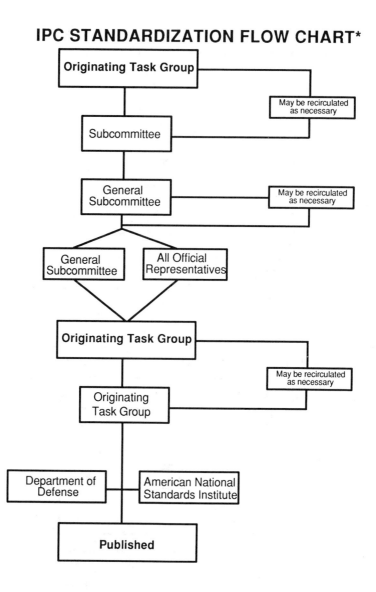

*SOURCE: SURFACE MOUNT COUNCIL SMT STATUS AND ACTION
PLAN, JANUARY 1989, Available from IPC and EIA.

**Figure A-1**

# EIA APPROVAL PROCESS CHART*
Procedure for Registration/Standardization of Outlines

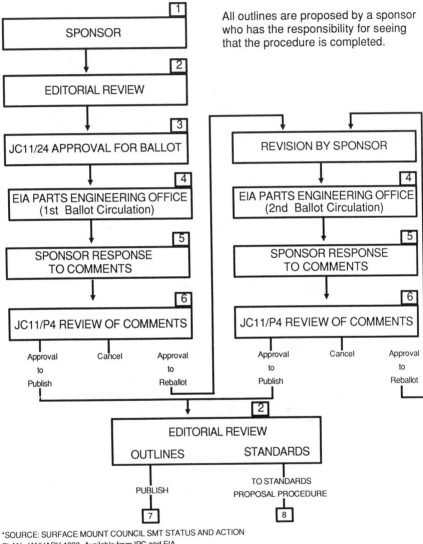

All outlines are proposed by a sponsor who has the responsibility for seeing that the procedure is completed.

*SOURCE: SURFACE MOUNT COUNCIL SMT STATUS AND ACTION PLAN, JANUARY 1989, Available from IPC and EIA.

**Figure A-2**

## 1.2    RESISTORS (EIA-Specifications)

IS-30    Resistor, Fixed, Surface Mount

IS-34    Resistor Networks


## 1.3    JEDEC (EIA) STANDARDS FOR SURFACE MOUNT (JC-11: MECHANICAL STANDARDIZATION COMMITTEE)

*Microelectronic Ceramics*

MS-002-009 Ceramic Chip Carriers
MO-044 50 mil Ceramic J-Lead
MS-014 40 mil Ceramic Chip Carrier
MO-056 45 mil Ceramic Chip Carrier
MO-057 20 mil Ceramic Chip Carrier
MO-062 25 mil Ceramic Chip Carrier

*Microelectronic Plastics*

SOIC
MO-046 200 mil Gull Wing
MS-012 150 mil Gull Wing
MS-013 300 mil Gull Wing
MS-059 330 mil Gull Wing
PLCC
MO-053 25 mil Leaded Square
MO-052 50 mil J-Lead Rectangular
Mo-047 50 mil J-Lead Square
MO-071 20 mil Gull Wing Chip Carrier
Chip Carrier
MO-041 Type E Rectangular, Leadless
MO-042 Type F Rectangular, Leadless
Small outline J lead (SOJ) plastic body
MO-061 0.400 inch body SOJ family
MO-063 0.350 inch body family
MO-065 0.300 inch body family

Discrete Devices

Transistors
TO-236—similar to SOT-23
TO-243—similar to SOT-89
TO-253—similar to SOT-143
Diodes
DO-207 Leadless
DO-213 Leadless
DO-214 C bend plastic
DO-215 Gull Wing plastic

## 2.0 SURFACE MOUNT SUBSTRATES AND SOLDER MASK

IPC-CF-152    Metallic Foil Specification for Copper/Invar/Copper for Printed Wiring and Other Related Applications

IPC-D-949    Design Standard for Rigid Multilayer Printed Boards

IPC-ML-950    Performance Specification for Multilayer Printed Boards

IPC-SM-840    Qualification and Performances of Permanent Polymer Coating (Solder Mask) for Printed Boards

IPC-FC-250    Performance Specification for Single and Double Sided Flexible Printed Boards

IPC-SD-320    Performance Specification for Rigid Single and Double Sided Printed Boards

MIL-P-50884    Military Specification Printed Wiring, Flexible and Rigid Flex

MIL-P-55110    Military Specification Printed Wiring Boards, General Specification Form

## 3.0 SURFACE MOUNT LAND PATTERN DESIGN

IPC-SM-782    Surface Mount Land Patterns (Configuration and Design Rules)

## 4.0 SURFACE MOUNT DESIGN FOR MANUFACTURABILITY

IPC-SM-780    Electronic Component Packaging and Interconnection with Emphasis on Surface Mounting

## 5.0 SURFACE MOUNT ADHESIVE

IPC-SM-817    General Requirements for Surface Mount Adhesive

## 6.0 SURFACE MOUNT SOLDER PASTE

IPC-SM-819    General Requirements and Test Methods for Electronic Grade Solder Paste

## 7.0 SOLDERABILITY

IPC-S-804    Solderability Test Method for Printed Wiring Boards

IPC-S-805    Solderability Test for Component Leads and Terminations

EIA-IS-49    Solderability Test Procedures

EIA-IS-46    Test Procedure for Resistance to Soldering (Vapor Phase Technique) for Surface Mount Devices

EIA-JEDEC Method B 102    Surface Mount Solderability Test (JESD22-B)

MIL-STD-883C Method 202    Wetting Balance for Leads

MIL-STD-202F Method 208F    Dip and Look Test

IEC-68-2-20
-Test T          Globule Test for Leads
-RS-319A         Dip and Look Method

## 8.0  SURFACE MOUNT COMPONENT MOUNTING

IPC-CM-770          Guidelines for Printed Board Component Mounting

EIA Bulletin CB11   Guidelines for the Surface Mounting of Multilayer Ceramic Chip Capacitors

DOD-2000-2          Part and Component Mounting for High Quality/ High Reliability Soldered Electrical and Electronic Assemblies

## 9.0  SURFACE MOUNT SOLDERING

IPC-S-815           General Requirements for Soldering Electronic Interconnections

IPC-S-8XX*          SMT Process Guidelines and Checklist

DOD-STD-2000-1      Soldering Technology, High Quality/High Reliability

DOD-STD-2000-4      General Soldering Requirement for Electrical and Electronic Equipment

## 10.0  SURFACE MOUNT FLUX AND CLEANING

IPC-SF-818          General Requirements for Electronic Soldering Fluxes

IPC-SC-60           Post Solder Solvent Cleaning Handbook

IPC-AC-62           Post Solder Aqueous Cleaning Handbook

MIL-F-14256-E       Flux, Soldering, Liquid (Rosin base)

## 11.0   SURFACE MOUNT INSPECTION

IPC-A-610            Acceptability of Printed Board Assemblies

IPC-AI-641           User Guidelines for Automated Solder Joint In-
                     spection Systems

IPC-AI-643           User Guidelines for Automatic Optical Inspection
                     of Populated Packaging and Interconnection As-
                     semblies

DOD-STD-2000-3       Criteria for High Quality/High Reliability Solder-
                     ing Technology

## 12.0   SURFACE MOUNT REPAIR/REWORK

IPC-R-700      Guidelines for Repair and Modification of Printed Boards
               and Assemblies

## 13.0   RELIABILITY

IPC-TR-579           Round Robin Evaluation for Small Diameter Plated
                     Through Holes

IPC-SM-782-RR*       Round Robin Test Plan for Surface Mount Land
                     Pattern Reliability

IPC-SM-XXX*          Guidelines on Accelerated Surface Mount Attach-
_785_                ment Reliability Testing

IPC-SM-XXX*          Testing and Handling of Surface Mount Plastic
_786_                Packages Susceptible to Moisture Induced Cracking

## 14.0   FUTURE SURFACE MOUNT TECHNOLOGIES

IPC-SM-784     Guidelines for Direct Chip Attachment

---

*New Surface Mount document under development by the IPC.

# Appendix B*
# Detailed Questionnaire for Evaluating Pick-and-Place Equipment for Surface Mounting

Name of company/machine: ———————————— Model # ————

Contact name: ——————————— Telephone number: ———————

Date: ————————————————

Does the equipment comply with Surface Mount Equipment Manufacturers

Association (SMEMA) standards? Yes ——————— No ———————

COMMENTS: ————————————————————————————

BOARD SIZE

Standard machine:
Maximum board size? ———————————— Minimum board size? ————

Maximum board thickness? ———— Minimum board thickness? ————

Maximum board warpage allowable? ————————————————

Larger board size capability available? Yes ——————— No ———————

If yes: Maximum board size? ———— Minimum board size? ————

*(Courtesy of M. Christensen, Intel Corporation)

If the larger board size capability is forecasted but not yet available, when will the size above be available? _____

COMMENTS: _____

## COMPONENT/PACKAGING

What size ranges of components can be placed? (part type and min/max dimensions)

Chips                    _____

MELFs                    _____

SOT                      _____

PLCC                     _____

SOIC (wide body)         _____

SOIC (narrow body)       _____

LCCC                     _____

SOJ                      _____

Other                    _____

Part feeders available: Tape _____ Tube _____ Bulk _____

Maximum reel diameter that can be mounted on the machine: _____

Sizes of tape that can be handled (8 mm, 12 mm, etc.): _____

_____

Maximum number of 8 mm feeders that can be mounted on the machine:

_____

Number of input slots that each feeder requires (8 mm feeder = 1):

12 mm _____ 16 mm _____ 24 mm _____ 32 mm _____

Tape types handled: Paper _____ Plastic _____ Other _____

COMMENTS: _____

## SPEED

Placement rate for 0805 and 1206 chips (including adhesive)? _____
parts/hr

Placement rate for 0805 and 1206 chips (without adhesive)? _____
parts/hr

Typical placement rate for SOICs and PLCCs (without adhesive)? _____
parts/hr

Conditions in which the speeds above are valid: _____

_____

COMMENTS: _____

## ADHESIVE AND COMPONENT PLACEMENT

Is on-line component test/verification available? _____ If yes, what
brand of tester is used?_____ Passive parts only? _____

Does the system have an autorecovery system for repairing misplaced

parts? _____

If yes, describe? _____

Placement reliability (with autorecovery): _____

Placement reliability (without autorecovery): _____

Placement head details:
Type (i.e. drum, XY, etc.): _____

Resolution (min increments): X _____  Y _____  Z _____

Repeatability (±0.000): X _____  Y _____  Z _____

Rotational resolution (Theta) _____

Minimum part spacing: _____

Does the head rotate the parts? _____

Are centering fingers used to center the parts? _____

Are chuck changes required for different parts? _____

Adhesive head details:
Type (i.e. syringe, etc.): _____  Quantity of dispensers? _____

Programmable location: _____  Programmable amount (time,

pressure): _____

Adhesive viscosity required or common adhesives supported? _____

_____

Programmable patterns of adhesive: _____

Maximum of patterns quantity: _____

Is dispensing unit integral to the machine or is it a separate unit?

_____

Maximum speed of the adhesive dispenser: _____

COMMENTS: _____

## BOARD HANDLING

Direction of machine flow (e.g., right to left): _____

Is track width adjustable? _____

If yes, how is it done: Manual (crank): _____ Fully automatic: _____

Does the board ever get rotated? _____ If so, why? _____

Is manual load/unload available as an option? _____

Are centering stations available that locate using:
Tooling pins? _____ Edge of board? _____

Other? _____

Can the machine handle palletes that carry circuit boards? _____

Does your company design the pallets? _____

What types of pallet? _____

Will your board handler and centering station work if leaded components

are autoinserted before SMT placement: _____


COMMENTS: _____


## PRODUCTION CHANGEOVER TIMES

Estimated time to load a tape feeder: _____

Estimated time to replace a reel on the input carriage: _____

Estimated time to download a program: _____

Estimated time to change from one board width to another (assume tool-

ing holes in same locations): _____

COMMENTS: _____

## PROGRAMMING

Methods available to program:

CAD download: _____ Manual coordinate entry: _____

Teach mode _____

Is an off-line programming station available? _____

Estimated time to create a program consisting of 100 total components with adhesive, and 10 different types of part, using:

CAD download: _____ Manual coordinate entry: _____

Teach mode: _____

Maximum number of steps per program: _____

Step and repeat capability of multipack boards: _____

Maximum number of step and repeats per program: _____

NC tape code (i.e., ASCII parity, etc.): _____

Computer Information

Does each machine come with a separate computer or do they all link to

a host computer system? _____

Maximum number of machines per host computer: _____

Computer brand: _____ Model: _____ CPU: _____

Amount of RAM in computer: _____

How many programs at one time can be placed in RAM? _____

Type of host interface (i.e., RS-232, IEEE, etc.) _____

## MAINTENANCE/WARRANTY

Standard warranty on parts? _____

Standard warranty on labor? _____

What is your guaranteed lead time for spares and service on down equipment:

Spare parts? _____   Service? _____

Where are parts and service centers located?

Location (s): _____

Estimated frequency of repair: _____

Estimated average time per repair: _____

Estimated total up-time: _____

Power requirements: _____   Air requirements: _____


COMMENTS: _____


## TRAINING/DOCUMENTATION/TESTING

Documentation included: Training: _____   Maintenance: _____

Schematics: _____   Operation manual: _____

Programming manual: _____   Other: _____

How much training is included? At factory: _____

At installation: _____

Maximum number of people who participated in this training: _____

How long does installation team will stay at our [user's] factory?

How long does a typical acceptance test at [vendor's] factory take?

_____

Where is [vendor's] factory located? _____

COMMENTS: _____

How many installations completed? _____

Where? _____

Names and telephone numbers of users? _____

_____

_____

_____

_____

# Appendix C
# Glossary

**Aspect Ratio.** A ratio of the thickness of the board to its preplated diameter. A via hole with aspect ratio greater than 3 may be susceptible to cracking.

**Azeotrope:** A blend of two or more polar and nonpolar solvents that behaves as a single solvent to remove polar and nonpolar contaminants. It has one boiling point like any other single component solvent, but it boils at a lower temperature than either of its constituents. The constituents of the azeotrope cannot be separated by distillation.

**B. Stage.**—*See* Prepreg

**Blind Via.** A via extending from an inner layer to the surface. *See also* **Via Hole.**

**Blowhole.** A large void in a solder connection created by rapid outgassing during the soldering process.

**Buried Via.** A via hole connecting internal layers that does not extend to the board surface.

**Butt Joint.** A surface mount device lead that is sheared, so that the end of the lead contacts the board land pattern. (Also called "I-Lead").

**C-Stage Resin:** A resin in a final state of cure. *See also* **B-Stage** *and* **Prepreg**.

**Castellation.** Metallized semicircular radial features on the edges of LCCC's that interconnect conducting surfaces. Castellations are typically found on all four edges of a leadless chip carrier. Each lies within the termination area for direct attachment to the land patterns.

**Chip Component.** Generic term for any two-terminal leadless surface mount passive devices, such as resistors and capacitors.

**Chip-on-Board Technology.** Generic term for any component assembly technology in which an unpackaged silicon die is mounted directly on the printed wiring board. Connections to the board can be made by wire bonding, tape automated bonding (TAB), or flip-chip bonding.

**CFC.** Chlorinated flurocarbon, cause depletion of ozone layer and scheduled for restricted use by the environmental protection agency. CFCs are used in air conditioning, foam insulation and solvents, etc.

**CLCC.** Ceramic leaded chip carrier.

**Cold Solder Joint.** A solder connection exhibiting poor wetting and a grayish, porous appearance due to insufficient heat or excessive impurities in the solder.

**Component Side.** A term used in through-hole technology to indicate the component side of the PWB. *See also* **Primary Side** *and* **Secondary Side**.

**Constraining Core Substrate.** A composite printed wiring board consisting of epoxy-glass outer layers bound to a low thermal-expansion core material, such as copper-invar-copper, graphite-epoxy, and aramid fiber-epoxy. The core constraints the expansion of the outer layers to match the expansion coefficient of ceramic chip carriers.

**Contact Angle.** The angle of wetting between the solder fillet and the termination or land pattern. A contact angle is measured by constructing a line tangent to the solder fillet that passes through a point of origin located at the plane of intersection between the solder fillet and termination or land pattern. Contact angles of less than 90°C (positive wetting angles) are acceptable. Contact angles greater than 90°C (negative wetting angles) are unacceptable.

**Control Chart.** A chart that tracks process performance over time. Trends in the chart are used to identify process problems that may require corrective action to bring the process under control.

**Coplanarity.** The maximum distance between the lowest and the highest pin when the package rests on a perfectly flat surface. 0.004 inch maximum coplanarity is acceptable.

**Crazing.** An internal condition that occurs in the laminated base material in which the glass fibers are separated from the resin at the weave intersections. This condition manifests itself in the form of connected white spots, or "crosses," below the surface of the base material, and is usually related to mechanically induced stress.

**CTE (Coefficient of Thermal Expansion).** The ratio of the change in dimensions to a unit change in temperature. CTE is commonly expressed in ppm/°C.

**Delamination.** A separation between plies within the base material, or between the base material and the conductive foil, or both.

**Dendritic Growth.** Metallic filament growth between conductors in the presence of condensed moisture and electrical bias. (Also known as "whiskers.")

**Dewetting.** A condition that occurs when molten solder has coated a surface and then receded, leaving irregularly shaped mounds of solder separated by areas covered with a thin solder film. Voids may also be seen in the dewetted areas. Dewetting is difficult to identify since solder can be wetted at some locations and base metal may be exposed at other locations.

**Dielectric Constant.**   A property that is a measure of a material's ability to store electrical energy.

**DIP (Dual In-Line Package).**   A package intended for through-hole mounting that has two rows of leads extending at right angles from the base with standard spacing between leads and row.

**Disturbed Solder Joint.**   A condition that results from motion between the joined members during solder solidification. Disturbed solder joints exhibit an irregular surface appearance, although they may also appear lustrous.

**Drawbridging.**   A solder open condition during reflow in which chip resistors and capacitor resemble a draw bridge. *See also* **Tombstoning**.

**Dual-Wave Soldering.**   A wave soldering process that uses a turbulent wave with a subsequent laminar wave. The turbulent wave ensures complete solder coverage in tight areas and the laminar wave removes bridges and icicles. Designed for soldering surface mount devices glued to the bottom of the board.

**Electroless Copper.**   Copper plating deposited from a plating solution as a result of a chemical reaction and without the application of an electrical current.

**Electrolytic Copper.**   Copper plating deposited from a plating solution by the application of an electrical current.

**Eutectic.**   The alloy of two or more metals that has a lower melting point than either of its constituents. Eutectic alloys, when heated, transform directly from a solid to a liquid and do not show pasty regions.

**Fiducial Mark.**   A geometric shape incorporated into the artwork of a printed wiring board, and used by a vision system to identify the exact artwork location and orientation. Generally three fiducial marks are used per board. Fiducial marks are necessary for the accurate placement of fine pitch components.

**Fine Pitch.**   Surface Mount Packages with lead pitches less than 50 mil, more commonly 25 mils and under. Fine pitch packages generally require vision system for accurate placement and lower solder paste prints than 50 mil pitch packages.

**Flatpack.**   An integrated circuit package with gull wing or flat leads on two or four sides, with standard spacing between leads. Commonly the lead pitches are at 50 mil centers, but lower pitches may also be used. The packages with lower pitches are generally referred to as fine pitch packages.

**Flip-Chip Technology.**   A chip-on-board technology in which the silicon die is inverted and mounted directly to the printed wiring board. Solder is deposited on the bonding pads in vacuum. When inverted, they make contact with the corresponding board lands and the die rests directly above the board surface. It provides the ultimate in densification also known as **C4** (controlled collapse chip connection).

**Footprint.**   A nonpreferred term for Land Pattern. *See* Land Pattern.

**Functional Test.**   An electrical test of an entire assembly that simulates the intended function of the product.

**Glass Transition Temperature.**   The temperature at which a polymer changes from a hard and relatively brittle condition to a viscous or rubbery condition. This transition generally occurs over a relatively narrow temperature range. It is not a phase transition. In this temperature region, many physical properties undergo significant and rapid changes. Some of those properties are hardness, brittleness, thermal expansion, and specific heat.

**Gull Wing Lead.**   A lead configuration typically used on small outline packages where leads bend and out. An end view of the package resembles a gull in flight.

**Icicle (Solder).**   A sharp point of solder that protrudes out of a solder joint, but does not make contact with another conductor. Icicles are not acceptable.

**In-Circuit Test.**   An electrical test of an assembly in which each component is tested individually, even though many components are soldered to the board.

**J-Lead.**   A lead configuration typically used on plastic chip carrier packages which have leads that are bent underneath the package body. A side view of the formed lead resembles the shape of the letter "J."

**Laminar Wave.**   A smoothly flowing solder wave with no turbulence. *See* **Dual-Wave Soldering**.

**Land.**   A portion of a conductive pattern usually, but not exclusively, used for the connection, or attachment, or both of components. (Also called a "pad").

**Land Pattern.**   Component mounting sites located on the substrate that are intended for the interconnection of a compatible Surface Mount Component. Land patterns are also referred to as "lands" or "pads."

**LCC.**   A nonpreferred term for "leadless ceramic chip carrier."

**LCCC (Leadless Ceramic Chip Carrier).**   A ceramic, hermetically-sealed, integrated circuit package commonly used for military applications. The package has metallized castellations on four sides for interconnecting to the substrate. (Also known as LCC).

**Leaching.**   The dissolution of a metal coating, such as silver and gold, into liquid solder. Nickel barrier underplating is used to prevent leaching.

**Lead Configuration.**   The solid formed conductors that extend from a component and serve as a mechanical and electrical connection that is readily formed to a desired configuration. The gull wing and the J-lead are the most common surface mount lead configurations. Less common are butt leads formed by cutting standard DIP package leads at the knee. *See also* **Butt Joint**.

**Lead Pitch.** The distance between successive centers of the leads of a component package. The lower the lead pitch, the smaller the package area for a given pin count in a package. In DIP, the lead pitch is 100 mil; in surface mount packages it is 50 mil. For fine pitch, commonly used lead pitches are 33 (Japanese), 25, and 20 mils. In tape automated bonding (see TAB), generally 10 mil pitches are used.

**Legend.** A format of letters, numbers, symbols, and/or patterns on the printed board that are primarily used to identify component locations and orientation for convenience in assembly and replacement operations.

**Mass Lamination.** The simultaneous lamination of a number of pre-etched, multiple image, C-stage panels or sheets, sandwiched between layers of **prepreg** (B-stage) and copper foil.

**Mealing.** A condition at the interface of the conformal coating and base material, in the form of discrete spots or patches, which reveals a separation of the conformal coating from the surface of the printed board, or from the surfaces of attached components, or from both.

**Measling.** An internal condition that occurs in laminated base material in which the glass fibers are separated from the resin at the weave intersection. This condition manifests itself in the form of discrete white spots or "crosses" below the surface of the base material, and is usually related to thermally induced stress.

**MELF (Metal Electrode Leadless Face).** A component package that is cylindrical and has metallization on its two ends. The package is commonly used for diodes, capacitors, and resistors.

**Metallization.** A metallic deposited on substrates and component terminations by itself, or over a base metal, to enable electrical and mechanical interconnections.

**Multilayer Board.** A printed wiring board that uses more than two layers for conductor routing. Internal layers are connected to the outer layers by way of plated via holes.

**Neutralizer.** An alkaline chemical added to water to improve its ability to dissolve organic acid flux residues.

**Node.** An electrical junction connecting two or more component terminations.

**Nonwetting.** A condition whereby a surface has contacted molten solder, but the solder has not adhered to all of the surface. The base metal remains exposed. Nonwetting is caused by a physical barrier (intermetallic or oxide) between surfaces to be joined.

**Outgassing.** De-aeration or other gaseous emission from a printed circuit board or solder joint.

**PAD.** A portion of a conductive pattern usually, but not exclusively, used for the connection, attachment, or both of components. Also called a "Land."

**P/I Structure.** Packaging and interconnecting structure. *See* **Printed Circuit Board PCB/Printing Wiring Board (PWB).**

**PLCC (Plastic Leaded Chip Carrier).** A component package that has J-leads on four sides with standard spacing between leads.

**Prepreg.** Sheet material (e.g., glass fabric) impregnated with a resin cured to an intermediate stage (B-stage resin).

**Primary Side:** The side of the assembly that is commonly referred to as the component side in through-hole technology. In SMT, the primary side is reflow soldered.

**Printed Circuit Board (PCB)/Printed Wiring Board (PWB).** The general term for completely processed printed circuit configurations. It includes rigid or flexible, single, double, or multilayer boards. A substrate of epoxy glass, clad metal, or other material upon which a pattern of conductive traces is formed to interconnect components.

**Printed Wiring Assembly (PWA).** A printed wiring board on which separately manufactured components and parts have been added. The generic term for a printed wiring board after all electronic components have been completely attached. Also called "printed circuit assembly."

**PTH (Plated Through Hole).** A plated via used as an interconnection between top and bottom sides or inner layers of a PWB. Intended for mounting component leads into through-hole technology.

**Quadpack.** Generic term for SMT packages with leads on all four sides. Most commonly used to describe packages with gull wing leads. Also known as a flat pack, but flat packs may have gull wing leads on either two or four sides.

**Reflow Soldering.** A process of joining metallic surfaces (without the melting of base metals) through the mass heating of preplaced solder paste to solder fillets in the metallized areas.

**Resin Recession.** The presence of voids between the barrel of the plated through hole and the wall of the holes, seen in cross-sections of plated through holes in boards that have been exposed to high temperatures.

**Resin Smear.** Resin transferred from the base material to the wall of a drilled hole covering the exposed edge of the conductive pattern. Normally caused by drilling.

**Resist.** Coating material used to mask or protect selected areas of a pattern from the action of an etchant, solder, or plating.

**Saponifier.** An alkaline chemical added to water to improve its ability to dissolve rosin flux residues.

**Secondary Side.** The side of the assembly that is commonly referred to as the solder side in through hole technology. In SMT, the secondary side may be either reflow soldered (active component) or wave soldered (passive component).

**Self-Alignment.** Due to the surface tension of molten solder, the tendency of slightly misaligned components (during placement) to self align with respect to their land pattern during reflow soldering. Minor self-alignment is possible, but one should not count on it.

**Shadowing (Infrared Reflow).** A condition in which component bodies block radiated infrared energy from striking certain areas of the board directly. Shadowed areas receive less energy than their surroundings and may not reach a temperature sufficient to completely melt the solder paste.

**Shadowing (Solder).** A condition in which solder fails to wet the surface mount device leads during the wave soldering process. Generally the trailing terminations of a component are affected, because the component body blocks the proper flow of solder. Requires proper component orientation during wave soldering to correct the problem.

**Single-Layer Board.** A printed wiring board that contains metallized conductors on only one side of the board. Through-holes are unplated.

**Single-Wave Soldering.** A wave soldering process that uses only a single, laminar wave to form the solder joints. Generally not used for wave soldering.

**SMC.** A surface mount component.

**SMD.** A surface mount device. Registered service mark of North American Philips Corporation to denote resistor, capacitor, SOIC, and SOT.

**SMOBC (Solder Mask Over Bare Copper).** The technology of using solder mask to protect the external bare copper circuitry from oxidation, and for coating the exposed copper circuitry with tin-lead solder [usually by using the hot air level (HAL) manufacturing process].

**SMT (Surface Mount Technology).** A method of assembling printed wiring boards or hybrid circuits, where components are mounted onto the surface rather than inserted into through-holes.

**SOIC (Small Outline Integrated Circuit).** An integrated circuit surface mount package with two parallel rows of gull wing leads, with standard spacing between leads and rows.

**SOJ (Small Outline J-Leaded).** An integrated circuit surface mount package with two parallel rows of J-Leads, with standard spacing between leads and rows. Generally used for memory devices.

**Solder Balls.** Small spheres of solder adhering to laminate, mask, or conductors. Solder balls are most often associated with the use of solder paste containing oxides. Baking of paste may minimize formation of solder balls, but overbaking may cause excessive balling.

**Solder Bridging.** The undesirable formation of a conductive path by solder between conductors.

**Solder Cream.** *See* **Solder Paste**.

**Solder Fillet.** A general term used to describe the configuration of a solder joint that was formed with a component lead or termination and a PWB land pattern.

**Solder Paste.** A homogeneous combination of minute spherical solder particles, flux, solvent, and a gelling or suspension agent used in surface mount reflow soldering. Solder paste can be deposited on a substrate via solder dispensing and screen or stencil printing.

**Solder Side.** A term used in through-hole technology to indicate the soldered side of the PWB. *See* **Primary Side and Secondary Side**.

**Solder Wicking.** The capillary action of molten solder to a pad or component lead. In the case of leaded packages, excessive wicking can led to an insufficient amount of solder at the lead/pad interface. It is caused by rapid heating during reflow or excessive lead coplanarity, and is more common in the vapor phase than in IR soldering.

**Solvent.** Any solution capable of dissolving a solute. Within the electronic industry, the term is primarily used to describe solvents with chloro-fluorochlorocarbon.

**SOT (Small Outline Transistor).** A discrete semiconductor surface mount package that has two gull wing leads on one side of the package and one on the other.

**Surfactant.** Contraction of "surface active agent." A chemical added to water in order to lower surface tension and allow penetration of water under tighter spaces.

**TAB (Tape Automated Bonding).** The process of mounting the integrated circuit die directly to the surface of the substrate, and interconnecting the two together using a fine lead frame.

**Tenting.** A printed board fabrication method of covering over plated via holes and the surrounding conductive pattern with a resist, usually dry film. (See Via hole and PTH.)

**Termination.** The metallization surfaces, or in some cases, metal end clips on the ends of passive chip components.

**Tombstoning.** A condition during reflow in which chip resistors or capacitors stand on their end resembling a tombstone. *See also* **Drawbridging**.

**Type I.** SMT using surface mounted devices on one or both sides of the substrate. The assembly is reflow soldered in one or two passes depending on the configuration.

**Type II.** SMT using surface mounted devices on one or both sides of the substrate and through-hole devices on the primary side. The assembly is reflow soldered on the primary side and wave soldered on the secondary side.

**Type III.**   SMT using surface mounted devices on the secondary side of the PWB and through-hole devices on the primary side. The assembly in wave soldered in one pass.

**Via Hole.**   A plated through hole connecting two or more conductor layers of a multilayer printed board. There is no intention to insert a component lead inside a via hole.

**Void.**   The absence of material in a localized area. *See also* **Blowhole.**

**Wave Soldering.**   A process of joining metallic surfaces (without the melting of the base metals) through the introduction of molten solder to metallized areas. Surface mount devices are attached using adhesive and are mounted on the secondary side of the PWB.

**Weave Exposure.**   A surface condition of base material in which the unbroken fibers of woven glass cloth are not completely covered by resin.

**Wetting.**   The effect of molten solder spreading along the base metal/metallization surfaces to produce complete and uniform solder coverage. A good intermetallic bond between surfaces is formed. Good wetting is indicated by a low contact angle (positive wetting angle) between the solder fillet and the base metal/metallization.

**Wicking.**   Absorption of liquid by capillary action along the fibers of the base metal. *See also* **Solder Wickering.**

# Index